HORMONES AND CELL REGULATION

European Symposium Volume 8

HORMONES AND CELL REGULATION

Proceedings of the Eighth INSERM European Symposium on Hormones and Cell Regulation, held at Sainte Odile (France), 26-29 September, 1983.
Sponsored by Institut National de la Santé et de la Recherche Médicale.

Edited by:

J. E. DUMONT and J. NUNEZ

VOLUME 8

Scientific Committee:

E. Carafoli	B. Hamprecht	F. Morel
R. M. Denton	R. J. B. King	J. Nunez
J. E. Dumont	H. J. van der Molen	G. Schultz

1984

ELSEVIER SCIENCE PUBLISHERS
AMSTERDAM · NEW YORK · OXFORD

© Elsevier Science Publishers B.V., 1984

All rights reserved. No part of this publication may be reproduced, stored in a retrieval system or transmitted in any form or by any means, electronic, mechanical, photocopying, recording or otherwise, without the prior permission of the copyright owner. However, this book has been registered with the Copyright Clearance Center, Inc. Consent is given for copying pages for personal or internal use, or for the personal or internal use of specific clients. This consent is given on the condition that the copier pay through the Center the per-page fee stated below for copying beyond that permitted by the U.S. Copyright Law.
The appropriate fee should be forwarded with a copy of the front and back of the title page of the book to the Copyright Clearance Center, Salem, MA 01970. This consent does not extend to other kinds of copying such as for general distribution, resale, advertising and promotional purposes, or for creating new works. Special written permission must be obtained from the publisher for such copying.
The per-page fee code for this book is 0-444-80583-4:84/$0+.80.

Published by:
Elsevier Science Publishers B.V.
P.O. Box 211
1000 AE Amsterdam, The Netherlands

Sole distributors for the USA and Canada:
Elsevier Science Publishing Company Inc.
52 Vanderbilt Avenue
New York, N.Y. 10017, USA

ISBN for this volume: 0-444-80583-4
ISBN for the series: 0 7204 0657 9

The Library of Congress has cataloged this work as follows:

Inserm European Symposium on Hormones and Cell Regulation.
  Hormones and cell regulation : proceedings of the ... I.N.-S.E.R.M. European Symposium on Hormones and Cell Regulation / sponsored by Institut national de la santé et de la recherche médicale et Délégation générale à la recherche scientifique et technique. — 1st (Sept. 27-30, 1976)- — Amsterdam ; New York : North-Holland ; New York : distributed in the U.S. by Elsevier/North-Holland, 1977-

  v. : ill. ; 24 cm.

  Annual.
  Editors: J. Dumont and J. Nunez.
  1. Cellular control mechanisms—Congresses. 2. Hormones—Congresses. I. Dumont, Jacques E. 1931- II. Nunez, J. (Jacques) III. Institut national de la santé et de la recherche médicale. IV. France. Délégation générale à la recherche scientifique et technique. V. Title.
  [DNLM: W3 HO812]
  QH604.I55a   599.01'927   81-645291
                                                        AACR 2 MARC-S

Library of Congress   [8403]

Printed in the Netherlands

CONTENTS

Foreword
Avant-propos IX

Summary of the meeting
Sommaire des séances XI

List of participants
Liste des participants XVII

CONTROL OF GROWTH AND CELL PROLIFERATION
CONTROLE DE LA CROISSANCE ET DE LA PROLIFERATION CELLULAIRES

Structure and function of insulin-like growth factors
Structure et fonction des facteurs de croissance analogues à l'insuline
    R.E. Humbel 3

Platelet-derived growth factor
Facteur de croissance dérivé des plaquettes
    B. Westermark, A. Wasteson and C.-H. Heldin 9

Role of ion fluxes and cyclic nucleotides in signalling mitogenesis in 3T3 cells
Rôle des flux ioniques et des nucléotides cycliques comme signaux mitogènes dans les cellules 3T3
    E. Rozengurt 17

N-glycosylation of nascent proteins early in the prereplicative phase constitutes a process for controlling animal cell proliferation
La N-glycosylation de protéines nascentes au début de la phase préreplicative constitue un processus de contrôle de la prolifération des cellules animales
    L. Jimenez de Asua, S. Poskocil, M.K. Foecking and A. Otto 37

Insulin effects on the proliferation and the differentiation of OB17 cells into adipocyte-like cells
Effets de l'insuline sur la prolifération et la différenciation des cellules OB17 en cellules semblables aux cellules adipeuses
    G. Ailhaud, E.-Z. Amri, P. Djian, C. Forest, P. Grimaldi,
    R. Négrel and C. Vannier 53

INTRACELLULAR SIGNALS
SIGNAUX INTRACELLULAIRES

Kinetical and physicochemical properties of V1 and V2 vasopressin receptors: Relation to cyclic AMP dependent and calcium dependent activation processes
Etudes des propriétés cinétiques et physicochimiques des récepteurs vasopressiques du foie et du rein de rat
    G. Guillon and D. Butlen 69

The mechanism of control of cGMP phosphodiesterase by photoexcited rhodopsin in retinal cells. Analogies with hormone controlled systems
Mécanisme du contrôle de la phosphodiestérase cGMP dépendante par la rhodopsine photoexcitée dans les cellules de la rétine. Analogies avec les systèmes sous contrôle hormonal
    M. Chabre, C. Pfister, P. Deterre and H. Kühn   87

Cyclic nucleotides and calcium in Paramecium: A neurobiological model organism
Les nucléotides cycliques et le calcium dans le Paramecium: un modèle neurobiologique
    J.E. Schultz, G. Boheim, D. Gierlich, W. Hanke, R. von Hirschhausen, G. Kleefeld, S. Klumpp, M.K. Otto and U. Schönefeld   99

CONTROL OF INTERMEDIARY METABOLISM
CONTROLE DU METABOLISME INTERMEDIAIRE

Regulation of fatty acid synthesis by insulin, glucagon and catecholamines
Régulation de la synthèse des acides gras par l'insuline, le glucagon et les catécholamines
    D.G. Hardie, R. Holland and M.R. Munday   117

Molecular mechanisms regulating glucose oxidation in insulin deficient animals
Mécanismes moléculaires régulant l'oxydation du glucose dans des animaux déficients en insuline
    P.J. Randle, S.J. Fuller, A.L. Kerbey, G.J. Sale and T.C. Vary   139

Reversible phosphorylation of hormone-sensitive lipase in the hormonal control of adipose tissue lipolysis
Phosphorylation réversible de la lipase sensible aux hormones dans la régulation hormonale de la lipolyse dans le tissu adipeux
    P. Strålfors, G. Fredrikson, H. Olsson and P. Belfrage   153

Kinase F_A mediated modulation of protein phosphatase activity
Modulation de l'activité de la protéine phosphatase médiée par la kinase F_A
    J.R. Vandenheede, S.-D. Yang and W. Merlevede   163

Deiodination and conjugation of thyroid hormone in rat liver
Désiodation et conjugaison des hormones thyroidiennes dans le foie de rat
    T.J. Visser, D. Fekkes, M.H. Otten, J.A. Mol, R. Docter and G. Hennemann   179

NEUROBIOLOGICAL CONTROLS
CONTROLES NEUROBIOLOGIQUES

Functional aspects of the coexistence of classical neurotransmitters and peptide neurotransmitters
Aspects fonctionnels de la coexistence de neurotransmetteurs classiques et peptidiques
    A. Westlind, J. Abens and T. Bartfai   195

Histamine in brain: Actions and functions
L'histamine dans le cerveau: son action et sa fonction
    M. Garbarg, J.-M. Arrang and J.-C. Schwartz   213

Role of specific neuro-neuronal and neuro-glial interactions in the *in vitro* development of dopaminergic neurons from the mouse mesencephalon
Rôle d'interactions spécifiques neuro-neuronales et neuro-gliales dans le développement, in vitro, de neurones dopaminergiques du mesencephale de la souris
    A. Prochiantz   223

Biosynthesis and degradation of carnosine and related dipeptides by brain cells in primary culture and purified enzyme preparations
Biosynthèse de la carnosine et de peptides apparentes par des cultures primaires de cellules
    K. Bauer, M. Schulz, N. Kunze, H. Kleinkauf, K. Hallermayer and
    B. Hamprecht   231

CONTROLS AT THE GENE LEVEL
CONTROLES AU NIVEAU DU GENE

Genes involved in development and differentiation control in plants
Gènes impliqués dans le contrôle du développement et de la différenciation des plantes
    J. Schell, M. Van Montagu, J. Schröder, G. Schröder, D. Inze,
    R. Deblaere, J.-P. Hernalsteens and M. De Block   245

Amplification of mouse mammary tumor virus DNA in mammary tumors of GR/A mice, a model for hormone sensitive mammary tumors
Amplification du DNA du virus induisant des tumeurs mammaires chez la souris dans des tumeurs mammaires de souris GR/A: un modèle de tumeurs mammaires hormono-sensibles
    R. Michalides   255

Prolactin receptors and intracellular mediator for prolactin action on the mammary cells
Récepteur de la prolactine et transfert de la stimulation hormonale au noyau de la cellule mammaire
    J. Djiane, L.-M. Houdebine, P.A. Kelly, I. Dusanter-Fourt,
    B. Teyssot and M. Katoh   269

Telomeric DNA rearrangements and antigenic variation in African trypanosomes
Réarrangements d'ADN télomérique et variation antigénique chez les trypanosomes africains
    E. Pays and M. Steinert   289

Electron microscopy and gene morphology
Microscopie électronique et morphologie du gène
    V. Pohl   309

Author index
Index des auteurs   325

Subject index
Index des sujets   327

FOREWORD

During the last few years, substantial progresses have been achieved in the elucidation of systems of cell regulation in superior organisms. In the field of hormone and neurotransmitter action, at least three major breakthroughs have greatly stimulated our interest and understanding : first, these problems can now be studied at the molecular level; second, several hormones are now revealed to act as neurotransmitters; third, all these agents act in different cells on the same fundamental regulation systems. It seems clear now that a limited number of regulation models can account for the main characteristics of such different extracellular signals as hormones, neurotransmitters, ions, etc. All these models imply a primary interaction of the signal with a specific protein, the receptor. Such receptors may belong to the plasma membrane, but also to other subcellular structures. Several types of receptors may correspond to one signal; each type of cell is submitted to several controls; the cell itself may modulate its response to extracellular signals. In any case, the primary interaction of the signal with its receptor leads to a cascade of intracellular events taking place at the level of the plasma membrane, of specific intracellular enzymes, of protein synthesis machinery, etc. The known physiological responses to the signals are the more or less distal consequences of these cascades.

Those ideas led a group of researchers to organize each year, since 1976, a four day International Symposium in a small village on the Alsatian side of the Vosges, first at the Bischenberg Centre, then at Sainte-Odile. They believed that it would be of great interest to compare results and concepts derived from studies in widely different areas but bearing all on the mechanisms of cell regulation. Moreover, the need to organize such a meeting at a European level was obvious, as a simple inventory demonstrated the existence of many groups of international reputation which had no regular direct contacts, such as the Laurentian or Gordon Conferences in America. For these groups we wished to promote the regular cross fertilization of ideas and techniques. This organization was made possible by the help of the "Institut National de la Recherche Médicale" and of the "Délégation Générale à la Recherche Scientifique et Technique" (France), and, in 1983, also of the Wellcome Trust.

The meetings have been successfull as they allowed, in a relaxed and informal atmosphere, to review and discuss recent advances in various fields of cell regulation, which are presented either as scattered and partial communications in general congresses or at length in separate specialized colloquia. The interdisciplinary character of the symposium obliges everybody to remain comprehensible for a general audience. Thus, it has become a very useful forum for European researchers in the field, who attend regularly, not only to talk but also to listen and learn. It is the hope of the organizers that they will, in time, constitute a nucleus for European collaboration in the field. The symposium has therefore been organized each year.

The success of a meeting should not necessarily imply the publication of its proceedings. In this case, the periodicity of the meetings allows to review regularly recent advances in the various fields of cell regulation by extracellular signals. All authors have been asked to present a rather brief synthetic view of their subject and their research. The length and depth of these reviews should place them between reviews in Molecular and Cellular Endocrinology and those of Physiological Reviews. This should benefit researchers in the same field but also non specialists and students. This book should give them brief and authoritative introductions and syntheses of the state of the art in various fields without having to scan the very dispersed specialized articles. The very rapid publication allows the reviews to be up to date; the predominantly European participation will ensure a European flavor and a fair consideration of European litterature, which, for various reasons, is sometimes conveniently forgotten. The conferences on fundamental biological problems (eg. membranes, DNA organization, etc.), which have always constituted a significant part of the Bischenberg meetings, have not in general been printed in a book which is dedicated to cell regulation. The editors have discussed the possibility of editing the discussion of each presentation. However, full publication of the discussions was thought to be both financially and materially difficult and might stiffen the informal character of these discussions. Summaries of the discussions were prepared for Volume 3, but these are open to the criticism that they may reflect more the opinion of the commentator than a true summary. They have therefore been omitted from later volumes. In this volume, a summary of the meeting, prepared originally for TIPS, has been included.

Briefly, we would like this book to provide to those who are unable to attend the meeting a good reflection of its scientific information. As to the series, when anybody in need of a short, synthetic and up to date review on cell regulation by hormones or neurotransmitters will turn first to Hormones and Cell Regulation, the editors will think that the endeavour was worthwhile. This year, we wish to thank Mrs. G. Wilmes who carried out with much enthusiasm and proficiency the bulk of the work of the organizing of the meeting, Mrs. J. Van Sande and F. Lamy, who compiled the index and, with B. Van Heuverswyn and P. Roger, helped for slide projections during the meeting.

SUMMARY OF THE MEETING

(From Trends in Pharmacological Sciences, with permission of the Editor)

The 8th Symposium on Hormones and Cell Regulation, held each year in Alsace in a secluded monastery on top of the Vosges mountains, purports to present, in an informal atmosphere and to a small size audience, recent major developments on the mechanisms of action of extracellular signal molecules on their target cells. A signal for a cell may be defined as any type of physical event or chemical agent for which a receptor, i.e. a specific sensor coupled to an effector, exists in the cell. Extracellular signals may be hormones, neurotransmitters, growth factors or even photons or oxygen pressure.

The analysis of the action of extracellular signals on the cell membranes to generate intracellular signal molecules was analyzed on three different models. The retinal rhodopsin phosphodiesterase system is very analogous to adenylate cyclase : it involves a receptor for photons (rhodopsin with its retinal moiety), and a transducin controlled by GTP, which then activates a cyclic GMP phosphodiesterase. Its reduction of cyclic GMP concentration in turn shuts off Na^+ channels. (Chabre - Grenoble). The concepts of floating receptor, several steps reaction, collision coupling and amplification which are still sometimes discussed in the cyclase field are here established. The analogy goes even further as the retinal transducer can activate cyclase in reconstituted systems and is ADP ribosylated by choleratoxin (like NS in cyclase) and islet activating protein (like Ni), which turn it on and off respectively. Thus, this transducin holds a role similar to a combination of Ni and NS in adenylate cyclase. The remarkable persistence of these proteins and functions in such different cell types bears testimony to the universality of the system (Bourne - San Francisco, Stryier - Stanford). The fact that rhodopsin is turned off by a phosphorylation suggests that such a mechanism should be looked for in desensitization of hormonal receptors.

The detailed biochemical dissection of adenylate cyclase was not considered at this meeting. However, several aspects of its function were analyzed. Four types of desensitization of the β adrenergic receptor controlled adenylate cyclase system were considered : a rapid agonist specific, cyclic AMP independent process, which is rapidly reversible and presumably bears on the receptor; a slower non specific cyclic AMP dependent process, which is reversible and involves an uncoupling (by vesicle sequestration) of the receptor; a still slower induction of phosphodiesterase; a delayed permanent loss of receptors which can only be compensated by new protein synthesis (Perkins, Chapel Hill).

It is becoming clear that, like norepinephrine, many hormones and neurotransmitters act by way of at least 2 mechanisms. Guillon (Montpellier) demonstrated definitely for a peptide hormone such a duality of receptors similar to the duality of the adrenergic (α, β) the histamine (H_1, H_2), and the dopamine receptors. V_1 receptors, which predominate in the kidney, activate adenylate cyclase; V_2 receptors found in the liver are coupled to phosphatidylinositol turnover and calcium fluxes. Direct measurement of cytosolic free Ca^{++} by the fluorescence indicator quin 2/AM demonstrates a vasopressin induced increase of this concentration (Berthon - Orsay).

To complicate matters even further, the once classical concept of one cell-one signal molecule or more precisely one neurone-one neurotransmitter has now been abandoned. In fact, most neurones may secrete several neurotransmitters, the secretion being differently affected by the rate of firing of the neurone (eg. vasointestinal peptide VIP released at the high rate and acetylcholine at the low rate). This arrangement thus provides a means to translate a frequency code into a chemical code. The different neurotransmitters may act in parallel or opposite, modulate each other's action at presynaptic or postsynaptic receptors, exert short or long term effects. For example in the submandibular gland VIP potentiates acetylcholine action (Bartfai - Stockholm). Histamine in the nervous system provides another example of this complexity. It is secreted by one category of neurones; it inhibits its secretion by a presynaptic receptor of a special type and activates the following neurone by postsynaptic H_1 (Ca^{++} channel linked) and H_2 (linked to adenylate cyclase) receptors (Garbarg - Paris). Thus, the possibilities of various behaviours and of modulation at the level of one synapse are numerous.

Another major topic was the hormonal regulation of carbohydrate and fat metabolism through the reversible phosphorylation of key enzymes. Considerable emphasis was placed on the action of adrenaline (through both alpha and beta-receptors) and insulin. Dr. P. Stralfors (Lund) presented evidence that the stimulation of lipolysis by adrenaline resulted from the increased phosphorylation of a single serine on triglyceride lipase by cyclic AMP dependent protein kinase. The reversal of this stimulation by insulin could not be accounted for by simply a decrease in cyclic AMP concentration. The action of insulin on skeletal muscle glycogen metabolism probably involves activation of the major protein phosphatase activity. Dr. J. Vandenheede (Leuven) argued that this might involve a stable conformational change, probably initiated by the phosphorylation of an inhibitory component (inhibitor I_2) by a separate factor (known as FA or GSK3) independent of calcium and cyclic nucleotides. I_2 would be an inhibitor binding to the catalytic unit of the enzyme but would become an activator of this unit upon phosphorylation (Villa-Moruzzi - Seatle).

Professor P.J. Randle (Oxford) described how long-term insulin deficiency, such as in starvation or diabetes, leads to increases in the inactive, multiply-phosphorylated form of pyruvate dehydrogenase through the elevated synthesis of a mitochondrial protein (perhaps the kinase itself) which stimulates phosphorylation. Inactivation of pyruvate dehydrogenase in these conditions leads to a decreased glucose oxidation. Dr. G. Hardie (Dundee) reviewed the considerable progress that he and his group have made in investigating the regulation of acetyl CoA carboxylase, the rate limiting enzyme in lipogenesis, by a phosphorylation which may occur at six or more different sites on the enzyme. As with pyruvate dehydrogenase, it appears that starvation can cause profound effects on enzyme activity by mechanisms which involve changes in phosphorylation quite distinct from those associated with acute changes in insulin concentration. Liver acetyl CoA carboxylase is inactivated by cyclic AMP dependent protein kinase. This effect correlates with the phosphorylation of a specific site of the enzyme. A similar effect is caused by α_1 adrenergic agents, vasopressin and angiotensin II, which through the release of diacylglycerol but not of inositol 1,4,5 phosphate (the 2 putative intracellular signals arising from phosphatidylinositol biphosphate hydrolysis) also phosphorylate this enzyme.

The role of calmodulin in calcium action was discussed at several stages. One particularly interesting case is its direct regulation of Paramecium guanylate cyclase, a membrane enzyme to which it is tightly bound. This previously unreported calmodulin can only be detached from membranes by washing with lanthanum solutions (Schultz - Tubingen).

Three types of control of gene expression were discussed. The first involves a known intracellular signal molecule as intermediate. Plasminogen activator synthesis by a pig kidney cell line is controlled at the transcription level, positively by calcitonin through cyclic AMP, negatively by hydrocortisone. The latter control overrides the former. However, there is no parallelism between mRNA levels and synthesis rates which suggests additional control at the translation level (Reich - Basel).

Prolactin interacts with membrane receptors and is later internalized with them. The first but not the second process is necessary and sufficient to elicit hormone action, i.e. mammary cell proliferation and casein synthesis. It is reported that these effects would be mediated by a heat stable, trypsin sensitive intracellular signal molecule which can be formed by isolated membranes and can act directly on isolated nuclei (Djiane -Jouy-En-Josas).

Steroid hormones act as both extra and intracellular signal molecules. Estrogens after entering the cell, bind to an intracellular receptor which then binds itself directly to a specific DNA site. Study of this receptor by monoclonal antibodies suggests that it involves separate DNA and estrogen domains and that it may be permanently located in the nucleus (Greene - Chicago). In MMTV transformed cells, a 200 nucleotide sequence from the long terminal repeat (LTR) of viral genome, integrated in cell DNA binds the cortisol receptor complex, which causes transcription and MMTV formation. When located proximal or distal to a cell oncogene, this sequence (similar to enhancer sequences) will induce hormone dependent tumors (Michalides - Amsterdam).

In Trypanosoma Brucei, the sequential expression of surface antigens involves the replication of the corresponding gene and the insertion of this replica at an expression site (Pays - Brussels).

In the case of Agrobacterium Tumefaciens, the information is transfered from a bacteria to the plant cell by a plasmid. The sequence which induces tumor formation has in effect the structure of a eukaryotic gene sequence. It codes for factors having the same functions as plant hormones including hormone(s) inducing proliferation of normal cells (Schell - Köln).

The control of cell proliferation and differentiation was a major topic of the meeting. B. Westermark (Upsala) has isolated from "buckets of Finnish platelet lysates" the factor which, in platelet rich serum, activates the proliferation of fibroblasts : PDGF (platelet derived growth factor). The primary structure of this factor presents great homologies with the transforming protein coded by simian sarcoma virus $p26^{SIS}$. It is also very similar to a osteoma growth factor secreted by osteomas. It confers to cells in vitro some of the characteristics of transformed cells (growth in soft agar). Within the concept of autocrine regulation of growth (cells secrete a growth factor that activates their own proliferation) the hypothesis is proposed that constitutive transformation that would derepress the process at any level (growth factor secretion, GF receptor, intracellular GF action) could cause cancer transformation.

The action of the growth factors discussed at the meeting, PDGF, EGF (epidermal growth factor), EDGF (eye derived growth factor) (Courty - Paris) seems to involve similar steps : activation of the phosphorylation of their own receptor at the inside of membrane, capping in coated pits, endocytosis and hydrolysis; immediate effects on ion transport and delayed stimulation of proliferation. The biochemical mechanism of action of growth factors was analyzed by Rozengurt (London), using Swiss 3T3 cell lines as a model. The decision of cell division would result from the interaction of a network of factors : mainly activated ion fluxes (with increased Na^+/H^+ exchange resulting in an enhancement of cytosolic pH, K^+ uptake, intracellular Na^+ concentration, and thus Ca^{++} release from mitochondria) and increased cyclic AMP levels would be synergistic. The latter would be caused by prostaglandins released through phosphatidylinositol hydrolysis. This same concept of 2 synergistic actions necessary for mitogenesis also applies to different extracellular signal factors acting on 3T3 cells : mitogens (EGF, $PGF_{2\alpha}$, PDGF, vasopressin) and modulators (PGE_1, insulin, hydrocortisone) (Jimenez de Asua - Basel). How these essentially membrane phenomena are related to the decision to divide remains unknown. However, one of the known biochemical steps in this action, the increase of ribosomal S_6 phosphorylation, is caused by activation of a specific S_6 kinase (Thomas - Basel). N glycosylation of proteins also appears to be involved (Jimenez de Asua). The use of microcarrier culture for such studies was demonstrated (Agius - Newcastle).

Detailed genetic and biochemical studies on tumors induced by the plasmid Ti of Agrobacterium Tumefaciens in plants, show that the genes involved in transformation act, like plant hormones, auxins and cytokinins to repress specific differentiation pathways (eg. root or shoot formation) (Schell - Köln). It is therefore remarkable that the EGF action on the thyroid cells in culture (Roger, Van Heuverzwijn - Brussels) is also to inhibit the differentiation program of the cells (eg. iodide trapping and thyroglobulin formation) and to promote growth. On the other hand, glucocorticoids, i.e. differentiating hormones, inhibit the proliferation of NHIK 3025 cells (Bakke - Trondheim).

A hormone of great physiological and clinical importance is also a growth factor. Insulin growth factor I, a somatomedin, is induced by growth hormone and its level parallels body growth in acromegaly and pituitary dwarfism. IGF_I and insulin activate each other's receptors but at concentrations several orders of magnitude higher than their respective physiological levels (Humbel - Zurich). In the differentiation of preadipocytes to adipocytes, several sequential hormonal treatments are necessary : first for proliferation, IGF_I; then for conversion to adipocytes, thyroid hormone is permissive; finally insulin is necessary for full expresion of the program (Ailhaud - Nice). In another more complex type of differentiation, some experiments suggest that in embryogenesis neurones are attracted by their target cells (Prochiantz - Paris).

LIST OF PARTICIPANTS

AGIUS, L.
Univ. Newcastle U/Tyne
Dept. Clin. Biochem.
& Metab. Medicine
The Royal Victoria Infirmary
NEWCASTLE UPON TYNE
NE1 4LP ENGLAND

AILHAUD, G.
Centre de Biochimie
Université de Nice
Parc Valrose
06034 NICE CEDEX FRANCE

ARATAN-SPIRE, S.
INSERM U-30
Hôpital des Enfants malades
149 rue de Sèvres
PARIS 15è FRANCE

ASSIMACOPOULOS-JEANNET, F.
Lab. de Rech. Métaboliques
Faculté de Médecine
64, av. de la Roseraie
1211 GENEVE 4 SUISSE

BAKKE, O.
The University of Trondheim
Div. of Biophysics
Sem Saelands vei 9
N-7034 TRONDHEIM-NTH NORWAY

BARTFAI, T.
Dept. of Biochemistry
Aarheniuslaboratory
University of Stockholm
S-106 91 STOCKHOLM, SWEDEN

BAUER, K.
Technische Univ. Berlin
Abteilung Biochemie
Franklinstrasse, 29
D-1 BERLIN 10, F.R.G.

BEECROFT, L.
University of Leicester
Dept. of Biochemistry
University Road
LEICESTER, ENGLAND

BISHOP, J.S.
Univ. of Minnesota
Dept. of Pharmacology
Medical School
3-260 Millard Hall
435 Delaware Str. S.E.
MINNEAPOLIS, MINNESOTA 55455

BORSODI, A.
Institute of Biochemistry
Biological Res. Center
Hungarian Acad. of Sciences
6701 SZEGED, P.O.B. 521
HUNGARY

BOTTARI, S.
Dept. Gynéco. & Obstét.
Hôpital Univ. Brugmann
Place A. Van Gehuchten, 4
B-1020 BRUXELLES, BELGIQUE

BOULANGER, Ch.
Faculté de Pharmacie,
INSERM U-243
74, route du Rhin
67400 ILLKIRCH, FRANCE

BURGER, A.,
Hôp. Univ. de Genève,
Lab. d'Invest. Clinique,
1211 GENEVE 4, SUISSE

BUTCHER, R.W.,
The University of texas,
Health Science Center,
P.O.Box 20334
Astrodome Station,
HOUSTON, TEXAS, 77025

CAPPONI, A.M.,
Division d'Endocrinologie,
Hôpital Universitaire,
CH-1211 GENEVE 4, SUISSE

CHABARDES, D.,
Physiologie Cellulaire,
Collège de France,
11, place M. Berthelot,
75231 PARIS CEDEX 05, FRANCE

CHABRE, M.,
Lab. Biol. Mol. & Cell.
F38041 GRENOBLE FRANCE

CLARET, M.,
Lab. Physiologie Comparée,
Univ. Paris-Sud
91405 ORSAY CEDEX, FRANCE

COOKE, B.A.,
Dept. of Biochemistry,
Royal Free Hospital,
School of Medicine
Rowland Hill Street,
LONDON, NW3 2PF, ENGLAND

COURTY, J.,
Centre de Gérontologie,
INSERM U-118,
29, rue Wilhem
75016 PARIS, FRANCE

DEGENHART, H.,
Sophia Children's Hospital,
Pediatric laboratory,
Gordelweg,160,
3038 GE ROTTERDAM,
THE NETHERLANDS

DENTON, R.M.,
Dept. of Biochemistry,
University of Bristol,
Medical School,
University Walk,
BRISTOL BS8 1TD ENGLAND

DJIANE, J.,
Physiologie de la Lactation,
Inst. Nat. Rech. Agronom.
78350 JOUY-EN-JOSAS, FRANCE

DUMONT, J.E.,
I.R.I.B.H.N.,
Faculty of Medicine,
University of Brussels,
Campus Erasme,
808 route de Lennik,
B-1070 BRUSSELS, BELGIUM

EDELMANN, A.,
Inst. of Clinical Biochem.
University of Bonn,
Sigmund-Freud-Str., 25,
D-5300 BONN, F.R.G.

EL MERNISSI, G.,
Collège de France,
Lab. de Physiol. Cellulaire,
11 pl. Marcelin Berthelot,
75690 PARIS CEDEX 14, FRANCE

FRANKLIN, T.J.,
I.C.I.,
Mereside Alderley Park,
Macclesfield Cheshire,
SK10, 4TG ENGLAND

GANELLIN, C.R.,
Smith Kline & French Res. LTD,
The Frythe,
WELWYN, HERTS, ENGLAND

GARBARG, M.,
INSERM U-109
2, rue d'Alésia,
PARIS 75014 FRANCE

GREENE, G.L.,
University of Chicago,
The Ben May Lab. for Cancer Res.,
950 East, 59th Street,
CHICAGO, ILLINOIS 60637, U.S.A.

GUILLON, G.,
Centre CNRS,
INSERM de Pharmacologie,
Rue de la Cardonille,
34033 MONTPELLIER CEDEX FRANCE

HAMPRECHT, B.,
Physiol-Chemisches Inst.
der Universität,
Koellikerstrasse, 2,
WURZBURG 87, GERMANY

HARDIE, D.G.,
Department of Biochemistry,
University of Dundee,
DUNDEE DD1, 4HN SCOTLAND

HENDERSON, D.,
Schering AG,
Biochem. Pharmakologie,
Müllerstr. 170-178,
D-1000 BERLIN 65, GERMANY

HOUDEBINE, L.M.,
Inst. National Rech. Agronom.,
Lab. Physiol. de la Lactation,
C.N.R.Z.,
78350 JOUY-EN-JOSAS FRANCE

HUMBEL, R.E.,
Biochemisches Institut
der Universität Zürich,
Zürichbergstrasse, 4,
CH-8028 ZURICH, SWITZERLAND

JIMENEZ DE ASUA, L.,
Friedrich-Mischer-Institut,
P.O.Box 273,
CH-4002 BASEL SWITZERLAND

KATTENBURG, D.,
Faculté de Pharmacie,
Lab. de Pharmacologie,
74, route du Rhin
67400 ILLKIRCH, FRANCE

KING, R.J.B.,
Imperial Cancer Res.
Fund Laboratories,
Hormone Biochem. Dept,
Lincoln's Inn Fields,
LONDON WC2A 3PX, ENGLAND

KUNG, W.,
Kantonsspital Basel,
University Women's Clinic,
Lab. Biochem. & Endocrinol.,
Hebelstrasse, 20,
4031 BASEL, SWITZERLAND

LAMPRECHT, S.A.,
Division of Gastroenterology,
Soroka Medical Center,
BEER-SHEVA 84105, ISRAEL

LAMY, F.,
I.R.I.B.H.N.,
Faculty of Medicine,
University of Brussels,
Campus Hôpital Erasme,
Route de Lennik, 808,
B-1070 BRUSSELS, BELGIUM

LAWSON, R.,
Univ. Newcastle Upon Tyne,
Dept. of Biochemistry,
NEWCASTLE UPON TYNE NE1 7RU, ENGLAND

LEWIN, M.J.M.,
INSERM U-10,
Hôpital Bichat,
179, bld. Ney,
75877 PARIS, CEDEX 18 FRANCE

MANY, M.C.,
Laboratoire d'Histologie,
Avenue Mounier, 52,
UCL-5229,
B-1200, BRUXELLES, BELGIQUE

MARCHETTI, J.,
Physiologie Cellulaire,
Collège de France,
11, pl. M. Berthelot,
75231 PARIS CEDEX 05, FRANCE

MARTIN, M.,
Laboratoires P.O.S.,
68240 KAYSERBERG, FRANCE

MELCHIORRI, P.,
Istituto di Farmacol. Medica,
Universita di Roma,
Citta Universitaria,
I-100100 ROMA, ITALIA

MICHALIDES, R.,
Netherlands Cancer Institute,
Plesmanlaan, 121,
1066 CX AMSTERDAM,
THE NETHERLANDS

MOREL, F.,
Collège de France,
11, pl. Marcelin Berthelot,
75231 PARIS CEDEX 05, FRANCE

MORGAN, M.J.,
The Wellcome Trust,
1, Park Square West,
LONDON NW1 4LJ, ENGLAND

MORUZZI, E.,
Dept. of Biochemistry,
Univ. of Washington
SEATTLE WA 98195, U.S.A.

MULLINS, J.,
University of Leicester,
ICI Joint Lab.,
Dept. of Biochemistry
University Road,
LEICESTER, ENGLAND

MUNIR, M.I.,
Dept. of Biochemistry,
University of Glasgow,
GLASGOW G12, 8QQ, ENGLAND

NAKHLA, A.,
The Population Council,
Rockefeller University,
Center for Biomedical Res.,
1230 York Avenue,
NEW YORK, N.Y. 10021 U.S.A.

NOVAK-HOFER, I.,
Friedrich Miescher-Institut,
P.O.Box 2543,
CH-4002 BASEL, SWITZERLAND

NUNEZ, J.,
I.R.I.B.H.N.
Faculty of Medicine,
University of Brussels,
Campus Erasme,
808 route de Lennik,
B-1070 BRUSSELS, BELGIUM

PAVLOVIC-HOURNAC, M.,
Unité de Recherche sur
la Glande Thyroïde,
INSERM U-96
79, av. du Général Leclerc,
94 BICETRE, FRANCE

PAYS, E.,
Cytol. & Embryol. Moléculaires,
Faculté des Sciences
Rue des Chevaux, 67,
RHODE-ST-GENESE, BELGIQUE

PERKINS, J.P.,
University North Carolina,
at Chapel Hill,
Dept. of Pharmacology,
Faculty Laboratory,
CHAPEL HILL, N.C. 27514, USA

V., POHL,
Lab. d'Histologie,
Faculté de Médecine,
Université Libre de Bruxelles,
2, rue Evers,
B-1000 BRUXELLES, BELGIQUE

PROCHIANTZ, A.,
Laboratoire de Neurobiologie,
Coll. de France, INSERM U-114
13, pl. Marcelin Berthelot,
PARIS 5è FRANCE

PUTA, Ch.,
Dept. of Biochemistry,
Royal Holloway College,
Egham Hill,
EGHAM, SURREY TW20 0EX,
ENGLAND

RANDLE, P.J.,
Dept. of Clinical Biochemistry,
University of Oxford,
OXFORD, ENGLAND

RATAJCZAK, T.,
Division of Endocrinology,
P.O.Box 134,
SUBIACO 6008,
WESTERN AUSTRALIA

REICH, E.,
Friedrich-Miescher-Institut,
Postfach 273,
CH-4002 BASEL, SWITZERLAND

RICKETTS, T.,
PFIZER Central Research,
SANDWICH, KENT CT13 9NJ, ENGLAND

RIZACK, M.A.,
The Rockefeller University,
1230 York Avenue,
NEW YORK, N.Y., 10021, U.S.A.

ROGER, P.,
I.R.I.B.H.N.,
School of Medicine,
University of Brussels,
Campus Hôpital Erasme,
808 route de Lennik,
B-1070 BRUSSELS, BELGIUM

ROZENGURT, E.,
Imperial Cancer Research
Fund Laboratories,
P.O.Box 123,
Lincoln's Inn Fields,
LONDON WC2A 3PX ENGLAND

SAEZ, J.M.,
U. Rech. Endocriniennes,
Hôpital Debrousse
INSERM U-34
29, Chemin Soeur Bouvier,
69322 LYON CEDEX FRANCE

SCHELL, J.,
Max-Planck Institut
für Züchtungsforschung,
Egelsplad,
5000 KOLN 30 R.F.A.

SCHNACKERZ, K.D.,
University of Wurzburg,
Physiological Chemistry,
Koellikerstr. 2
D-8700 WURZBURG, GERMANY, F.R.G.

SCHULTZ, J.E.,
Pharmazeutisches Institut
der Universität,
Auf der Morgenstelle, 8,
D-7400 TUBINGEN, F.R.G.

SCHULTZ, G.,
Pharmakologisches Inst.
der Universität Heidelberg,
Im Neuenheimer Feld, 366,
6900 HEIDELBERG 1, GERMANY

SENSENBRENNER, M.,
Centre de Neurochimie du CNRS,
5 rue Blaise Pascal,
67034 STRASBOURG CEDEX, FRANCE

SIMANTOV, R.,
Weizmann Inst. of Science,
Rehovot 76100,
ISRAEL P.O.B. 26

STRALFORS, P.,
Dept. Physiol. Chemistry,
University of Lund,
P.O.Box 750,
S-220 07 LUND 7 SWEDEN

VANDENHEEDE, J.R.,
Faculteit Geneeskunde,
Dept. Humane Biologie,
Afdeling Biochemie,
K.U.L.,
Herestraat, 49,
3000 LEUVEN

VAN DER MOLEN, H.J.,
Afdeling Biochemie,
Erasmus Universiteit,
P.O.Box 1738,
ROTTERDAM, THE NETHERLANDS

VAN HEUVERSWYN, B.,
I.R.I.B.H.N.,
Faculty of Medicine,
University of Brussels,
Campus Hôpital Erasme,
808, route de Lennik,
B-1070 BRUSSELS, BELGIUM

VAN SANDE, J.,
I.R.I.B.H.N.,
Faculty of Medicine,
University of Brussels,
Campus Hôpital Erasme,
808, route de Lennik,
B-1070 BRUSSELS, BELGIUM

VISSER, T.J.,
Afdeling Biochemie,
Erasmus Universiteit,
P.O.Box 1738,
ROTTERDAM, THE NETHERLANDS

WESTERMARK, B.,
The Wallenberg Laboratory,
Upsala University,
UPSALA, SWEDEN

WILMES, G.
I.R.I.B.H.N.,
Faculty of Medicine,
University of Brussels,
Campus Hôpital Erasme,
808 route de Lennik,
B-1070 BRUSSELS, BELGIUM

YANG, S.D.,
K.U.L. Leuven,
Human Biology,
Biochemistry Division,
Herestraat,
Campus Gasthuisberg,
3000 LEUVEN, BELGIUM

ZONEFRATI, R.,
Metabolic Unit,
Clinica Medica III,
Viale Morgagni, 85,
50134 FIRENZE, ITALY

CONTROL OF GROWTH AND CELL PROLIFERATION

CONTRÔLE DE LA CROISSANCE ET DE LA PROLIFÉRATION CELLULAIRES

STRUCTURE AND FUNCTION OF INSULIN-LIKE GROWTH FACTORS

RENE E. HUMBEL

Biochemisches Institut, Universität Zürich, Winterthurerstr. 190, CH-8057 Zürich (Switzerland)

INTRODUCTION

The term insulin-like growth factor (IGF) was introduced when it was realized that two polypeptides with growth promoting properties isolated from human serum had amino acid sequences homologous to insulin (1, 2, 3). Recently, chemical identity between IGF I and somatomedin C (SmC) has been established (4) so that either term can be used. Somatomedins have been defined as growth factors under control of and mediating the effects of growth hormone (5). IGF I fulfills all criteria of a somatomedin, whereas the physiologic function of IGF II is far from being clear.

A related growth factor described originally as multiplication stimulating activity (MSA) (6) has been isolated from conditioned medium of a rat liver tumor line and shown to be homologous to human IGF II (7). This MSA is now called rat IGF II.

STRUCTURE OF IGF

Chemically, IGF I and II have very similar amino acid sequences which in turn are similar to the one of proinsulin (3). These similarities suggest that the three peptides are homologous, i.e. that the IGF's and proinsulin have diverged from a common ancestor by gene duplication (3).

Although the three-dimensional structure of the IGF's has not been determined directly by x-ray crystallography, extensive conservation of the hydrophobic core and of all three disulphide bridges present in insulin has allowed to build models of IGF (8). These models of the three-dimensional structures of IGF I and II support the evolutionary relationship to proinsulin and are furthermore compatible with the known data of the interaction of IGF with antibodies against and receptors for insulin and IGF's (9).

IGF RECEPTORS

Earlier studies on the insulin-like and growth promoting activities of IGF had suggested that a weak crossreaction of IGF with the insulin receptor is responsible for the insulin-like effects such as stimulation of glucose transport or glycogen deposition, whereas direct interaction of IGF with its own receptor(s) is responsible for its growth promoting effects (10). Conversely, the growth promoting effects of large doses of insulin have been explained by a weak interaction with the IGF receptor(s). Recent investigations on insulin and IGF receptors have largely confirmed these notions. Briefly, three types of receptors have been characterized: the insulin receptor with a heterotetrameric structure, the IGF receptor type I with a structure very similar to the insulin receptor, and the monomeric IGF receptor type II which apparently bears no resemblance to the insulin receptor (11). Insulin crossreacts with the type I, but not with the type II receptor. IGF I has a high affinity to the type I receptor and crossreacts with the insulin receptor. IGF II reacts preferentially with the type II receptor (12).

However, several data are not explained by this simple model: Evidence for insulin-like effects in muscle due to IGF receptor occupancy (13) and in adipose tissue for IGF receptor-insulin receptor interactions (12, 14) point to a situation which might well be more complex than anticipated.

PHYSIOLOGIC FUNCTION

Most of the data so far obtained are in accord with the notion that IGF I is indeed the long-sought somatomedin. Specific radioimmunoassays for IGF I have shown that serum levels in man and some experimental animals are growth hormone dependent, IGF I being low in hypopituitary dwarfs and high in acromegaly or after GH injections (15, 16). Continuous infusion of IGF I into hypophysectomized rats leads to widening of the tibial epiphyseal growth plate (17). Furthermore, IGF I acts as a feedback inhibitor of growth hormone release at the hypothalamic and/or pituitary levels (18, 19).

IGF II is under less strict control of growth hormone (16), is much less active on growth in vivo (E. Schoenle, private communication) and is less active as feedback control on pituitary GH secretion (19). There is evidence that IGF II might fulfill the role of a fetal somatomedin, at least in rodents (20). However, in man, high levels of IGF II persist into adulthood and it is at present difficult to assign it a plausible physiologic role. The recently demonstrated occurrence of IGF II in granular form in the brain at widely varying concentrations in different brain regions (21) may point to a regulatory role in the central nervous system.

CONTRIBUTION OF IGF TO THE GROWTH PROMOTING EFFECT OF SERUM

The fact that supraphysiologic concentrations of insulin promote cell growth in serum-free media has led Temin (23) to postulate that serum contains an insulin-like growth factor which was later identified as MSA (6), somatomedin C (24) and IGF (25), respectively. IGF has been found to stimulate DNA synthesis with or without subsequent mitosis in cells such as fibroblasts, chondrocytes, myoblasts, mesenchymal cells and lens epithelium (26). However, in most cases, the effect of IGF in serum-free media is inferior to the maximal mitogenic effect of serum (27). Obviously, most cells that are arrested in G_0/G_1 by serum deprivation need in addition to IGF some other growth factor(s) to traverse through G_1 into S. The minimal requirement for many cells in serum-free media to enter into S are IGF and PDGF and/or EGF. In BALB c/3T3 cells, the progression of PDGF treated cells into S phase was found to be proportional to the amount of IGF I added (28). It has been postulated that progression through G_1 is dependent on two growth factors, a "competence" factor such as PDGF and a "progression" factor such as IGF I (28). In another study, EGF has been shown to substitute for PDGF and that IGF I is required for traverse of the last 6 hours of G_1 (29).

Further experiments are certainly needed before this apparently simple interplay between two growth factors can be validated and generalized. Much progress can also be expected from ongoing experiments employing recombinant DNA technology. The structure of the genes for PDGF and IGF may well shed light not only on the mechanisms of physiological growth control, but also on mechanisms of transformation.

ACKNOWLEDGEMENTS

The author wishes to acknowledge the help of his present collaborators Gisela Haselbacher, Annemarie Honegger, Urs Läubli and Peter Zumstein and the secretarial help of Anna-Maria Mosca. The author's own work was funded by the Swiss National Science Foundation (3.719.80).

REFERENCES

1. Rinderknecht, E. and Humbel, R.E. (1976) Proc. Natl. Acad. Sci. U.S.A., 73, 4379.

2. Rinderknecht, E. and Humbel, R.E. (1978) J. Biol. Chem., 253, 2769.

3. Rinderknecht, E. and Humbel, R.E. (1978) FEBS Letters, 89, 283.

4. Klapper, D.G., Svoboda, M.E. and Van Wyk, J.J. (1983) Endocrinology, 112, 2215.

5. Daughaday, W.H., Hall, K., Raben, M.S., Salmon, W.D., Jr., Van den Brande, J.L. and Van Wyk, J.J. (1972) Nature, 235, 107.

6. Pierson, R.W., Jr. and Temin, H.M. (1972) J. Cell Physiol., 79, 319.

7. Marquardt, H., Todaro, G.J., Henderson, L.E. and Oroszlan, S. (1981) J. Biol. Chem., 256, 6859.

8. Blundell, T.L., Bedarkar, S., Rinderknecht, E. and Humbel, R.E. (1978) Proc. Natl. Acad. Sci. U.S.A., 75, 180.

9. Blundell, T.L., Bedarkar, S. and Humbel, R.E. (1983) Fed. Proc., 42, 2592.

10. Zapf, J., Froesch, E.R. and Humbel, R.E. (1981) Curr. Top. Cell. Reg., 19, 257.

11. Massague, J. and Czech, M.P. (1982) J. Biol. Chem., 257, 5038.

12. Czech, M.P. (1982) Cell. 31,8.

13. Meuli, C. and Froesch, E.R. (1976) Arch. Biochem. Biophys., 177, 31.

14. Schoenle, E., Zapf, J. and Froesch, E.R. (1976) FEBS Letters, 67, 175.

15. Clemmons, D.R., Van Wyk, J.J., Ridgway, E.C., Kliman, B., Kjellberg, R.N. and Underwood, L.E. (1979) New Engl. J. Med., 301, 1138.

16. Zapf, J., Walter, H. and Froesch, E.R. (1981) J. Clin. Invest., 68, 1321.

17. Schoenle, E., Zapf, J., Humbel, R.E. and Froesch, E.R. (1982) Nature, 296, 252.

18. Berelowitz, M., Szabo, M., Frohmann, L.A., Firestone, S., Chu, L. and Hintz, R.L. (1981) Science, 212, 1279.

19. Brazeau, P., Guillemin, R., Ling, N., Van Wyk, J.J. and Humbel, R.E. (1982) CR Acad. Sci. Paris t 295, Série III, 651.

20. Adams, S.O., Nissley, S.P., Handwerger, S., and Rechler, M.M. (1983) Nature, 302, 150.

21. Haselbacher, G. and Humbel, R. (1983) in preparation.

22. Gey, G.O. and Thalhimer, W. (1924) J. Am. Med. Assoc., 82, 1609.

23. Temin, H.M. (1967) in: V. Defendi and M. Stoker (Eds.), Wistar Inst. Symp., Monogr. No. 7, Wistar Inst. Press, Philadelphia, pp. 103-114.

24. Van Wyk, J.J., Underwood, L.E., Baseman, J.B., Hintz, R.L. Clemmons, D.R. and Marshall, R.N. (1975) Adv. Metab. Disord., 8, 127.

25. Rinderknecht, E. and Humbel, R.E. (1976) Proc. Natl. Acad. Sci. U.S.A., 73, 2365.

26. Van Wyk, J.J. and Underwood, L.E. (1978) Biochemical Actions of Hormones, 5, 101.

27. Morell, B. and Froesch, E.R. (1973) Eur. J. Clin. Invest., 3, 119.

28. Stiles, C.D., Capone, G.T., Scher, C.D., Antoniades, H.N., Van Wyk, J.J. and Pledger, W.J. (1979) Proc. Natl. Acad. Sci. U.S.A., 76, 1279.

29. Leof, E.G., Wharton, W., Van Wyk, J.J. and Pledger, W.J. (1982) Exp. Cell Res., 141, 107.

Résumé

Des facteurs de croissance analogues à l'insuline (IGF) sont des polypeptides stimulant la croissance. Ils peuvent être isolés à partir de sérum. Leurs structures primaires sont homologues à celles de la pro-insuline. IGF_1 est identique à la somatomédine C, MSA chez le rat est homologue au facteur IGF II humain. Ces IGF agissent via leurs propres récepteurs comme des facteurs de croissance. L'insuline, utilisée à des concentrations supraphysiologiques interagit avec le récepteur de IGF_I. IGF_I est une vraie somatomédine c'est-à-dire il est sous contrôle de l'hormone de croissance dont il médie les effets sur la croissance. Le rôle physiologique de l'IGF_{II} est moins clair : chez les rongeurs on a trouvé des arguments en faveur d'un rôle en tant que somatomédine foetale, et la présence d'IGF_{II} en quantité variables dans différentes régions du cerveau suggère, pour ce facteur, un rôle dans le système nerveux central. IGF agit probablement sur le passage de G1 en phase S en coordination avec d'autres facteurs de croissance tels que PDGF et EGF.

PLATELET-DERIVED GROWTH FACTOR

BENGT WESTERMARK[1], ÅKE WASTESON[2] AND CARL-HENRIK HELDIN[2]
[1]Department of Pathology and [2]Institute of Medical and Physiological Chemistry, University of Uppsala, Uppsala, Sweden

INTRODUCTION

The first observation that platelets harbor a growth factor was made by Balk (1), who noted that chicken fibroblasts at low calcium concentration were incapable of growing in medium supplemented with plasma. Balk's finding was later confirmed in other laboratories including our own (2-4). After the initial observations, several laboratories have been involved in the purification of the platelet mitogen, termed platelet-derived growth factor (PDGF) (see refs.5,6 for recent reviews). The physiological function of PDGF is still not clear. Its localization in the platelet α-granule has led to the suggestion that it may act as a wound hormone, stimulating migration, and matrix production of cells at the site of an injury. PDGF has also been given a role in certain pathological conditions, e.g. atherosclerosis (7,8). In this brief review, we will highlight some of the important features of PDGF concerning its structure, mechanism of action, relation to oncogenes and its possible implication in the growth of human glioma and sarcoma.

PURIFICATION AND STRUCTURE OF PDGF

PDGF is present in platelets in minute amounts; in fact, it has been calculated that each platelet contains only less than a thousand PDGF molecules (9). Hence, the major difficulty in the purification of PDGF is to find enough starting material. Through the courtesy of the Finnish Red Cross in Helsinki and the Dept. of Virology, University of Uppsala, we have been able to collect large quantities of fresh platelet pellets. After freeze-thawing of the pellets, PDGF is purified from the resulting lysate by a four step procedure, utilizing the physicochemical properties of PDGF (basic pI, hydrophobicity and molecular weight). The purification protocol involves ion exchange chromatography on CM-Sephadex, chromatography on Blue Sepharose, gel chromatography on Bio-Gel P 150 followed by high performance liquid chromatography on a hydrophobic matrix (RP-8) (10-12).

Gel electrophoresis of PDGF in SDS under non-reducing conditions gives rise to a broad band in the molecular weight range of 28,000-33,000 (12). Mitogenic activity can be recovered from this entire region of the gel. The heterogeneity in molecular weight is probably a result of a limited proteolytic cleveage of

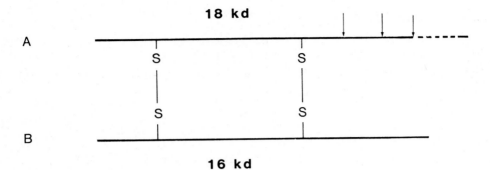

Fig. 1 Tentative molecular model of PDGF. The molecule consists of two polypeptide chains, denoted A and B, linked by disulphide bonds. Note, however, that the exact number of intra and inter molecular bonds is not known. After reduction and alkylation the two chains can be separated on HPLC. On SDS gels, the B chain migrates as a single 16 kDa species whereas the A chain appears as multiple forms, possibly due to a limited proteolysis within the platelet or during preparation.

the factor. Available data suggests that PDGF consists of two polypeptide chains, designated A and B, which are linked by disulphide bonds (Fig. 1). The biological activity of PDGF seems to require the two chain structure; after reduction and alkylation PDGF loses its mitogenic activity.

PDGF AS A MITOGEN FOR CULTURED CELLS

PDGF is a potent mitogen in various cell culture systems. However, PDGF seems to have a remarkably narrow range of tissue specificity since only mesenchymal and glial cells have been shown to be responsive. Various epithelial, hematopoietic and endothelial cells do not respond to PDGF with an increased growth rate. Responsiveness to PDGF in terms of cellular growth correlates well with the presence of PDGF receptors (see below). A maximal growth response is generally elicited by 1-5 ng of pure PDGF per ml, independent of the cell type. In cultures of human foreskin fibroblasts, PDGF is about as potent as EGF both on a molar basis and in terms of maximal growth response (per cent ^{3}H-thymidine labeled cells at the optimal concentration of the growth factor) [13].

In cultures of normal human glial cells [14] or fibroblasts [13], PDGF is mitogenic in a completely defined medium (MCDB 105) with no other additive than PDGF. In this respect, these cells differ from mouse 3T3 cells which, in addition to PDGF, require either plasma or the sequential addition of EGF and somatomedin C (15,16). The nature of this difference on PDGF responsiveness has not been elucidated.

THE PDGF RECEPTOR

Specific PDGF receptors have been demonstrated on responsive cells such as dermal fibroblasts, normal and neoplastic glial cells, vascular smooth muscle cells and 3T3 cells (17-19). Various non-responsive cells do not display any specific PDGF-binding. At $37^{\circ}C$, bound ^{125}I-PDGF is metabolically processed by the cell. Recently obtained EM autoradiographical data suggests that bound PDGF is internalized by coated pits/coated vesicles which subsequently fuse with lysosomes whereupon PDGF is degraded (20). Degradation is blocked various inhibitors of lysosomal activity such as chloroquine, methylamine and ammonium ions (21). Binding and internalization of PDGF is accompanied by a decrease in binding capacity, i.e. down regulation of the receptor occurs. The restoration of PDGF binding requires **de novo** protein synthesis. This may indicate that the receptor is degraded together with PDGF after internalization and not recycled to any significant extent.

Binding of PDGF to its receptor on fibroblast membranes leads to a rapid phosphorylation of two membrane proteins of molecular weights 185,000 and 130,000, respectively (22,23). Available data strongly suggest that the 185,000 dalton component is the PDGF receptor itself whereas the 130,000 dalton species is a proteolytic cleveage product of the 185,000 dalton component (24). The PDGF receptor appears to have at least two functional domains: an extracellular PDGF binding site, and an intracellular kinase activity (Fig. 2). Further, the exterior part of the receptor is glycosylated; certain lectins such as wheat germ agglutinin inhibit the binding of ^{125}I-PDGF to the receptor. The interior part of the receptor also contains as phosphoacceptor site which becomes phosphorylated on tyrosine residues following PDGF stimulation. The enzymatic activity of the PDGF receptor is thus analogous to that of the EGF (25) and insulin (26) receptors, both of which have have an inherent tyrosine-specific kinase activity which is stimulated by binding of the specific ligand.

A search for secondary substrates for the PDGF receptor kinase was started (27) and is currently carried out in several laboratories with the object to elucidate the post-receptor pathway that transmits the growth stimulatory signal from the receptor to the cell interior.

Phosphorylation on tyrosine residues is a rare event in normal cells. However, it has recently been found that a number of retroviral oncogenes encode tyrosine specific protein kinases, the expression of which is required for the maintenanace of the transformed phenotype (28). The best known of these transforming kinases is pp60^{v-src} the product of the transforming gene of Rous

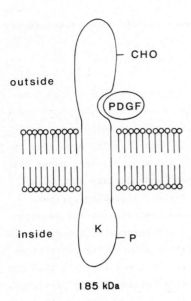

Fig. 2 Tentative model of the PDGF receptor. See text for explanation.

sarcoma virus (29). Nucleotide sequence analysis of molecular clones of retroviral kinase oncogenes has revealed a marked structural homology, suggesting a common evolutionary origin (30). Future work will show if the kinase domains of the growth factor receptors are similarly homologous with each other and with the oncogene products.

STRUCTURAL HOMOLOGY BETWEEN PDGF AND $p28^{sis}$

A partial amino acid sequence for PDGF has recently been obtained in two independent laboratories (31,32). In the study of Waterfield at al. (31) it was found that a stretch of 104 contiguous amino acids was shared by PDGF and the amino acid sequence of the transforming protein $p28^{sis}$ of simian sarcoma virus, the latter sequence predicted from the nucleotide sequence of the v-**sis** region of the SSV genome (33). Essentially the same data was reported by Doolittle et al. (32) Moreover, sequence data on isolated PDGF A and B chains revealed a considerable homology (31). One interpretation of this finding is that the two chains are encoded by gene sequences of a common ancestral origin.

It is well established that RNA tumor viruses have evolved by a recombination by retrovirus particles and cellular genes. The expression of the latter gene sequences (v-**onc**) is required for the maintenance of the transformed phenotype. Structural analysis of the viral oncogenes show that they are not isogenic to their cellular ancestors (c-**onc** or proto-**onc** genes) (see ref. 34 for a recent review). Rather, v-**onc** genes have undergone several

structural changes compared to their cellular homologues, including point
mutations, deletions and recombinations. Considering the two chain structure of
PDGF (Fig. 1) and the partial amino acid sequence of the two chains in
relation to $p28^{sis}$ amino acid sequence, one may assume that simian sarcoma
virus has aquired sequences for only one of the two chains (31). Alternative
interpretations are, however, not to be excluded.

The structural homology between PDGF and $p28^{sis}$ suggests a similar mechanism
of action, i.e. that the transforming activity of $p28^{sis}$ is exerted via the
PDGF receptor. However, one might argue that this may not be so if only one of
the two PDGF chains is expressed in SSV transformed cells, since the mitogenic
activity of PDGF requires the two chain structure of the growth factor.
Clearly, more information on the structural and functional properties of PDGF
in relation to $p28^{sis}$ is required before this type of question can be
answered.

SYNTHESIS OF PDGF OR RELATED FACTORS BY HUMAN TUMOR CELLS

Neoplastic transformation of cultured cells is in most cases correlated to a
partial or complete loss of dependence of exogenous growth factors. Howard
Temin suggested that tumor cells may synthesize and release growth factors
which trigger the cell cycle by activating normal growth factor receptors (35)
This hypothesis of "autocrine" activation has in recent years been revived
mainly through the work of Todaro and collaborators (36). We reported several
years ago (4) that human osteosarcoma cells release a growth factor which we
termed osteosarcoma-derived growth factor (ODGF). As we made progress in the
purification of this factor we became aware of the striking resemblance of ODGF
and PDGF and it now appears that the two factors are very similar or even
identical (37). The availability of specific PDGF radioligand assays (5) has
recently made it possible to screen a number of conditioned tumor cell media
for factors related to PDGF. In addition to osteosarcoma cells we now have
evidence that human rhabdomyoma (RD) and glioma (U-343 MGa Cl2) cells release
factors which are structurally, immunologically and functionally closely
related to PDGF (38).

To date, PDGF related factor(s) have only been found in sarcomas and gliomas
and not in any carcinoma or neuroblastoma cell lines. Moreover, glioma and
sarcoma (but not carcinoma cell) lines are known to express, at a high
frequency, a 4.2-kilobase transcript which hybridizes with a v-**sis** probe (39).
Since glial cells and mesenchymal cells normally respond to PDGF, it is an
interesting aspect that abnormal expression of the PDGF gene, which may be
identical to the proto-**sis** gene, leads to an uncontrolled growth of these
cells.

ACKNOWLEDGEMENTS

This work was supported by grants from the Swedish Cancer Society (Nos. 689, 786 and 1794), the Swedish Medical Research Council (No. 4486), Konung Gustaf V:s 80-årsfond, Försöksdjursnämnden, Dept. of Agriculture and the University of Uppsala.

REFERENCES

1. Balk, S.D. (1971) Proc. Natl. Acad. Sci. USA 68, 271

2. Kohler, N. and Lipton, A. (1974) Exp. Cell Res. 87, 297

3. Ross, R., Glomset, J.A., Kariya, B., Harker, L. (1974) Proc. Natl. Acad. Sci. USA 71, 1207

4. Westermark, B. and Wasteson, Å. (1975) Adv. Metabolic Disorders, 8, 85

5. Westermark, B., Heldin, C.-H., Ek, B., Johnsson, A., Mellström, K., Nistér, M. and Wasteson, Å. (1983) in: Guroff, G. (Ed.), Growth and Maturation Factors, John Wiley and Sons, New York, pp. 73-115

6. Heldin, C.-H., Westermark, B., Mellström, K., Johnsson, A., Ek, B., Nistér, M, Betsholtz, C. Rönnstrand, L. and Wasteson, Å. (1983) Survey and Synthesis of Pathology Research 1, 153

7. Ross, R. and Glomset, J.A. (1976) N. Engl. J. Med. 295, 369

8. Ross, R. and Glomset, J.A. (1976) N. Engl. J. Med. 295, 420

9. Singh, J.P., Chainkin, M.A. and Stiles, C.D. (1982) J. Cell Biol. 95, 667

10. Heldin, C.-H., Westermark, B. and Wasteson, Å. (1979) Proc. Natl. Acad. Sci. USA 76, 3722

11. Heldin, C.-H., Westermark, B. and Wasteson, Å (1981) Biochem. J. 193, 907

12. Johnsson, A., Heldin, C.-H., Westermark, B. and Wasteson, Å. (1982) Biochem. Biophys. Res. Commun. 104, 66

13. Betsholtz, C. and Westermark, B. (1984) J. Cell. Physiol. In press

14. Heldin, C.-H., Wasteson, Å. and Westermark, B. (1980) Proc. Natl. Acad. Sci. USA 77, 6611

15. Stiles, C.D., Capone, G.T., Scher, C.D., Antoniades, H.N., Van Wyk, J.J. and Pledger, W.J. (1979) Proc. Natl. Acad. Sci., USA 76, 1279

16. Leof, E.B., Wharton, W., Van Wyk, J.J. and Pledger, W.J. (1982). Exp. Cell Res. 141, 107

17. Heldin, C.-H., Westermark, B. and Wasteson, Å. (1981) Proc. Natl. Acad. Sci. USA 78, 3664

18. Huang, J.S., Huang, S.S., Kennedy, B. and Deuel, T.F. (1982) J. Biol. Chem. 257, 8130

19. Bowen-Pope, D.F. and Ross, R. (1982) J. Biol. Chem. 257, 5161

20. Nilsson, J., Thyberg, J., Heldin, C.-H., Westermark, B. and Wasteson, Å. (1983) Proc. Natl. Acad. Sci. USA in press

21. Heldin, C.-H., Wasteson, Å. and Westermark, B. (1982) J. Biol. Chem. 257, 4216

22. Ek, B., Westermark, B., Wasteson, Å. and Heldin, C.-H. (1982) Nature 295, 419

23. Ek, B. and Heldin, C.-H. (1982) J. Biol. Chem. 257, 10486

24. Heldin, C.-H., Ek, B. and Rönnstrand, L. (1983) J. Biol. Chem. 258, 10054

25. Cohen, S., Carpenter, G. and King, L., Jr. (1980) J. Biol. Chem. 255, 4834

26. Kasuga, M., Zick, Y., Blithe, D.L., Crettaz, M. and Kahn, C.R. (1982) Nature 298, 667

27. Cooper, J.A., Bowen-Pope, D.F., Raines, E., Ross, R. and Hunter, T. (1982) Cell 30, 263

28. Hunter, T. and Sefton, B.M. (1982) In: Cohen, P. and Van Heyningen, S. (Eds.) The Molecular Aspects of Cellular Regulation Elsevier/North-Holland Biomedical Press, Amsterdam, Vol. 2., pp. 337-370

29. Erikson, R.L., Purchio, A.F., Erikson, E., Collett, M.S. and Brugge, J.S. (1980) J. Cell Biol. 87, 319

30. Wang, J.Y.J. (1983) Nature 304, 400

31. Waterfield, M.D., Scrace, G.T., Whittle, N., Stroobant, P., Johnsson, A., Wasteson, Å., Westermark, B., Heldin, C.-H., Huang, J.S. and Deuel, T.F. (1983) Nature 304, 35

32. Doolittle, R.F., Hunkapiller, M.W., Hood, L.E., Devare, S.G., Robbins, K.C., Aaronson, S.A. and Antoniades, H.N. (1983) Science 221, 275

33. Devare, S.G., Reddy, P., Law, J.D., Robbins, K. and Aaronson, S.A. (1983) Proc. Natl. Acad. Sci. USA 80, 731

34. Duesberg, P.H. (1983) Nature 304, 219

35. Temin, H.M., Pierson, R.W., Jr. and Dulak, N.C. (1972) In: Rothblat.,G.H. and Cristofalo, V. (Eds.) Growth, Nutrition and Metabolism of Cells in Culture, Academic Press, New York, pp. 50-81

36. Todaro, G.J., De Larco, J.E., Fryling, C., Johnson, P.A. and Sporn, M.B. (1981) J. Supramol. Struct. Cell. Biochem. 15, 287

37. Heldin, C.-H., Westermark, B. and Wasteson, Å. (1980) J. Cell. Physiol. 105, 235

38. Betsholtz, C., Heldin, C.-H., Nistér, M., Ek. B., Wasteson, Å. and Westermark, B. In preparation

39. Eva, A., Robbins, K.C., Andersen, P.R., Srinivasan, A., Tronick, S.R., Reddy, E.P., Ellmore, N.W., Galen, A., Lautenberger, J.A., Papas, T.S., Westin, E.H., Wong-Staal, F., Gallo, R.C. and Aaronson, S.A. Nature 295, 116

Résumé.

Le facteur de croissance dérivé des plaquettes (PDGF) est un puissant mitogène pour les cellules gliales et du mésenchyme en culture. Ce facteur a un poids moléculaire d'environ 30.000 et est composé de deux chaînes polypeptidiques liées par des ponts disulfure. Le PDGF se lie à des récepteurs spécifiques sur les cellules sensibles. Après liaison à 37°C, il est internalisé et dégradé dans les lysosomes, probablement en même temps que le récepteur. Le récepteur du PDGF a été identifié comme étant une glycoprotéine membranaire de poids moléculaire 185.000 possédant au moins deux domaines fonctionnels : un site extracellulaire de liaison au PDGF et une activité kinase intracellulaire qui catalyse la phosphorylation de résidus tyrosine dans les protéines substrats. Une séquence partielle des acides aminés du PDGF révèle une grande homologie structurelle avec la protéine transformante p28^{sis} du simian sarcoma virus. Ceci suggère que ce virus a acquis une activité transformante en acquérant les séquences du gène cellulaire PDGF. L'implication du gène PDGF dans les tumeurs malignes humaines est suggérée par le fait que les cellules en culture de sarcome et de gliome peuvent synthétiser et libérer un facteur qui est proche ou même identique au PDGF.

ROLE OF ION FLUXES AND CYCLIC NUCLEOTIDES IN SIGNALLING MITOGENESIS IN 3T3 CELLS

ENRIQUE ROZENGURT
Imperial Cancer Research Fund, P O Box 123, Lincoln's Inn Fields, London WC2A 3PX, U.K.

INTRODUCTION

The cells of many tissues and organs exist in a non-proliferating state in G_0/G_1 in which they remain viable and metabolically active. However, they retain the capacity to respond to extracellular signals such as hormones, peptide factors and antigens by increasing their rate of proliferation. The salient features of the phenomenon of cell stimulation can be studied in cell culture (1,2). Thus, normal fibroblasts in general and mouse 3T3 cells in particular (3) become quiescent in the G_0/G_1 phase of the cell cycle when they deplete the serum present in the medium of an essential growth factor(s). The arrest of growth is reversible; addition of fresh serum to such quiescent cultures stimulates reinitiation of DNA synthesis and cell division (1,2,4-6).

In recent years it has become apparent that the proliferation of normal cells can be regulated by a variety of extracellular factors (6). Thus, quiescent 3T3 cells can be stimulated to reinitiate DNA synthesis by a variety of exogenous agents including the platelet-derived growth factor (PDGF), a potent mitogen present in serum but not in plasma (7), the peptides epidermal growth factor (EGF) and insulin (6), the neurohypophyseal hormone vasopressin and its related peptides, the tumour promoting agents of the phorbol ester family and certain modulators of membrane permeability such as melittin (see section IIB), compounds that disrupt the microtubule organization (8-10), vitamin A derivatives (6,11) and agents that elevate the intracellular level of cAMP (see section III). Studies carried out with combinations of defined growth-promoting molecules have revealed an important aspect of their action: the existence of synergistic interactions (6). By virtue of synergistic effects, specific combinations of mitogenic molecules can be as effective as whole serum in eliciting a complex set of biochemical events (12) and in stimulating DNA synthesis (6). Tumour cell lines display a marked reduction in their dependence on exogenous growth factors for proliferation and produce growth factors which could contribute to their autonomous growth (see 13 for ref.). For example, a virus-transformed cell line releases into the medium a polypeptide (fibroblast-derived growth factor, FDGF) which is a potent mitogen for 3T3 cells (14) and exhibits many properties in common with PDGF (15). A link between production of certain growth factors and the expression of malignant transformation has been strikingly reinforced by the finding of a remarkable homology between the amino acid sequence of PDGF and the transforming protein of the simian sarcoma virus encoded by the cys^{28} oncogene (16,17). All these findings suggest that growth factors play a critical role in modulating normal and abnormal cell proliferation. In accord with this, our effort has been directed towards the understanding of the mechanism by which these diverse external signals modulate cell proliferation of fibroblastic cells. In particular, our attention has been focused on the initial cellular responses associated with the interaction of mitogenic factors and hormones with the cell in the expectation that the early events will provide useful clues to primary regulatory mechanisms (6,12,13).

The first step in the action of many growth factors is to bind specific receptors located at the cell surface (18). A central problem in understanding the mechanism of action of growth factors is to elucidate how, after binding to specific surface receptors, such factors trigger the generation of internal signals capable of eliciting a mitogenic response. We are currently evaluating whether changes in the opening of ion permeability pathways through the plasma membrane and/or in the intracellular concentration of cyclic nucleotides, play a role as internal signals in the regulation of the proliferative response. The purpose of this paper is to review our evidence suggesting that increases in ion fluxes and elevations in the cellular level of cAMP can synergistically signal the initiation of DNA synthesis in quiescent fibroblastic cells.

I. REGULATION OF ION FLUXES BY SERUM AND GROWTH FACTORS IN QUIESCENT CELLS

A. <u>Monovalent ion fluxes</u>. One of the earliest events elicited by addition of serum to quiescent cultures of 3T3 fibroblasts is an increase in the activity of the ouabain-sensitive Na^+/K^+ pump (19). In addition to serum (19-22), a specific group of growth-promoting agents including PDGF (20), FDGF (23), vasopressin (24-26) and phorbol esters (27,28) stimulate Na^+/K^+ pump activity in quiescent 3T3 cells. Because the activity of the Na^+/K^+ pump in intact fibroblasts is limited and regulated by the supply of Na^+ (29,30) and because serum (20,29) and growth factors stimulate the entry of Na^+ into quiescent cells (20,22,25,26,28-30), it has been suggested that growth-promoting factors stimulate the pump by increasing Na^+ entry into the cells. These findings which have been previously reviewed (see 25,30,31) lead to the hypothesis that one of the initial events that occurs in fibroblastic cells stimulated to proliferate is an increase in the rate of Na^+ influx into the cells. Indeed, recent reports further demonstrate the ubiquity of Na^+ fluxes as an early event in a variety of quiescent cultured cells stimulated to proliferate, including human fibroblasts (20,32,33,34) hamster fibroblasts (20,35), rat liver cells (36), neuroblastoma (37,38), lymphocytes (39,40), BSC-1 epithelial cells (41), rat pheochromocytoma cells (42) and glial cells (43).

The translocation of Na^+ across the plasma membrane of a variety of cell types is mediated, at least in part, by an amiloride-sensitive electroneutral, Na^+/H^+ antiport system which is driven by the Na^+ electrochemical gradient across the plasma membrane (44-45). Recently, Schuldiner and Rozengurt (46) have produced several lines of evidence indicating the presence of a functional Na^+/H^+ antiport system in substratum attached Swiss 3T3 cells: 1. The maintenance of a physiological pH requires the presence of Na^+ in the extracellular medium; 2. Addition of Na^+ to Na^+-depleted cells causes a rapid increase in intracellular pH; 3. The effect of Na^+ is dose-dependent and it can be replaced by Li^+ but not by choline Cl or KCl; 4. The Na^+-induced increase in intracellular pH is blocked by amiloride which also inhibits the entry of radiolabelled Na^+ into the cells at a comparable half-maximal concentration; 5. Finally, increased extracellular pH leads to a significant enhancement in Na^+ entry which is sufficient to stimulate the activity of the Na-K pump (46). All these observations are consistent with the presence of a functional Na^+/H^+ antiport in Swiss 3T3 cells. Recent reports from other laboratories also suggest the presence of an amiloride-sensitive Na^+/H^+ antiport in other cultured cell types (33,47-51).

Since Na^+/H^+ exchange represents a significant pathway for the entry of Na^+ into 3T3 cells (46) it would be expected that a concomitant rise in intracellular pH should occur simultaneously with the increase in Na^+ influx triggered by the addition of mitogenic agents to cultures of quiescent cells (see above). Recently this prediction has been verified experimentally: incubation of quiescent Swiss 3T3 cells in the presence of PDGF, vasopressin and insulin, a potent mitogenic combination, increases the intracellular pH by 0.16 ± 0.04 pH units as derived from the uptake of the weak acids (46). Burns and Rozengurt (Ms in preparation) also showed that mitogenic stimulation leads to cytoplasmic alkalinization using a different combination of factors; stimulation of 3T3 cells with EGF, vasopressin and insulin increased intracellular pH by 0.18 units, an effect blocked by the removal of Na^+ from the medium. Significantly, Moolenar et al. (52) have recently reported that

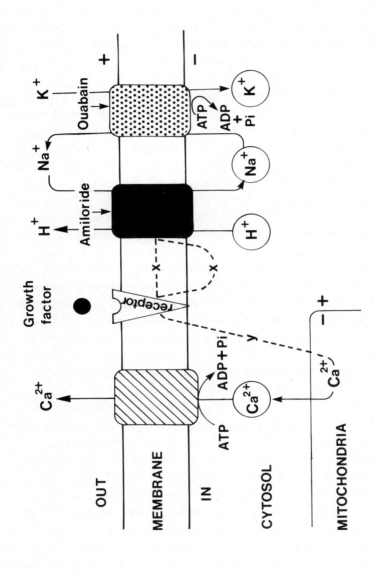

Fig. 1. Ion fluxes stimulated by growth factors in quiescent cells. The mitogenic enhancement of Na^+, K^+ and H^+ fluxes in quiescent cells can be envisaged as a "Na^+ cycle" composed by Na^+ influx via an amiloride-sensitive Na^+/H^+ antiport which modulates intracellular pH and Na^+ efflux via the ouabain-sensitive Na^+/K^+ pump which leads to K^+ accumulation. Growth factors also cause a rapid mobilization of Ca^{2+} from the mitochondria. The molecular mechanism by which the binding of the mitogenic ligand elicits the stimulation of ion fluxes remains unknown and is represented by broken lines.

addition of fresh serum to quiescent cultures of human fibroblasts causes an increase in intracellular pH by 0.2 units as revealed by using an internalized fluorescent indicator. The ability of growth factors to induce cytoplasmic alkalinization suggest that the activation of Na^+/H^+ exchange is a primary effect of the mitogens rather than a secondary mechanism for the extrusion of protons resulting from a growth factor-induced acceleration of cellular metabolism.

In summary, the stimulation by mitogens of Na^+, K^+ and H^+ fluxes in intact 3T3 cells can be envisaged as part of a "Na^+ cycle" composed of Na^+ influx via an amiloride-sensitive Na^+/H^+ antiport which results in pH modulation, and Na^+ efflux via the ouabain-sensitive Na/K pump which leads to K^+ accumulation and to the restoration of the Na^+ electrochemical gradient across the plasma membrane. (see Fig. 1.)

B. <u>Divalent cation fluxes</u>. In addition to rapid changes in monovalent ion fluxes, Lopez-Rivas and Rozengurt (53) have recently reported that addition of dialysed fetal bovine serum to quiescent cultures of Swiss 3T3 cells and other fibroblast cell lines induces the mobilization of Ca^{2+} from an intracellular store (Fig 1).This is one of the earliest events (15 sec) which takes place in quiescent fibroblasts after the stimulation with serum (53). Changes in Ca^{2+} distribution are not restricted to serum-stimulated cells since the neurohypophyseal nonapeptide vasopressin, which is a potent mitogen for Swiss 3T3 cells (24,25) stimulates the efflux of $^{45}Ca^{2+}$ from $^{45}Ca^{2+}$-loaded cells and causes a net decrease in the cellular Ca^{2+} content (Lopez-Rivas and Rozengurt, Ms in preparation). In contrast, neither insulin nor EGF affected $^{45}Ca^{2+}$ efflux from 3T3 cells. Since various inhibitors of oxidative phosphorylation prevent the stimulation of Ca^{2+} efflux by serum, it appears that mitogenic stimulation releases Ca^{2+} from a mitochondrial store (53). It is plausible that as a result of this Ca^{2+} flux there is an increase in the cytosolic concentration of this cation which, in turn, leads to the observed efflux of Ca^{2+} from the cell mediated by the plasma membrane Ca^{2+} ATPase pump (54). In view of the wide variety of cellular processes regulated by cytosolic Ca^{2+} (54), an analysis of the cause-effect relationship among Ca^{2+} movements, monovalent ion fluxes and other events associated with mitogenesis as well as further studies on the effect of other pure growth-promoting agents on the cellular distribution of Ca^{2+} are warranted.

II. ION FLUXES AND INITIATION OF DNA SYNTHESIS

Changes in ion fluxes and redistributions have been proposed to play a central role in the initiation of cell proliferation (30,31,56). The recent findings on the coupled movements of Na^+, K^+, H^+ and Ca^{2+} in mitogen-stimulated 3T3 cells described in the preceding section, provide further support for the proposition that changes in ion fluxes may play a role as internal signals in the stimulation of quiescent cells since such ionic events could have multiple and profound effects on cell metabolism and organization (30,54,55). For example, changes in intracellular pH have been demonstrated in other biological systems and in some of these the increase in intracellular pH has been implicated as a crucial part of the control mechanism (see 55 for review). However, to prove that a given early event in growth stimulation is causally related to the subsequent events leading to DNA replication is a very difficult task. The possibility that increased ionic fluxes may play a significant role in mitogenesis is supported by the following lines of evidence: a) as discussed above, considerable evidence indicates the generality of increased Na^+, K^+ and H^+ fluxes as an early event in a variety of quiescent cells stimulated to proliferate; b) the concentration of Na^+ and K^+ in the culture medium markedly influences the development of the mitogenic response; c) stimulation of Na^+ influx by hormonal peptides, tumour promoters or membrane permeability modulators initiates DNA synthesis acting synergistically with other growth-promoting factors. Points b) and c) will be the subjects of the following section.

A. <u>Inhibition of ion fluxes prevents the initiation of DNA synthesis.</u> The stimulation of DNA synthesis by serum in quiescent cultures of 3T3 cells depends on the concentration of Na^+ and K^+ in the nutrient medium. A decrease in the K^+ concentration in the medium inhibits the stimulation of DNA synthesis by serum (19). Recently, Lopez-Rivas et al. (57) found that the stimulation of DNA synthesis by insulin, EGF and vasopressin in serum-free medium is steeply dependent on the intracellular K^+ content or K^+ concentration. The relationship between these parameters is sigmoid; an increase in the intracellular K^+ above a certain threshold level (o.56 µmoles/mg protein; 90 mM) is required to sustain the proliferative response of quiescent fibroblasts to peptide growth factors (57). The effects of K^+ on the G_1-S transition are, at least in part, exerted via its control of protein synthesis (57). The findings suggest that a relatively small change in the intracellular K^+ level can influence the ability of 3T3 cells to initiate DNA synthesis in response to peptide factors in serum free medium.

Reducing external Na^+ below 100 mM inhibits serum stimulated DNA synthesis until at 20 mM it is completely blocked (29). Recently, Burns and Rozengurt (Ms in preparation) using Swiss 3T3 cells stimulated by EGF, vasopressin and insulin demonstrated that there is a striking dependence of the initiation of DNA synthesis on the Na^+ concentration of the medium. The relationship was approximately sigmoidal with a plateau at 70 mM and with half-maximal point at 35 mM Na^+. Thus, removal of Na^+ from the medium blocks the stimulation of DNA synthesis induced by peptide growth factors in serum-free medium.

However, the possibility that these treatments interfere with the proliferative response by mechanisms other than their effects on Na^+ influx is difficult to completely rule out (30,58,59). Because of this consideration, we focused our attention on the cellular effect of hormones, ionophores and toxins which promote Na^+ influx in cultured cells and investigated whether such substances can be mitogenic to quiescent cells.

B. <u>Increased ion fluxes promotes initiation of DNA synthesis</u>. If ionic events at the plasma membrane mediate the effect of growth-promoting factors, substances capable of promoting such events should be mitogenic for quiescent cells. This constitutes an important prediction of the hypothesis that ion fluxes play a signalling role on mitogenesis. However, a major problem in determining whether an ion flux modulator has mitogenic activity is that the assay for initiation of DNA synthesis requires a long (about 20-40 h) exposure of the cells to the agent, a time during which side effects of the agent can inhibit the proliferative response. An ideal substance should confine its effects to the plasma membrane, rather than modifying the permeability of other cellular membranes. Furthermore, in view of the existence of synergistic effects, it is likely that promoters of ion fluxes would induce DNA synthesis only when added with other growth promoting factors which could regulate cell metabolism in a complementary fashion acting through other mechanisms. Pursuing the possibility that an increased ionic flux could play a role in the stimulation of DNA synthesis by quiescent cells we have found that peptides which induce cation fluxes in their target cells such as vasopressin or bombesin or the membrane permeability modulators melittin and amphotericin B can act as mitogens in serum-free medium.

B1. <u>Vasopressin, phorbol esters and bombesin</u>. The addition of vasopressin to quiescent cultures of Swiss 3T3 cells rapidly stimulates ion movements across the membrane (25,26) and acts synergistically with insulin (12,24,60), EGF (19,61), FDGF (25) and PDGF (15) and cAMP-elevating agents (see next sections), to stimulate DNA synthesis in Swiss 3T3 cells. These cells exhibit a striking specificity in their response to neurohypophyseal hormones. Vasopressin is 10^2-10^3 fold more potent than oxytocin in stimulating ion fluxes and DNA synthesis in quiescent cultures of 3T3 cells (24,26,60).

Dicker and Rozengurt proposed that the mitogenic actions of vasopressin and the potent tumour promoting agents of the phorbol ester family (61-63) are mediated via a common mechanism (61). Thus, vasopressin and phorbol esters can substitute for each other in synergistically stimulating DNA synthesis when added in the presence of other growth factors but show neither synergistic nor additive effects with each other (61). Since these agents bind to different receptors (64), the convergence of their action must occur at a post-receptor step. Interestingly, phorbol esters, like vasopressin, cause an increased rate of Na^+ entry into ouabain-blocked 3T3 cells (28). The stimulation of Na^+ influx is produced by phorbol ester derivatives (28), which are potent inducers of DNA synthesis (64,65), whereas biologically inactive analogs were not effective in stimulating Na^+ flux (28).

Recently, several reports indicate that biologically active phorbol esters can substitute for diacylglycerol (66-69) in stimulating a new type of cyclic nucleotide-independent protein kinase which is activated by association with membrane phospholipids in the presence of Ca^{2+} (70-72). Similar observations were made with extracts prepared from Swiss 3T3 cells (Rodriguez-Pena, Smith and Rozengurt, unpublished results). Whether or not these observations "in vitro" are relevant to the mechanism whereby phorbol esters modulate ion fluxes and cell proliferation in the intact cells remained difficult to assess because little attention has been given to the early effects of phorbol esters on protein phosphorylation in intact, responsive cells. For these reasons, recently we focused our analysis on phorbol esters and protein phosphorylation in intact cells.

Rozengurt et al. (73) found that addition of biologically active phorbol esters to intact, quiescent 3T3 cells stimulates an extremely rapid (detectable within seconds) phosphorylation of a Mr 80,000 cellular protein (termed 80K). The possibility that this phosphorylation is related to the activation of Ca^{2+}-activated phospholipid-dependent protein kinase is suggested by the fact that phospholipid breakdown and generation of diacylglycerol induced by exogenous phospholipase C from Clostridium perfinges or by PDGF, a potent activator of endogenous phospholipase C, also causes a rapid enhancement of 80K phosphorylation (73). Prolonged pretreatment of the cells with phorbol esters, which leads to a marked decrease in the number of specific phorbol ester binding sites in Swiss 3T3 cells (64,74) prevents the phosphorylation of 80K stimulated by all these diverse agents, namely, phorbol esters, phospholipase C and PDGF. Preliminary experiments also indicate that both vasopressin and bombesin (see below) rapidly enhance the phosphorylation of 80K. These findings provide evidence that implicate the stimulation of Ca^{2+}-activated, phospholipid-dependent protein kinase in the action of phorbol esters and other growth factors in intact cells. The possibility that this protein kinase regulates the rate of ion fluxes by phosphorylating the Na^+/H^+ antiport or a regulatory subunit of this transport system is an intriguing possibility that warrants further experimental work.

Recently, we found that the regulatory tetradecapeptide bombesin is a potent mitogen for Swiss 3T3 cells (75). At low (nanomolar) concentrations, bombesin stimulates initiation of DNA synthesis in Swiss 3T3 cells maintained in serum-free medium. The mitogenic effect is dose and time dependent, specific and markedly enhanced by insulin and other growth promoting agents (75). Neither vasopressin nor phorbol esters enhance the maximal level of DNA synthesis induced by bombesin. These findings suggest that vasopressin, phorbol esters and bombesin share common pathways in their mechanism of action in Swiss 3T3 cells (75). Current experiments demonstrate that bombesin causes a potent stimulation of Na^+ entry and of Na^+/K^+ pump activity in Swiss 3T3 cells (unpublished results). All these findings provide further support to the proposition that ion fluxes may be involved in signalling the initiation of DNA synthesis induced by regulatory peptides, hormones and tumour promoters. However, these agents stimulate a complex array of cellular responses (6,61,64) some of which may be elicited by other, as yet undiscovered, signals.

B2. <u>Permeability modulators</u>. Melittin, the principal component of bee venom, is a water-soluble, strongly amphipathic, 26-amino acid polypeptide (76). At micromolar concentrations, this toxin binds to phospholipids (77) and increases permeability to ions in liposomes and artificial bilayer membranes (76). This amphipathic polypeptide provides a useful tool to perturb selectively the ion permeability of the plasma membrane (rather than intracellular membranes) and to test further whether ion fluxes may signal the initiation of mitogenesis.

Rozengurt et al. (78) found that addition of melittin to quiescent 3T3 cells stimulates Na^+ entry and causes a several-fold increase in the initial rate of ouabain-sensitive $^{86}Rb^+$ uptake. The toxin at concentrations which promote ion fluxes, stimulates DNA synthesis in serum-free medium when added in the presence of insulin, EGF or FDGF (78). The concentration-dependence of the melittin stimulation of DNA synthesis parallels that of increases of $^{86}Rb^+$ uptake (79). Interestingly, melittin did not act synergistically with either vasopressin or phorbol esters in stimulating DNA synthesis (78), further suggesting that the toxin, the hormone and the tumour promoter might act through a common mechanism.

Polyene antibiotics such as amphotericin B interact with both artificial and biological sterol-containing membranes to form channels or pores which permit the movement of monovalent cations (80-82). A subtoxic concentration of amphotericin B increases Na^+ influx and enhances the activity of the Na-K pump in 3T3 cells (25). Interestingly, the addition of amphotericin B stimulates DNA synthesis in quiescent cultures of 3T3 cells acting synergistically with insulin (31). The dose-response relationship for the stimulation of DNA synthesis was very similar to that found for the stimulation of $^{86}Rb^+$ uptake in quiescent 3T3 cells (25,31). The level of DNA synthesis induced by amphotericin B is considerably less pronounced than that induced by serum or combination of growth factors. Since the transmembrane channel formed by amphotericin B allows the inward movement of Na^+ as well as the exit of K^+ from the cell (25), it would not be expected to reproduce perfectly the selective increase in Na^+ ion permeability caused by serum, hormones and growth factors in 3T3 cells.

In summary, considerable evidence supports the hypothesis that cation fluxes may be involved in signalling mitogenesis in cultured cells. This hypothesis poses a considerable number of important questions. In particular, the critical initial steps leading to the stimulation of these ionic fluxes by growth factors and the mechanism by which this ionic signal is translated into the initiation of DNA synthesis constitute important areas of future research.

III. CYCLIC NUCLEOTIDES AND INITIATION OF DNA SYNTHESIS

The possibility that cyclic nucleotides, cyclic AMP (cAMP) and cyclic GMP, may regulate the proliferative response of quiescent fibroblastic cells has been the subject of a large and controversial literature (2,83-87). In 3T3 cells and other fibroblastic cells, increased levels of cAMP were widely thought to reduce the rate of growth and inhibit the stimulation of DNA synthesis promoted by adding serum to quiescent cells (88-91). An objection to many of these studies has been that these effects were elicited by high concentrations of analogues of cAMP and could be regarded as non-specific (83,85,87). Because these objections precluded a definitive conclusion on the effects of cAMP on the initiation of DNA synthesis of 3T3 cells we decided to evaluate further the effects of cAMP-elevating agents on the initiation of DNA synthesis of 3T3 cells. In contrast to previous reports (88-91), we found that increased cellular concentrations of cAMP act synergistically with growth-promoting agents to stimulate DNA synthesis in quiescent cultures of 3T3 cells.

A. <u>Cholera toxin stimulates initiation of DNA synthesis by 3T3 cells</u>. Recently, we found that cholera toxin, at concentrations which increase cAMP levels in intact cells, promote (rather than inhibits) initiation of DNA synthesis in serum-stimulated Swiss 3T3 cells (87,92). Furthermore, cholera toxin added with insulin, EGF, vasopressin, TPA or FDGF (92), or with the recently discovered tumour promoter teleocidin (93), synergistically stimulated DNA synthesis in cultures of Swiss 3T3 cells. Cholera toxin elevated cyclic AMP and stimulated DNA synthesis in concentration-dependent fashion. The shape of the dose-response curves for inducing DNA synthesis (in the presence of insulin) and for increasing cyclic AMP was similar (92).

If the primary effect of cholera toxin on the initiation of DNA synthesis by 3T3 cells is due to its activation of the adenylate cyclase and cellular accumulation of cyclic AMP, inhibitors of cyclic nucleotide phosphodiesterase activity should be expected to potentiate the stimulation of DNA synthesis and the cellular accumulation of cyclic AMP produced by cholera toxin. We found that addition of IBMX or Ro 20-1724, potent inhibitors of phosphodiesterase activity, markedly potentiated the stimulation of DNA synthesis and the increase in cellular cyclic AMP levels promoted by cholera toxin (92). These findings indicate that an increase in the intracellular levels of cyclic AMP acts synergistically with other mitogenic agents to stimulate DNA synthesis in Swiss 3T3 cells. This conclusion has been substantiated by further experiments showing that other agents that activate the adenylate cyclase and increase the intracellular level of cyclic AMP are also mitogenic for Swiss 3T3 cells maintained in serum-free medium.

B. <u>Enhancement of cAMP levels and stimulation of DNA synthesis by adenosine agonists, prostaglandin E_1 and forskolin</u>. Adenosine binds to cell surface receptors of a variety of cell types and either stimulates or inhibits the activity of the adenylate cyclase (94). The most potent of the adenosine analogs that binds selectively to stimulatory receptors is 5'N-ethylcarboxamideadenosine (NECA) (95). Recently we investigated whether NECA could alter cAMP levels and promote the initiation of DNA synthesis in quiescent 3T3 cells. We found that addition of NECA causes a rapid increase in the levels of cAMP, an effect markedly potentiated by Ro 20-1724, a potent and selective non-methylxanthine inhibitor of cAMP phosphodiesterase (96). Subsequently, the adenosine analog stimulated (^{3}H)-thymidine incorporation into acid-insoluble material in cultures of Swiss 3T3 cells maintained in the presence of Ro 20-1724 and insulin (96). The increase in cAMP levels and the stimulation of DNA synthesis produced by addition of NECA to Swiss 3T3 cells were completely and selectively blocked by aminophylline (96). Thus, the mitogenic activity of NECA closely parallels its ability to increase the intracellular levels of cAMP.

The prostaglandins of the E series (97,98) stimulate adenylate cyclase activity and increase the cellular levels of cAMP in a variety of cultured cells (83) including Swiss 3T3 cells (87,99). Several early reports have indicated that PGE_1, at high concentrations, inhibits the initiation of DNA synthesis in Swiss 3T3 cells (97). In view of our recent findings indicating that increased cAMP levels acts as a mitogenic stimulus for 3T3 cells, it was of interest to re-evaluate the influence of E-type prostaglandins on cAMP levels and DNA synthesis in these cells. In the presence of insulin and inhibitors of phosphodiesterase acitivty, PGE_1 induced DNA synthesis at concentrations which were orders of magnitude lower than those used in previous studies to elicit an inhibitory effect (99). We verified that PGE_1 at mitogenic concentrations, increases the intracellular level of cAMP in Swiss 3T3 cells (99). Our results indicate that the inhibitory effects produced by high concentrations of PGE_1 on the initiation of Swiss 3T3 DNA synthesis are not mediated by cAMP and should be regarded as non-specific. In contrast, the growth-promoting effects elicited by PGE_1 at low concentrations (5-50 ng/ml) provide further support for our proposal that cAMP acts as a positive signal for Swiss 3T3 cells proliferation.

Recently we found that the diterpene forskolin, which directly activates adenylate cyclase in membrane preparations and intact cells from a variety of tissues (100,101), causes a marked increase in cAMP levels and stimulates DNA synthesis acting synergistically with insulin in quiescent cultures of 3T3 cells (unpublished results). These cellular effects of adenosine agonists, prostaglandin E_1 and forskolin, clearly support the proposition that cAMP acts as a positive signal for initiation of DNA synthesis in Swiss 3T3 cells.

C. <u>Effect of cyclic nucleotide derivatives on the initiation of DNA synthesis in 3T3 cells</u>. The preceding findings prompted us to examine the effect of exogenously added cAMP and cGMP derivatives on the initiation of DNA synthesis by Swiss 3T3 cells. We found that the cAMP derivatives butcAMP or 8BrcAMP stimulate DNA synthesis when added with insulin, phorbol esters, vasopressin, EGF or fetal bovine serum (102,103). The mitogenic effect of these agents is specific because 8Br5'AMP, 5'AMP or butyrate, fail to stimulate DNA synthesis and because their mitogenic effects were markedly potentiated by the inhibitors of cyclic nucleotide phosphodiesterase activity (102,103). Furthermore, 8BrcGMP fails to initiate DNA synthesis in 3T3 cells under conditions in which 8BrcAMP is effective. These findings with exogenously added cyclic nucleotide derivatives support the proposition that an increase in the cellular levels of cAMP (but not cGMP) acts as a mitogenic stimulus for confluent and quiescent Swiss 3T3 cells.

D. <u>Time-courses of the stimulation of DNA synthesis produced by cAMP analogues and insulin</u>. In certain cell types, cyclic AMP promotes initiation of DNA synthesis at the G_1/S boundary of the cell cycle (104,105). In contrast, stimulation of DNA synthesis by cyclic AMP derivatives and insulin in 3T3 cells occurs after a lag period of 17 h, which is similar to the lag period observed in serum-stimulated 3T3 cells (102). Further, the same lag (17 h) was found when the cyclic AMP elevating agents or insulin were added to cells pre-incubated for 24 h with the hormone or the cyclic nucleotide respectively (102). These findings indicate that neither the cAMP analogues nor the insulin (added individually) stimulate 3T3 cells resting in the G_0 phase of the cell cycle to exit from this state and to become arrested in a subsequent point in G_1 which is positioned closer to S. The results suggest that insulin and cAMP-elevating compounds must act simultaneously in G_0 or early G_1 to stimulate initiation of DNA synthesis in Swiss 3T3 cells.

E. <u>The role of cAMP in the action of growth factors</u>. Since the data described above indicate that a sustained increase in the cellular level of cAMP constitutes one of the growth-promoting signals for Swiss 3T3 cells, it was important to evaluate whether physiological growth factors added to serum-free medium could alter cAMP metabolism in 3T3 cells.

In the course of studies designed to determine the effect of various mitogenic factors on the cellular level of cAMP, we found that addition of PDGF one of the most potent mitogens for untransformed fibroblastic cells (7), induces a striking accumulation of cAMP in confluent and quiescent cultures of 3T3 cells incubated in the presence of inhibitors of cyclic nucleotide degradation (106). The effect was dose and time-dependent and revealed using purified preparations of PDGF isolated either from porcine or human platelets. In contrast, other growth-promoting factors including EGF, vasopressin or insulin failed to increase the level of cAMP.

Several lines of evidence indicate that the accumulation of cAMP elicited by PDGF is mediated by increased synthesis of E-type prostaglandins, which in turn stimulate cAMP production by 3T3 cells. Thus, the enhancement of cAMP levels promoted by PDGF in the presence of phosphodiesterase inhibitors was completely blocked by indomethacin. The ability of this agent to block PDGF-induced accumulation of cAMP diminished with its time of addition after PDGF, with only a slight effect when added 120 min after PDGF (106). These results suggested that a stable fatty acid cyclooxygenase product formed largely during the initial 60 min of exposure to PDGF is responsible for cAMP production. This was confirmed directly; addition of PDGF to cultures of Swiss 3T3 cells stimulated a striking increase in the production of E-type prostaglandins which reached a concentration in the medium of 26 ng/ml 1 h after treatment with the growth factor. This concentration of PGE produced a similar level of cAMP to that found with PDGF, suggesting that the PDGF-induced increase in cAMP is mediated by E-type prostaglandins released in the culture medium (106). These findings suggest that cAMP may be one of the signals utilized by PDGF to stimulate initiation of cell proliferation in Swiss 3T3 cells.

F. <u>cAMP and ion fluxes</u>. Which, if any, is the interrelationship between these signals? To answer this question we have studied whether cAMP-elevating agents can modulate monovalent ion fluxes. Paris and Rozengurt (107) found that an increased cellular level of cAMP stimulates the Na-K pump mediated uptake of $^{86}Rb^+$ into Swiss 3T3 cells. The increase in Na-K pump activity occurs whether cAMP was generated endogenously by stimulation of the adenylate cyclase activity by cholera toxin, NECA, or PGE_1 or added exogenously as 8BrcAMP. The stimulatory effect of these compounds on $^{86}Rb^+$ uptake was potentiated by the inhibitors of cyclic nucleotide phosphodiesterase activity (107).

Several lines of evidence suggest that the mechanism by which cAMP regulates Na-K pump activity is fundamentally different from that of other mitogenic agents. In contrast to the rapid stimulation of the Na-K pump caused by addition of Na^+ flux modulators (serum, PDGF, vasopressin, phorbol esters, melittin), the stimulation of ouabain-sensitive $^{86}Rb^+$ uptake by cAMP-elevating agents reached a maximal effect after hours of incubation. Further, increased cAMP failed to augment Na^+ influx into 3T3 cells whereas under identical experimental conditions, serum markedly increased Na^+ entry into 3T3 cells (103,107). Further evidence suggesting that cAMP and Na^+ fluxes regulate the Na-K pump by independent mechanisms is furnished by the fact that serum reduces the level of cAMP (84) and phorbol esters, vasopressin and melittin did not affect the cellular content of cAMP under conditions which caused rapid stimulation of the Na-K pump (106,108). These findings indicate that the time-dependent stimulation of Na-K pump activity caused by increased cAMP levels contrasts mechanistically with the rapid control of pump activity by serum which is primarily mediated by increased Na^+ entry into the cells.

CONCLUSION

Quiescent cells resting in the G_1/G_0 phase of the cell cycle can be stimulated to reinitiate DNA synthesis by combinations of chemically diverse agents which act synergistically when added to cultures maintained in serum-free medium. The purpose of this paper was to summarize our recent evidence indicating that increases in ion fluxes and in cAMP levels can act as mitogenic signals for Swiss 3T3 cells. An attractive hypothesis is that cAMP and ion fluxes constitute separate mitogenic signals which are elicited by different sets of extracellular agents (56,103). In Swiss 3T3 cells neither of these signals appears sufficient to elicit a proliferative response in serum-free medium. However, when these signals are induced simultaneously in quiescent 3T3 cells by an appropriate combination of extracellular factors, they act synergistically to stimulate entry into DNA synthesis. Two main lines of evidence support this working hypothesis. First, addition of a combination of agents which only elevate cAMP or a combination of agents such as phorbol esters, vasopressin and melittin which stimulate ion fluxes does not lead to DNA synthesis (61,78,92), whereas an appropriate combination (e.g. cholera toxin and vasopressin) is mitogenic (61,108). Insulin, which can synergise with both cAMP-increasing agents and factors which stimulate ion fluxes cannot act identically to either group of agents. Second, the pure polypeptides PDGF and bombesin, which induce DNA synthesis without any other exogenously added growth factor, are capable of elevating cAMP (106, and unpublished results) and of stimulating ion fluxes (20, and unpublished results); these mitogenic ligands may elicit the generation of two complementary signals which synergistically lead to initiation of DNA synthesis. These synergistic effects can be further enhanced by disruption of the microtubule network (109,110). Although many aspects of this hypothesis require experimental verification and other growth-promoting signals may remain as yet undiscovered, this multiple control of the initiation of DNA synthesis in 3T3 cells by identifiable intracellular signals (i.e. ion fluxes, cAMP and cytoskeletal organisation) provides a model for understanding the organisation and strategy of the mechanisms whereby extracellular agents may regulate cell proliferation.

REFERENCES

1. Pardee, A.B., Dubrow R., Hamlin, J.L. and Kletzein, R.F. (1978) Ann. Rev. Biochem. 47, 715.
2. Rozengurt, E. (1979) in: Hynes, R. (Ed.), Surfaces of Normal and Malignant Cells, John Wiley and Sons, England, p. 323.
3. Todaro, G.J. and Green, H. (1963) J. Cell. Biol. 17, 299.
4. Holley, R.S. (1975) Nature 258, 487.
5. Todaro, G.J., Matsuya, U., Bloom, S., Robbin, A. and Green, H. (1967) in: Defendi, A. and Stoker, M. (Eds.) Growth Regulating Substances for Animal Cells in Culture, Monograph No. 7, Wistar Institute Press, Philadelphia, p. 87.
6. Rozengurt, E. (1980) Curr. Topics Cell. Regul. 17, 59.
7. Ross, R. and Vogel, A. (1978) Cell 14, 203.
8. Friedkin, M., Legg, A. and Rozengurt, E. (1979) Proc. Natl. Acad. Sci. USA 76, 3909.
9. Friedkin, M., Legg, A. and Rozengurt, E. (1980) Exp. Cell Res. 129, 23.
10. Friedkin, M. and Rozengurt, E. (1981) in:Weber, G. (Ed.) Advances in Enzyme Regulation, Pergamon Press, p. 39.
11. Dicker, P. and Rozengurt, E. (1979) Biochem. Biophys. Res. Commun. 91, 1203.
12. Rozengurt, E. (1979) Cold Spring Harbor Conferences on Cell Proliferation 6, 773.
13. Rozengurt, E. (1983) Mol. Biol. Med. 1, 169.
14. Rozengurt, E., Collins, M., Brown, K. and Pettican, P. (1982) J. Biol. Chem. 257, 3680.
15. Dicker, P., Pohjanpelto, P., Pettican, P. and Rozengurt, E. (1981) Exp. Cell Res. 135, 221.
16. Doolittle, R.F., Hunkapiller, M.W., Hood, L.E., Devare, S.G., Robbins, K.C., Aaronson, S.A. and Antoniades, H.N. (1983) Science 221, 275.
17. Waterfield, M.D., Scrace, G.T., Whittle, N., Stroobant, P., Johnsson, A., Wasteson, A., Westermark, B., Heldin, C.H., Huang, J.S. and Deuel, T.F. (1983) Nature 304, 35.
18. Rozengurt, E. and Collins, M.K.L. (1983) J. Pathol. (in press).
19. Rozengurt, E. and Heppel, L.A. (1975) Proc. Natl. Acad. Sci. USA 72, 4492.
20. Mendoza, S.A., Wigglesworth, N.M., Pohjanpelto, P. and Rozengurt, E. (1980) J. Cell. Physiol. 103, 17.
21. Tupper, J.T., Zorgniotti, F. and Mills, B. (1977) J. Cell. Physiol. 91, 429.
22. Smith, J.B. and Rozengurt, E. (1978) J. Cell. Physiol. 97, 441.
23. Bourne, H.R. and Rozengurt, E. (1976) Proc. Natl. Acad. Sci. USA 73, 4555.

24. Rozengurt, E., Legg, A. and Pettican, P. (1979) Proc. Natl. Acad. Sci. USA 76, 1284.
25. Rozengurt, E. and Mendoza, S.A. (1980) Ann. N.Y. Acad. Sci. 339, 175.
26. Mendoza, S.A., Wigglesworth, N.M. and Rozengurt, E. (1980) J. Cell. Physiol. 105, 153.
27. Moroney, J.A., Smith, A., Thoei, L.D. and Wenner, C.E. (1978) J. Cell. Physiol. 95, 287.
28. Dicker, P. and Rozengurt, E. (1981) Biochem. Biophys. Res. Commun. 100, 433.
29. Smith, J.B. and Rozengurt, E. (1978) Proc. Natl. Acad. Sci. USA 75, 5560.
30. Rozengurt, E. (1981) Adv. Enzyme Regul. 19, 61.
31. Rozengurt, E. (1981) in: Cellular Responses to Molecular Modulators, Miami Winter Symposia, 18, 149.
32. Villereal, M.L. (1981) J. Cell. Physiol. 108, 251.
33. Moolenaar, W.H., Yarden, Y., de Laat, S.W. and Schlessinger, J. (1982) J. Biol. Chem. 257, 8502.
34. Owen, N.E. and Villereal, M.L. (1983) Cell 32, 979.
35. Pouyssegur, J., Chambard, J.C., Franchi, A., Paris, S. and Van Obberghen-Schilling, E. (1982) Proc. Natl. Acad. Sci. USA 79, 3935.
36. Koch, K.S. and Leffert, H.L. (1979) Cell 18, 153
37. Moolenaar, W.H., Delaat, S.W. and van der Saag, P.T. (1979) Nature 279, 721.
38. Moolenaar, W.H., Munnery, C.L., van der Saag, P.T. and DeLaat, S.W. (1981) Cell 23, 789.
39. Segal, G.B., Simon, W. and Lichtman, M.A. (1979) J. Clin. Invest. 64, 834.
40. Felber, S.M. and Brand, M.D. (1983) Biochem. J. 210, 893.
41. Rothenberg, P., Reuss, L. and Glaser, L. (1982) Proc. Natl. Acad. Sci. USA 79, 7787.
42. Boonstra, J., Moolenaar, W.H., Harrison, P.H., Moed, P., van der Saag, P.T. and DeLaat, S.W. (1983) J. Cell Biol. 97, 92.
43. Benos, D.J. and Sapirstein, Y.S. (1983) J. Cell. Physiol. 116, 213.
44. Roos, A. and Boron, W.F. (1981) Physiol. Rev. 61, 297.
45. Boron, W.F. (1983) J. Memb. Biol. 72, 1.
46. Schuldiner, S. and Rozengurt, E. (1982) Proc. Natl. Acad. Sci. USA 79, 7778.
47. Rindler, M.J. and Saier, M.H. (1981) J. Biol. Chem. 256, 10820.
48. Moolenaar, W.H., Boonstra, J., van der Saag, P.T. and de Laat, S.W. (1981) J. Biol. Chem. 256, 12883.
49. Paris, S. and Pouyssegur, J. (1983) J. Biol. Chem. 258, 3503.
50. Frelin, C., Vigne, P. and Lazdunski, M. (1983) J. Biol. Chem. 258, 6272.
51. Rothenberg, P., Glaser, L., Schlesinger, P. and Cassel, D. (1983) J. Biol. Chem. 258, 4883.

52. Moolenaar, W.H., Tsien, R.Y., van der Saag, P.T. and de Laat, S.W. (1983) Nature 304, 645.
53. Lopez-Rivas, A. and Rozengurt, E. (1983) Biochem. Biophys. Res. Commun. 114, 240.
54. Carafoli, E. and Crompton, M. (1978) in: Current Topics in Membranes and Transport 10, 151.
55. Intracellular pH, its Measurement, Regulation and Utilization in Cellular Functions, Nuccitelli, C.R. and Deamer, D.W. (Eds.) Alan R. Liss, Inc., New York (1982).
56. Rozengurt, E. (1982) in: Boynton, A.L., McKeehan, W.L. and Whitfield, J.F. (Eds.) Ions, Cell Proliferation and Cancer, Academic Press, New York, pp. 259-281.
57. Lopez-Rivas, A., Adelberg, E. and Rozengurt, E. (1982) Proc. Natl. Acad. Sci. USA 79, 6275.
58. Balk, S.D. and Polimeni, P.I. (1982) J. Cell. Physiol. 112, 251.
59. Lubin, M. (1982) in: Boynton, A.L., McKeehan, W.L. and Whitfield, J.F. (Eds.) Ions, Cell Proliferation and Cancer, Academic Press, New York, pp. 131-150.
60. Rozengurt, E., Brown, K.D. and Pettican, P. (1981) J. Biol. Chem. 256, 716.
61. Dicker, P. and Rozengurt, E. (1980) Nature 287, 607.
62. Dicker, P. and Rozengurt, E. (1978) Nature 276, 723.
63. Dicker, P. and Rozengurt, E. (1979) J. Supramol. Struct. 11, 79.
64. Collins, M. and Rozengurt, E. (1982) J. Cell. Physiol. 112, 42.
65. Dicker, P. and Rozengurt, E. (1981) J. Cell. Physiol. 109, 99.
66. Castagna, M., Takai, Y., Kaibuchi, K., Sano, K., Kikkawa, U. and Nishizuka, Y. (1982) J. Biol. Chem. 257, 7847.
67. Niedel, J.E., Kuhn, L.J. and Vandenbark, G.R. (1983) Proc. Natl. Acad. Sci. USA 80, 36.
68. Kraft, A.S. and Anderson, W.B. (1983) Nature 301, 621.
69. Ashendel, C.L., Staller, J.M. and Boutwell, R.K. (1983) Biochem. Biophys. Res. Commun. 111, 340.
70. Takai, Y., Kishimoto, A., Iwasa, Y., Kawahara, Y., Mori, T. and Nishizuka, Y. (1979) J. Biol. Chem. 254, 3692.
71. Kishimoto, A., Takai, Y., Mori, T., Kikkawa, U. and Nishizuka, Y. (1980) J. Biol. Chem. 255, 2273.
72. Kuo, J.F., Anderson, R.G.G., Wise, B.C., Mackerlova, L., Salomonsson, I., Brackett, N.L., Katoh, N., Shoji, M. and Wrenn, R.W. (1980) Proc. Natl. Acad. Sci. USA 77, 7039.
73. Rozengurt, E., Rodriguez-Pena, M. and Smith, K. (1983) Proc. Natl. Acad. Sci. USA (in press).
74. Collins, M.K.L. and Rozengurt, E. (1984) J. Cell. Physiol. (in press).
75. Rozengurt, E. and Sinnett-Smith, J. (1983) Proc. Natl. Acad. Sci. USA 80, 2936.
76. Habermann, E. (1972) Science 177, 314.

77. Knoppel, E., Eisenberg, D. and Wickner, W. (1979) Biochem. J. 18, 4177.
78. Rozengurt, E., Gelehrter, T.D., Legg, A. and Pettican, P. (1981) Cell 23, 781.
79. Gelehrter, T.D. and Rozengurt, E. (1980) Biochem. Biophys. Res. Commun. 97, 716.
80. Kobayaski, G. and Medoff, G. (1977) Ann. Rev. Microbiol. 31, 291.
81. De Kruijff, B., Gerritsen, W.J., Oerlemans, A., Demel, R.A. and Van Deenen, L.L. (1974) Biochem. Biophys. Acta 339, 30.
82. Van Hoogevest, P. and De Kruijss, B. (1978) Biochem. Biophys. Acta 511, 397.
83. Chlapowski, F.J., Kelly, L.A. and Butcher, R.W. (1975) Adv. Cyclic Nuc. Res. 6, 245.
84. Pastan, I.H., Johnson, G.S. and Anderson, W.B. (1975) Ann Rev. Biochem. 44, 491.
85. Friedman, D.L., Johnson, R.A. and Zeilig, C.E. (1976) Adv. Cyclic Nuc. Res. 7, 69.
86. Johnson, G.S. and Pastan, I. (1972) J. Natl. Cancer Inst. 48, 1377.
87. Rozengurt, E. (1981) Adv. Cyclic Nuc. Res. 14, 429.
88. Willingham, M.C., Johnson, G.S. and Pastan, I. (1972) Biochem. Biophys. Res. Commun. 48, 743.
89. Burger, M.M., Bombik, B.M., Breckenridge, B. McL and Seppard, J.R. (1972) Nature 239, 161.
90. Bombik, B.M. and Burger, M.M. (1973) Exp. Cell Res. 80, 88.
91. Kram, R., Mamont, P. and Tomkins, G.M. (1973) Proc. Natl. Acad. Sci. USA 70, 1432.
92. Rozengurt, E., Legg, A., Strang, G. and Courtenay-Luck, N. (1981) Proc. Natl. Acad. Sci. USA 78, 4392.
93. Collins, M. and Rozengurt, E. (1982) Biochem. Biophys. Res. Commun. 104, 1159.
94. Fain, J.N. and Malbon, C.C. (1979) Mol. Cell. Biochem. 25, 143.
95. Londos, C., Cooper, D.M.F. and Wolff, J. (1980) Proc. Natl. Acad. Sci. USA 77, 2551.
96. Rozengurt, E. (1982) Exp. Cell Res. 139, 71.
97. Hammerstrom, S. (1982) Arch. Biochem. Biophys. 214, 431.
98. Samuelsson, B., Goldyne, M., Granstrom, E., Hamberg, M., Hammerstrom, S. and Malmsten, C. (1978) Ann. Rev. Biochem. 47, 997.
99. Rozengurt, E., Collins, M. and Keehan, M. (1983) J. Cell. Physiol. (in press).
100. Seamon, K.B., Padgett, W., and Daly, J.W. (1981) Proc. Natl. Acad. Sci. USA 78, 3363.
101. Seamon, K., Padgett, W. and Daly, J.W. (1981) J. Biol. Chem. 256, 9799.
102. Rozengurt, E. (1982) J. Cell. Physiol. 112, 243.
103. Rozengurt, E. and Courtenay-Luck, N. (1982) Bioscience Rep. 2, 589.
104. Whitfield, J.F., Boynton, A.L., MacManus, J.P., Siroska, M. and

Tsang, B.K. (1979) Mol. Cell. Bioche. 27, 155.
105. Boynton, A.L. and Whitfield, J.F. (1979) J. Cell Physiol. 101, 139.
106. Rozengurt, E., Stroobant, P., Waterfield, M.D., Deuel, T.D. and Keehan, M. (1983) Cell 34, 265.
107. Paris, S. and Rozengurt, E. (1982) J. Cell. Physiol. 112, 273.
108. Collins, M.K.L. and Rozengurt, E. (1983) Proc. Natl. Acad. Sci. USA 80, 1924.
109. Friedkin, M.E. and Rozengurt, E. (1981) Adv. Enzyme Reg. 19, 39.
110. Wang, Z.W. and Rozengurt, E. (1983) J. Cell Biol. 96, 1743.

Résumé

Les cellules quiescentes 3T3 au repos en phase Go du cycle cellulaire peuvent être stimulées à recommencer la synthèse du DNA par des combinaisons de divers agents chimiques qui, lorsqu'ils sont ajoutés à des cultures maintenues en milieu sans sérum, agissent en synergie. La compréhension du mécanisme(s) par lequel des agents extracellulaires interagissent pour moduler la prolifération cellulaire requiert l'identification des signaux intracellulaires importants dans l'initiation d'une réponse mitogène. Notre connaissance actuelle indique qu'une augmentation des flux ioniques faisant intervenir les ions Na^+, H^+, K^+ et Ca^{2+}, ainsi qu'une augmentation de la quantité d'AMP cyclique (produite par la cholératoxine, les agonistes de l'adénosine, la prostaglandine E_1 ou les dérivés de l'AMP cyclique) peuvent être des signaux mitogènes pour les cellules Swiss 3T3. Selon notre hypothèse, l'initiation de la synthèse de DNA pourrait être induite par l'interaction synergique de deux signaux identifiables, à savoir : une vitesse accrue des mouvements ioniques et une augmentation du taux cellulaire d'AMP cyclique.

N-GLYCOSYLATION OF NASCENT PROTEINS EARLY IN THE PREREPLICATIVE PHASE CONSTITUTES A PROCESS FOR CONTROLLING ANIMAL CELL PROLIFERATION

LUIS JIMENEZ DE ASUA, STANISLAVA POSKOCIL, M. KATHERINE FOECKING AND ANGELA M. OTTO
Friedrich Miescher-Institut, P.O. Box 2543, CH-4002 Basel, Switzerland.

INTRODUCTION

Normal animal cells in vivo as well as in vitro are able to regulate the replication of chromosomal DNA and cell division in response to changes in the extracellular environment (1-3). Cultured Swiss mouse 3T3 cells have provided a useful model system for studying the mechanisms that regulate cell proliferation (1,3). Different purified mitogens such us epidermal growth factor (EGF) (4), platelet derived growth factor (PDGF) (5,6), prostaglandin $F_{2\alpha}$ ($PGF_{2\alpha}$) (7) or vasopressin (8) when added to confluent resting Swiss 3T3 cells stimulate the initiation of DNA replication. The kinetics of this process consists of two different phenomena: 1) a constant prereplicative phase of 15 hours and 2) the initiation process, which occurs at a rate which is regulated by the concentration of mitogen present in the culture medium. This rate which follows apparent first order kinetics can be quantified by a rate constant k (3). Other hormones, which in confluent resting Swiss 3T3 cells are not mitogenic by themselves, can in combination with some of these growth factors enhance or reduce the final rate of entry into S phase (9,10). From these studies and from more recent research on the interaction of different mitogens such as EGF with $PGF_{2\alpha}$ (11), we have proposed that proliferation of animal cells is regulated by a multi-signalling system which is triggered by the interaction of a variety of mitogens at the cell surface and programmed by the genetic expression of the cell (9,12). Of central importance in the understanding of the control of normal and malignant cell proliferation is the elucidation of the cascade of events that link the signals originating at the surface membrane with the nuclear targets which ultimately are involved in initiation of DNA replication.

In recent years it has become clear that protein modifications play an important role in controlling many cellular functions, including those associated with or involved in the regulation of mammalian cell proliferation (13,14). Among these modifications, protein glycosylation plays an important role in controlling many cellular events. Glycoproteins are ubiquitous molecules with specific and important biological functions (15). In particular, there is evidence that N-glycosylation of newly synthesized proteins can play a crucial role in several cellular processes such as hormone-receptor functions (16,17), cell-cell interaction (18,19), immunology (20-22), cell differentiation (23) and embryonic development (24).

The biosynthesis of N-glycosylated proteins occurs through the synthesis of lipid-sugar intermediates (25,26). The lipid moiety is dolichol, a saturated polyprenol of 20 isoprene units, derived from mevalonic acid. A rate limiting step leading to the biosynthesis of dolichol is the formation of mevalonic acid, a reaction catalyzed by hydroxylmethylglutaryl-CoA-reductase (HMG-CoA-reductase) and taking place in the endoplasmic reticulum (27). The activity of the enzyme is regulated by hormones and by conformational and covalent modifications of the enzyme (27). Mevalonic acid is then converted to farnesylpyrophosphate, the precursor in the synthesis of dolichol as well as cholesterol and ubiquinone (coenzyme Q) (Fig. 1). Inhibiting the activity of HGM-CoA-reductase by compactin, mevinolin or analogous of cholesterol also

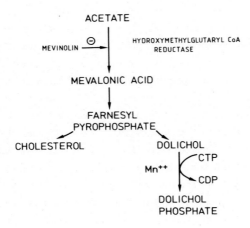

Fig. 1. Metabolic routes of mevalonic acid in animal cells.

blocks the synthesis of dolichol (28-30) and subsequently inhibits N-glycosylation of nascent polypeptides (31). Furthermore, the inhibition of mevalonic acid synthesis leads to a decrease in the levels of cholesterol and coenzyme Q, thereby affecting other cellular functions.

Lennarz and his co-workers have shown in sea urchin eggs that the phosphorylation of dolichol can also be a rate limiting step for the subsequent utilization of the dolichol as a lipid carrier molecule (32). The first step in the

N-glycosylation process is the transfer of N-acetylglucosamine from UDP-N-acetylglucosamine, resulting in dolichol-pyrophosphoryl-N-acetylglucosamine. Subsequent to this reaction, the lipid-sugar intermediate is elongated by the addition of a second N-acetylglucosamine and nine mannosyl residues, which are transferred from GDP-Mannose or from dolichol-phosphoryl-mannose (Fig. 2).

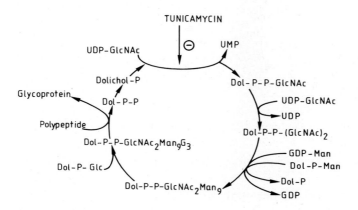

Fig. 2. The dolichol pathway for N-glycosylation. Readapted from Leloir (26).

Thereafter, three molecules of glucose are added from dolichol phosphoryl-glucose, which are again removed before the entire oligosacharide is transferred to the asparagine residue of the nascent polypeptide (Fig. 2). All these processes, which were elucidated by Leloir and his co-workers, occur in the microsomal fraction (25,26). The N-glycosylated protein is subjected to further sugar modifications, which occur in the Golgi apparatus before the N-glycosylated proteins are sorted and distributed to the extracellular matrix, to the intercellular fluid or they remain at different locations within the cell (25).

Tunicamycin is a lipophilic molecule, produced by Streptomyces lysosuperficus, containing N-acetylglucosamine, uracil, and a fatty acid (Fig. 3). It blocks the synthesis of dolichol-pyrophosphoryl-N-acetylglucosamine (33) and thereby inhibits the N-glycosylation process (15,25). These results lead to the following questions:

1) Does N-glycosylation of proteins play a role in regulating the initiation of DNA synthesis?

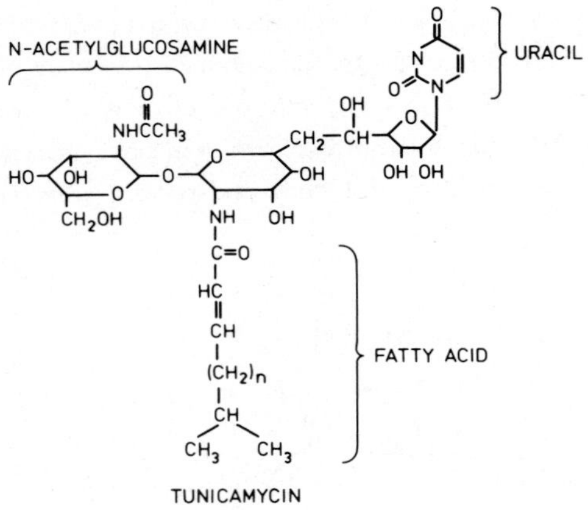

Fig. 3. Structure of tunicamycin. Readapted from Lennarz (15).

2) When during the prereplicative period does this process have to take place?
3) How do mitogens regulate N-glycosylation?
4) Are there specific N-glycosylated proteins required for the initiation of DNA synthesis, and where are they located?

Here we present evidence that in Swiss 3T3 cells mevinolin and tunicamycin inhibit the initiation of DNA synthesis stimulated by different mitogens. This indicates that events on the pathway leading to N-glycosylation of proteins are involved in regulating the rate of entry into S phase. We propose that N-glycosylated proteins play a role in controlling the rate of cell proliferation.

MATERIALS AND METHODS

Cell Cultures. Swiss mouse 3T3 cells (34) were maintained in Vogt-Dulbecco's Modified Eagle's Medium (VDMEM) containing 100 µg/ml streptomycin, 100 units/ml penicillin and supplemented with 10% fetal calf serum (FCS) as previously described (3).

Measurement of the initiation of DNA synthesis and determination of the rate constant for entry into S phase. Cells were plated at 10^5 cells per 35 mm dish in 2 ml of VDMEM supplemented with low molecular weight nutrients, 6% FCS and 1% newborn serum (VDMEMS) (3). Three days after seeding, the cells were given fresh VDMEMS and then allowed to become confluent and quiescent. They were used 3-4 days after the medium change, when no mitotic figures were observed. All additions were made directly to the conditioned medium. Cultures were radioactively labelled for autoradiography by exposing them to 1 µM [methyl^{3}H] thymidine (3 µCi/ml) from the time of the additions until the times indicated in each experiment. Pairs of cultures were then processed for autoradiography (3). The determinations of the rate constant for entry into S phase were calculated as described (3).

Measurement of [^{14}C]glucosamine and [^{3}H]mannose incorporation. Cultures were plated as described for the assay of initiation of DNA synthesis. Before labelling the cells with both radioactive substances, the medium was removed, and the cultures were washed twice with 3.0 ml of serum-free VDMEMS minus glucose pre-warmed at 37°C. Then the cultures received 2.0 ml of the same medium containing 10% dialyzed FCS or mitogens as indicated. Cultures were labelled with 1 µCi/ml of [^{14}C]glucosamine and 2.5 µCi/ml of [^{3}H]mannose from 0 to 5 hours after additions. Thereafter cells were washed, precipitated with trichloroacetic acid and prepared for scintillation counting as described before (7).

Total RNA and protein synthesis determination. The cells were plated as for the measurement of DNA synthesis. Addition of serum, mitogens and radioactive precursor were made directly to the conditioned medium. For total RNA synthesis cultures were labelled with 5.0 µCi/ml of [^{3}H]uridine for 5 hours and for protein synthesis with 2.5 µCi/ml of [^{3}H]leucine for 2 hours as indicated in each experiment. The cultures were processed as for sugar incorporation.

Materials. Prostaglandin F$_{2\alpha}$ and mevinolin were the generous gifts of Dr. John Pike, The Upjohn Company, Kalamazoo, Michigan, and of Dr. Alberts of Merck, Sharp and Dhome, Rahway, New Jersey, respectively. Radiochemicals were obtained from the Radiochemical Center, Amersham, England. Tunicamycin, insulin, N-acetylglucosamine, mannose, mevanolactone and thymidine as well as the low molecular nutrients used to supplement the culture medium were purchased from Sigma. Fetal calf serum, newborn bovine serum, and Vogt-Dulbecco's modified Eagle's medium were obtained from Gibco and Flow Laboratories.

RESULTS AND DISCUSSION

Effects of mevinolin and mevanolactone on the initiation of DNA synthesis stimulated by serum or mitogens. Addition of mevinolin at two different concentrations to cells stimulated by FCS or the mitogens EGF or PGF$_{2\alpha}$, alone or with insulin, markedly reduced the value of the labelling index (Table 1).

TABLE 1

INHIBITION OF THE INITIATION OF DNA SYNTHESIS STIMULATED BY EGF, $PGF_{2\alpha}$ OR FCS BY MEVINOLIN

Additions	Labelled Nuclei %
None	0.5
Mevinolin	0.9
FCS	95.0
FCS + mevinolin	85.0
EGF	15.0
EGF + mevinolin	3.0
EGF + insulin	40.0
EGF + insulin + mevinolin	20.0
$PGF_{2\alpha}$	18.0
$PGF_{2\alpha}$ + mevinolin	7.0
$PGF_{2\alpha}$ + insulin	47.0
$PGF_{2\alpha}$ + insulin + mevinolin	21.0

Cells were exposed to 1 µM [methyl^{3}H]thymidine (3 µCi/ml) from 0 to 28 hours after additions. Mitogens were added as follows: EGF(20 ng/ml), $PGF_{2\alpha}$ (300 ng/ml), insulin (50 ng/ml) and FCS (10% v/v). Mevinolin (20 µM) was dissolved in ethanol so that the final concentration was less than 0.01% in the culture medium.

Addition of mevinolin to cells stimulated with FCS inhibited the stimulation of DNA synthesis by about 10%, while DNA synthesis stimulated by EGF or $PGF_{2\alpha}$ was inhibited by 30-80%, the extent of inhibition being less in the presence of insulin. In order to test whether the inhibition of DNA synthesis by mevinolin is due to the inhibition of HMG-CoA-reductase or due to a non-specific effect, mevanolactone, a salt of mevalonic acid, the product of this reaction, was tested for the reversibility of this effect (Table 2). $PGF_{2\alpha}$ (300 ng/ml) added to quiescent Swiss 3T3 cells stimulated by ten-fold the incorporation of [methyl^{3}H]thymidine. Insulin, which at this concentration does not have any stimulatory effect on DNA synthesis, markedly enhanced the effect of $PGF_{2\alpha}$ (Table 2). Addition of mevinolin at 20 µM inhibited by about 65% the initiation of DNA synthesis stimulated by $PGF_{2\alpha}$ alone or with insulin (Table 2). Mevanolactone added at 20 µM or 40 µM reversed partially or completely the inhibitory effect of mevinolin. This suggests that the inhibition of DNA synthesis by mevinolin is exerted at the level of the HMG-CoA-reductase. However, we do not know yet whether the block in the initiation of DNA synthesis can also be reversed by any of the compounds derived from mevalonic acid, such as dolichol,

TABLE 2

REVERSION OF THE INHIBITORY EFFECT OF MEVINOLIN ON THE INITIATION OF DNA SYNTHESIS BY MEVANOLACTONE

Additions	[Methyl^{3}H]thymidine 10^4 cpm per 3.3×10^5 cells
None	0.37
Insulin (50 ng/ml)	0.42
PGF$_{2\alpha}$ (300 ng/ml)	3.99
+ mevinolin	1.55
+ mevinolin + mevanolactone (20 μM)	2.90
+ mevinolin + mevanolactone (40 μM)	3.95
PGF$_{2\alpha}$ + insulin	7.04
+ mevinolin	3.10
+ mevinolin + mevanolactone (20 μM)	4.10
+ mevinolin + mevanolactone (40 μM)	6.81

Mevinolin was added at a concentration of 20 μM. Cultures were exposed to 1 μM [methyl^{3}H]thymidine (3 μCi/ml) from 0 to 28 hours after additions. The radioactivity incorporated into DNA was measured as indicated in Materials and Methods.

coenzyme Q or cholesterol. It has been shown in Swiss 3T3 cells stimulated by PDGF that the inhibitory effect of compactin (an analogue of mevinolin) on DNA synthesis can not be reversed by the addition of cholesterol (35). Furthermore, in sea urchin eggs, addition of compactin causes abnormal gastrulation, and a decrease in the synthesis of dolichol-oligosaccharide. These inhibitory effects can only be reversed by dolichol or dolichol-phosphate but not by cholesterol, coenzyme Q, or both together (31). Thus, it is to be predicted that the inhibitory effect of mevinolin could be reversed by addition of dolichol or dolichol-phosphate to cells stimulated by different mitogens. The results so far suggest that the activity of HMG-CoA-reductase may play an important role in controlling biosynthetic steps involved in N-glycosylation and possibly in regulating events involved in DNA synthesis.

Tunicamycin inhibits the initiation of DNA synthesis stimulated by serum or mitogens. Tunicamycin added to confluent quiescent Swiss 3T3 cells stimulated by fetal calf serum inhibits the initiation of DNA synthesis. Addition of FCS (10% v/v) to such quiescent cultures stimulated 90% of the cells to initiate

DNA synthesis within 28 hours (Fig. 4A). Tunicamycin added from 50 to 1000 ng/ml to serum-stimulated cells reduced the value of the labelling index from 90% to about 30% at 1000 ng/ml. Added alone to quiescent cultures, tunicamycin had no apparent effect (Fig. 4A). Only at high concentrations (1 µg/ml or more) did tunicamycin produce some morphological changes characterized by the rounding of the cells.

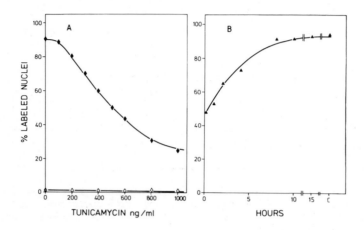

Fig. 4. Effect of tunicamycin on the initiation of DNA synthesis in confluent resting Swiss 3T3 cells stimulated by FCS. A. Tunicamycin added at different concentrations to (◇) quiescent cultures, or those stimulated by (◆) FCS (10% v/v). B. Time of addition of tunicamycin after stimulation by FCS (10% v/v). Tunicamycin was added at 500 ng/ml at the times indicated. Cultures were labelled from 0 to 28 hours after additions as indicated in Materials and Methods.

The inhibitory effect of tunicamycin depends on the time of addition after confluent quiescent Swiss 3T3 cells are stimulated by FCS. Addition of tunicamycin (500 ng/ml) together with FCS (10% v/v) markedly reduced the value of the labelling index from 96 to 50%. Adding tunicamycin at later times resulted in a less inhibitory effect, and additions at times later than 8 hours did not have any inhibitory effect on the initiation of DNA synthesis (Fig. 4B). Furthermore, addition of tunicamycin in the same concentration range to cultures stimulated by the mitogens EGF or $PGF_{2\alpha}$, with or without insulin, markedly inhibited the initiation of DNA synthesis (Jimenez de Asua, et al., manuscript in preparation).

What are the kinetic parameters of the initiation of DNA synthesis stimulated by FCS which are affected by tunicamycin? Addition of tunicamycin at 400 ng/ml or 800 ng/ml to cultures stimulated by FCS resulted in a decrease in the rate at which cells entered S phase without changing the length of prereplicative phase of 15 hours (Table 3). The decrease in the rate was dependent on the concentration of tunicamycin. With FCS alone the rate was 21.33×10^{-2}/h, while tunicamycin at 400 ng/ml or 800 ng/ml reduced the rate to 6.81 and 2.32, respectively. Later additions of tunicamycin at 9 or 15 hours did not have any effect on the rate, as would be expected from the data shown in Fig. 4B. Similar effects of tunicamycin were observed with cells stimulated by EGF or $PGF_{2\alpha}$ (Jimenez de Asua, et al., manuscript in preparation).

TABLE 3

EFFECT OF TUNICAMYCIN ON THE KINETICS OF INITIATION OF DNA SYNTHESIS IN CONFLUENT QUIESCENT SWISS 3T3 CELLS STIMULATED BY FCS

Additions	Lag Phase (hr)	Rate Constant ($\times 10^{-2}$/h)
None	–	0.06
Tunicamycin (800 ng/ml)	–	0.07
FCS	15	21.33
FCS + tunicamycin (400 ng/ml)	15	6.81
FCS + tunicamycin (800 ng/ml)	15	2.32

FCS was added at a concentration of 10% v/v. Cultures were labelled for autoradiography by exposing them to 1 μM [methyl-^{3}H]thymidine (3 μCi/ml) from the time of the additions. At intervals of 1 to 2 hours, dishes were processed as indicated in Materials and Methods. The value of the rate constant was calculated from the slope obtained between 16 and 21 hours as previously described (3).

These results strongly indicate that tunicamycin inhibits the initiation of initiation of DNA synthesis only in the early part of the prereplicative phase and suggest that N-glycosylation of proteins synthesized during this time may control subsequent biochemical events involved in the initiation of DNA replication.

Tunicamycin decreases the incorporation of [^{14}C]glucosamine and [^{3}H]mannose into acid precipitable material. Stimulation of confluent quiescent Swiss 3T3 cells by dialyzed FCS (10% v/v) resulted in a marked increase in the incorporation of [^{14}C]glucosamine and [^{3}H]mannose into acid precipitable material (Fig. 5). Addition of tunicamycin in a range from 100 to 1000 ng/ml decreased

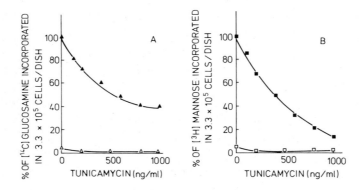

Fig. 5. Inhibition by tunicamycin of the incorporation of [^{14}C]glucosamine (A) and [^{3}H]mannose (B) into acid precipitable material from Swiss 3T3 cells stimulated by FCS. Tunicamycin was added to (△,□) quiescent cells, or to (▲,■) cells stimulated by dialyzed FCS (10% v/v). The maximal incorporation obtained with FCS is defined as 100%. Additions and labelling conditions were described in Materials and Methods.

by 50 and 70% the incorporation of [^{14}C]glucosamine and [^{3}H]mannose, respectively. The inhibitory effect was not due to an inhibition of sugar uptake since the level of radioactivity incorporated into the acid soluble material did not change with tunicamycin. These results indicate that there is a correlation between the inhibitory effect of tunicamycin on the initiation of DNA synthesis and the ability of this inhibitor to block the incorporation of precursor into acid precipitable material. Analysis of the total proteins labelled with [^{14}C]mannose from 0 to 4 h after stimulation by FCS by one dimensional gel electrophoresis showed that several proteins were glycosylated (not shown). Addition of tunicamycin at 500 ng/ml markedly reduced the labelling of these proteins. Similarly, when confluent resting 3T3 cells were stimulated by EGF or PGF$_{2\alpha}$ there was an increase in the incorporation of [^{14}C]glucosamine and [^{3}H]mannose into the acid precipitable material, as well as of [^{14}C]mannose into proteins, which was blocked in the presence of tunicamycin.

The specificity of the inhibitory effect of tunicamycin. Since the molecule of tunicamycin contains a uracil residue as well as a fatty acid, one could envisage that the inhibitory effect on the initiation of DNA synthesis may not be only a consequence of the block of N-glycosylation of newly synthesized proteins, but could also be due to inhibition of total RNA or protein synthesis occurring during the first 8 hrs of the prereplicative phase. In order to disclose between these possibilities, quiescent Swiss 3T3 cells stimulated by FCS were exposed to either [^{3}H]uridine, [^{3}H]leucine or to [^{14}C]glucosamine, in the presence or absence of tunicamycin, for different periods of time. Table 4 shows that adding tunicamycin at a concentration of 500 ng/ml, which blocks by 50% the incorporation of [^{14}C]glucosamine into acid precipitable material, did not block the incorporation of [^{3}H]uridine into RNA or the rate of protein synthesis measured by pulsing the cells with [^{3}H]leucine for a 2 hour period.

TABLE 4

EFFECT OF TUNICAMYCIN ON THE INCORPORATION OF [^{14}C]GLUCOSAMINE, [^{3}H]URIDINE AND [^{3}H]LEUCINE INTO ACID PRECIPITABLE MATERIAL IN CONFLUENT RESTING SWISS 3T3 CELLS STIMULATED BY FCS

Additions	[^{14}C]glucosamine	[^{3}H]uridine	[^{3}H]leucine
	Relative incorporation		
None	1.00	1.00	1.00
FCS	16.71	5.41	2.00
FCS + tunicamycin	8.01	5.21	1.95

FCS was added at 10% v/v. Prior to stimulation, confluent quiescent Swiss 3T3 cells were preincubated with 500 ng/ml of tunicamycin for 20 minutes. The cells were labelled with either [^{14}C]glucosamine (1.0 µCi/ml) or [^{3}H]uridine (5 µCi/ml) from 0 to 5 hours after stimulation. Labelling with [^{3}H]leucine (2.5 µCi/ml) was from 0 to 2 hours after FCS addition. Conditions for labelling were as described in Materials and Methods.

These results strongly suggest that the inhibition of initiation of DNA synthesis by tunicamycin may be a consequence of the inhibition of N-glycosylation processes rather than of an unspecific block of total RNA synthesis or inhibition of the rate of protein synthesis. Furthermore, other results from our laboratory indicate that cyclohexemide, an inhibitor of protein synthesis, added at any time after stimulation of resting cells by FCS or by EGF or PGF$_{2\alpha}$, decreased the rate of protein synthesis with concomitant decrease in the rate at which cells initiate DNA replication.

A WORKING HYPOTHESIS

The metabolic pathway leading from acetate to mevalonic acid, the formation of dolichol from mevalonic acid, the phosphorylation of dolichol, the linkage of sugars to dolichol-phosphate, and finally the transfer of the sugars to the free amino group of asparagine of nascent polypeptides constitutes a defined sequence of events with several regulatory steps. The knowledge of these steps provides a framework to test the hypothesis that signals generated at the surface membrane by specific mitogens stimulate the synthesis of certain N-glycosylated proteins, which are translocated to the nucleus, where they can affect events involved in DNA replication (Fig. 6). This constitutes a link

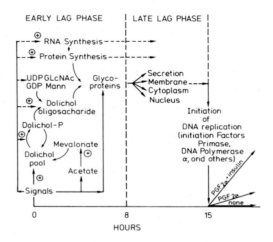

Fig. 6. Sequence of metabolic steps leading to N-glycosylation of newly synthesized proteins and their possible role in controlling the rate of initiation of DNA replication. We have represented by positive symbols the steps which can be influenced by the mitogenic signals.

between cell surface and nuclear events. Other types of studies support this hypothesis. It has been found that the receptors for EGF and insulin are N-glycosylated proteins (16,17,36). On the other hand, various components of the nuclear compartment have been described as mannose-containing proteins, suggesting their N-glycosylation, notably deoxyribonuclease I, some high mobility group proteins, and proteins of nucleosomes (36-38). In view of the present knowledge, the hypothesis would predict that: 1) specific mitogens regulate the different steps ultimately leading to N-glycosylation of proteins, and 2) there is a linkage of specific N-glycosylated proteins to events regulating the initiation of DNA replication. To challenge this hypothesis is the goal of our future research.

ACKNOWLEDGEMENTS

We thank Drs. Marit Nilsen-Hamilton, Iowa State University, Ames, Iowa, USA, and Armando J. Parodi, Instituto de Investigaciones Bioquimicas, Fundacion Campomar, Buenos Aires, Argentina, for encouraging us in this line of research. A.M.O. is a Special Fellow of the Leukemia Society of America, Inc.

REFERENCES

1. Holley, R.W. (1975) Nature (London) 258, 487-490.
2. Baserga, R. (1976) Multiplication and Division in Mammalian Cells, Dekker, New York.
3. Jimenez de Asua, L., Richmond, K.M.V., Otto, A.M., Kubler, A.M., O'Farrell, M.K. and Rudland, P.S. (1979) in: Sato, G.H. and Ross, R. (Eds.), Hormones and Cell Culture, Cold Spring Harbor Laboratory, Cold Spring Harbor, New York, vol. 6, pp. 403-424.
4. Carpenter, G. and Cohen, S. (1979) Annu. Rev. Biochem. 48, 193-216.
5. Heldin, C.H., Westermark, B. and Wasteson, A. (1981) Proc. Natl. Acad. Sci. USA 78, 3664-3667.
6. Ross, R., Nist, C., Kariya, B., Rivest, M.J., Raines, E. and Callis, J. (1978) J. Cell. Physiol. 97, 497-508.
7. Jimenez de Asua, L., Clingan, D. and Rudland, P.S. (1975) Proc. Natl. Acad. Sci. USA 72, 2724-2728.
8. Rozengurt, E., Legg, A., Pettican, P. (1979) Proc. Natl. Acad. Sci. USA 76, 1284-1287.
9. Otto, A.M. and Jimenez de Asua, L. (1982) in: Padilla, G.M. and McCarty, K.S. (Eds.) Genetic Expression in the Cell Cycle, Academic Press, New York, pp. 315-333.
10. Otto, A.M., Nilsen-Hamilton, M., Boss, B.D., Ulrich, M.O. and Jimenez de Asua, L. (1982) Proc. Natl. Acad. Sci. USA 79, 4992-4996.
11. Jimenez de Asua, L., Richmond, K.M.V. and Otto, A.M. (1981) Proc. Natl. Acad. Sci. USA 78, 1004-1008.
12. Jimenez de Asua, L. and Otto, A.M. (1983) in: Brunner, G. and Fischer, G. (Eds.) First European Conference on Serum-Free Cell Culture. Springer Verlag, in press.
13. Wood, F. (1981) Annu. Rev. Biochem. 50, 783-814.
14. Racker, E. (1983) Bioscience Rep. 3, 507-516.
15. Lennarz, W.J. (1980) The Biochemistry of Glycoproteins and Proteoglycans, Plenum Press, New York.
16. Carpenter, G. and Cohen, S. (1977) Biochem. Biophys. Res. Commun. 79, 545-552.

17. Rosen, O.M., Chia, G.H., Fung, C. and Rubin, C.S. (1979) J. Cell. Physiol. 99, 37-42.
18. Irimura, T. and Nicolson, G.L. (1982) J. Supramol. Struct. and Cell Biochem. 17, 325-336.
19. Olden, K., Bernard, B.A., White, S.L. and Parent, B. (1982) J. Cell Biochem. 18, 313-315.
20. Baenziger, J. and Kornfeld, S. (1974) J. Biol. Chem. 249, 7260-7269.
21. Baenziger, J. and Kornfeld, S. (1974) J. Biol. Chem. 249, 7270-7275.
22. Hickman, S. Kulczyckl, A., Lynch, R.G. and Kornfeld, S. (1977) J. Biol. Chem. 252, 4402-4408.
23. Grabel, L.B. and Martin, G.R. (1983) Dev. Biology 95, 115-125.
24. Heifetz, A. and Lennarz, W.S. (1979) J. Biol. Chem. 254, 6119-6127.
25. Parodi, A.J. and Leloir, L.F. (1979) Biochim. Biophys. Acta. 559, 1-37.
26. Leloir, L.F. (1981) Current Topics in Cell Regulation 18, 211-220.
27. Beytia, E.D. and Porter, J.W. (1976) Annu. Rev. Biochem. 45, 113-141.
28. Brown, M.S., Faust, J.R., Goldstein, J.L., Kanedo, I. and Endo, A. (1978) J. Biol. Chem. 253, 1121-1128.
29. Alberts, A.W., Chen, J., Kuron, G., Hunt, V., Huff, J., Hoffman, C., Rothrock, J., Lopez, M., Johua, H., Harris, E., Patchett, A., Monaghan, R., Curries, S., Stapley, E., Albers-Schonberg, G., Hensens, O., Hirshfield, J., Hoogsteen, K., Liesch, J. and Springer, J. (1980) Proc. Natl. Acad. Sci. USA 77, 3957-3961.
30. James, M.J. and Kandutsch, A.A. (1979) J. Biol. Chem. 254, 8442-8446.
31. Carson, D.D. and Lennarz, W.J. (1979) Proc. Natl. Acad. Sci. USA 76, 5709-5713.
32. Rossignol, D.P., Lennarz, W.J. and Waechter, C.J. (1981) J. Biol. Chem. 256, 10538-10542.
33. Takatsuki, A., Kohno, K. and Tamura, G. (1975) Agric. Biol. Chem. 39, 2089-2093.
34. Todaro, G.J. and Green, H. (1963) J. Cell Biol. 17, 299-313.
35. Habenicht, A.J.R., Glomset, J.A. and Ross, R. (1980) J. Biol. Chem. 255, 5134-5140.
36. Pratt, R.M. and Pastan, I.H. (1978) Nature (London) 272, 68-70.
37. Catley, B.J., Moore, S. and Stein, W. (1969) J. Biol. Chem. 244, 933-936.
38. Reeves, R., Chang, D. and Ching Chung, S. (1980) Proc. Natl. Acad. Sci. USA 78, 6704-6708.
39. Miki, B.L., Gurd, J.W. and Brown, M.S. (1980) Can. J. Biochem. 58, 1261-1269.

Résumé

L'initiation de la synthèse de DNA dans des cellules quiescentes en culture (Swiss 3T3) par des agents mitogènes est inhibée quand les premières étapes de la synthèse des glycoprotéines sont bloquées par des inhibiteurs spécifiques. Ceci suggère que la synthèse des glycoprotéines est essentielle pour la transition des cellules en phase S. Le dolichol joue le rôle de transporteur lipidique de la partie sucre pendant la biosynthèse des glycoprotéines. Lorsque la HMG CoA-réductase, enzyme qui intervient au début de la voie de synthèse de dolichol pour former de l'acide mévalonique, est inhibée par la mévinoline, l'initiation de la synthèse du DNA stimulée par l'ensemble PGF2α-insuline est inhibée. La spécificité de la mévilonine pour cet enzyme repose sur l'observation que la mévalolactone, un sel de l'acide mévalonique, annule l'inhibition par la mévinoline. L'inhibition par la tunicamycine de la première étape de la voie de synthèse des glycoprotéines liée au dolichol inhibe l'initiation de la synthèse de DNA due à l'effet du sérum ou d'autres mitogènes. La cinétique d'action de la tunicamycine suggère que la synthèse des glycoprotéines pendant les premières 8 heures de la phase de latence, est essentielle pour l'initiation plus tardive de la synthèse de DNA. Nous présentons une hypothèse de travail qui constitue une cascade potentielle qui pourrait relier les évènements mitogènes qui ont lieu au niveau de la membrane cellulaire à des évènements nucléaires plus tardifs.

INSULIN EFFECTS ON THE PROLIFERATION AND THE DIFFERENTIATION OF OB17 CELLS INTO ADIPOCYTE-LIKE CELLS

GERARD AILHAUD, EZ-ZOUBIR AMRI, PHILIPPE DJIAN, CLAUDE FOREST, PAUL GRIMALDI, RAYMOND NEGREL AND CHRISTIAN VANNIER
Centre de Biochimie du CNRS, Parc Valrose, Université de Nice, 06034 Nice cédex (FRANCE)

Extensive work has implicated insulin as the major hormone regulating, on a long-term basis, the availability of fatty acids in adipose tissue. Insulin appears to regulate endogenous fatty acid synthesis (1-3) and also the entry of exogenous fatty acids through the action of lipoprotein lipase (4,5). *In vivo* the adaptive changes induced by insulin in the content of fatty acid synthetase and of lipoprotein lipase of fat tissue have been well established. *In vitro* a direct stimulation by insulin of the cellular content of both enzymes has been shown to occur in ob17 cells (6,7). The ob17 preadipocyte clonal cell line has been established from the adipocyte fraction of epididymal fat pads of C57 BL/6J ob/ob mice (8), and a similar approach has been used to establish the preadipocyte cell line HGFu from the lean C57 BL/6J +/? mice (9).

The validity of using preadipocyte cell lines to study proliferation of adipose precursor cells, and their differentiation into adipocyte-like cells, relies primarily on the biochemical properties of the cells after adipose conversion. The comparable properties of ob17 (and HGFu) differentiated cells and of mature adipocytes isolated from rodents strongly support the validity of these cellular models for such studies. In both cases, the specific activities of key enzymes involved in lipogenesis and lipolysis are within the same range of magnitude (8,10). Differentiated ob17 cells possess β-receptors and show a lipolytic response to epinephrine added at physiological concentrations (8) or to isoproterenol (11). Ob17 cells have insulin receptors whose levels can be down-regulated after chronic exposure of confluent ob17 cells to the hormone (Table 1). The important but incomplete loss of "high-affinity" insulin binding sites ($K_d \simeq 1$ nM) is accompanied by complete disappearance of the stimulatory effect of insulin upon α-aminoisobutyrate uptake, whereas insulin removal leads to complete recovery both of the binding sites and of the stimulatory effect of the hormone (12,13).

Figure 1 describes the main features of adipose conversion of ob17 cells, which proceeds after confluence with the formation of fat cell clusters. These clusters are due to the co-existence of cells insusceptible or susceptible to such conversion, as shown in Fig.2. Stem cells (adipoblasts) can be committed,

TABLE 1

INSULIN BINDING TO OB17 CELLS AND INSULIN STIMULATION OF α-AMINOISOBUTYRATE (AIB) UPTAKE

Fetal calf serum was depleted of insulin by affinity chromatography on a column of anti-bovine insulin, guinea pig immunoglobulins coupled to Sepharose 4B, using [^{125}I]porcine insulin as a tracer. Cells were first grown to confluence as previously described (8) in insulin-depleted medium (<0.2 pM) and then treated at confluence as indicated. Assays were performed at least in triplicate. The basal (control) values of AIB uptake were 5.3 ± 0.4 pmol/10^6 cells/min under all conditions. The results are expressed as % stimulation in the presence of a maximally effective concentration of 35 nM insulin, above control values taken as 100%. The K_d values and the number of "high-affinity" insulin binding sites per cell were determined by linear regression of the high-affinity portion of the curvilinear Scatchard plots.

Culture conditions after confluence	Growth phase			Post-confluent phase		
	K_d (nM)	Sites/cell	% stimulation of AIB uptake	K_d (nM)	Sites/cell	% stimulation of AIB uptake
Insulin-depleted medium	0.8	6,200	176 ± 3.8	1.0	9,600	147 ± 1.4
Insulin-supplemented medium	-	-	-	0.5	1,600	100 ± 2.5
Insulin-supplemented medium followed by insulin removal	-	-	-	1.0	8,500	145 ± 1.3

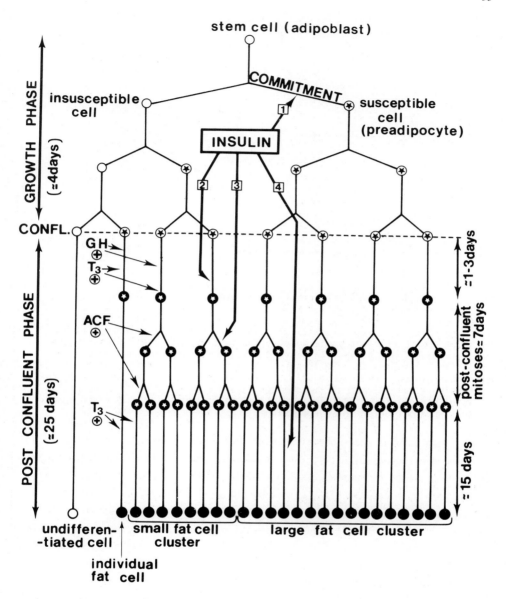

Fig. 1. Model for the differentiation of ob17 cells *in vitro*.

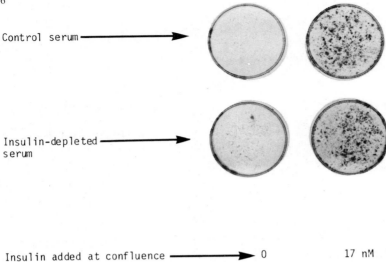

Fig. 2. Formation of fat cell clusters in insulin-depleted and insulin-supplemented medium.
Ob17 cells were treated under four conditions as indicated. Oil-Red O staining of the fat cell clusters was carried out 19 days after confluence. It should be pointed out that, despite the fact that fat cell clusters are hardly visible when no insulin was added at confluence, the number and the proportion of susceptible cells per dish, and therefore the number of clusters, remain unchanged (14). Since both the emergence of enzyme markers of adipose conversion (Table 2) and of lipid accumulation remain rather low in insulin-depleted medium, the stainable fat cell clusters are less visible and appear to be less numerous under these conditions.

that is can acquire the susceptible state, at any cell division occurring during the growth phase (10,14). At confluence, susceptible cells (preadipocytes) are present in low proportion and distributed in clones of variable sizes including single cells. After a resting phase of 1-3 days, post-confluent mitoses occur most preferentially in susceptible cells for a limited period of time (7-10 days). This limited proliferation leads to an increase in the number and in the proportion of susceptible cells, and thus leads to the formation of cell clusters of larger size which will develop 1-2 weeks later into fat cell clusters. According to this stochastic model (10), a variable but significant proportion of stem cells (about 10-30%) remain present undifferentiated at any time after confluence and remain "silent" with respect to the expression of enzyme markers such as lipoprotein lipase (6), glycerol-3-phosphate dehydrogenase and acetyl-CoA carboxylase (not shown). However, these insusceptible cells may become committed if they are allowed to divide either by making a "wound" in the dish or by reinoculating them selectively (14).

Hormonal requirements for growth and differentiation

Recent investigations have been focused on the characterization and the identification of extracellular signals which control growth and differentiation of various preadipocyte cell lines (15). Growth requirements of the ob17 cell line are qualitatively similar but quantitatively different from those of the 3T3-L1 cell line (16). Growth is highly dependent upon the presence of insulin, FGF and a partially purified fraction from sub-maxillary glands (SMGE), and is less dependent upon that of transferrin (17). This serum-free medium supports growth but is insufficient for differentiation, even when supplemented with triiodothyronine which is essential for adipose conversion (18). This observation shows indirectly the existence of additional factors present in medium supplemented with fetal calf serum. Recently Morikawa et al. have demonstrated that growth hormone at physiological concentrations acts as an adipogenic factor required for the adipose conversion of 3T3-F442A cells maintained in the presence of cat serum (19). The addition of growth hormone is also sufficient for the differentiation of ob17 cells in bovine serum-supplemented medium but remains insufficient in serum-free medium, suggesting indirectly that adipogenic factor(s) other than GH, present both in bovine and in fetal calf serum, are also required. In addition, an active fraction present in fetal calf serum (adipose conversion factor(s) or ACF), clearly different from GH, has been shown to control post-confluent mitoses of susceptible cells ; a fraction recovered from bovine pituitary extracts shows properties similar to those of ACF (18).

So far, insulin has been shown to be necessary for the adipose conversion of preadipocyte cell lines (8,20-24), with the exception of the preadipocyte cell lines isolated from bone marrow of mice which do not show this requirement but, instead, need corticosteroids for such conversion (25-27). Clearly insulin, usually added at confluence, could possibly be required at different levels of this multistep process of differentiation. As indicated in Fig.1, insulin could be involved in 1) the cell committment as already proposed (28) 3) the post-confluent mitoses in relation with its known mitogenic properties on cultured cells (14,16,17) 2) the onset and/or 4) the modulation of the expression of the differentiation program.

TABLE 2

DEVELOPMENT OF ENZYME MARKERS OF ADIPOSE CONVERSION IN INSULIN-DEPLETED OR IN INSULIN-SUPPLEMENTED MEDIUM

Ob17 cells were treated under four different conditions as indicated. Day 0 corresponds to confluence determined by microscopic examination. The activity levels of glycerol-3-phosphate dehydrogenase and of lipoprotein lipase are undetectable (und.) in exponentially-growing cells and in early confluent cells. The numbers in parenthesis correspond to values obtained at day 15 in insulin- and ACF-supplemented medium but which had been depleted of triiodothyronine as previously described (30).

Culture conditions		Glycerol-3-phosphate dehydrogenase (nmol/min/mg) at day			Acid:CoA ligase (nmol/min/mg) at day			Lipoprotein lipase (nmol/min/mg) at day			Lactate dehydrogenase (nmol/min/mg) at day		
Before confluence	After confluence	0	5	13	0	5	13	0	5	13	0	5	13
Control serum	Control serum	und.	8.3	77.3	0.19	0.61	1.39	und.	4.45	8.43	2034	2692	1992
Control serum	Control serum +17nM insulin	und.	150	887 (4)	0.19	1.20	10.25 (0.5)	und.	6.66	14.06 (und.)	2034	–	2870 (3050)
Insulin-depleted serum	Insulin-depleted serum	und.	7.8	66.1	0.21	0.57	1.43	und.	3.44	10.57	1962	2356	2307
Insulin-depleted serum	Insulin-depleted serum +17 nM insulin	und.	14.1	837	0.21	1.33	7.71	und.	4.95	9.95	1962	2644	2490

Development of enzyme markers and adipose conversion of ob17 cells in insulin-depleted and in insulin-supplemented medium

The formation of fat cell clusters is not affected when cells are first grown in insulin-depleted serum and then exposed to insulin after confluence in the presence of triiodothyronine (Fig.2). The emergence of enzyme markers of adipose conversion is also clearly not affected when ob17 cells are grown and maintained after confluence in insulin-depleted serum (Table 2). In this case, however, the specific activity values remain low but nonetheless significant, since lipoprotein lipase and glycerol-3-phosphate dehydrogenase are undetectable in cells which do not enter the differentiation program, that is i) in exponentially growing cells (6,8) ii) in post-confluent cells first grown in the presence of 5-bromodeoxyuridine (8,20) iii) in late post-confluent insusceptible cells (6). In contrast, ob17 cells maintained in medium supplemented with insulin but deprived of triiodothyronine do not express in a significant manner such enzyme markers (Table 2).

Altogether, these results indicate that insulin seems to be required neither for cell commitment nor for the onset of the differentiation program. The participation of insulin in post-confluent mitoses can also be excluded ; mitoses of susceptible cells, which occur during a limited period of time after confluence, were analyzed by [^{3}H]thymidine incorporation into DNA followed by autoradiography (14). The labelling pattern was found to be identical when confluent cells are maintained in insulin-depleted or in insulin-supplemented medium (not shown). Insusceptible cells were not significantly labelled under both conditions. As expected, post-confluent mitoses lead to an increase in the cell number per dish (Fig.3). This increase is approximately 2-fold under each condition, showing that insulin is neither a mitogen nor a permissive hormone required for post-confluent mitoses, despite its growth-promoting ability at higher concentrations (*vide infra*). Therefore, the number and the proportion of cells susceptible to adipose conversion remain quite similar in cells grown and maintained in insulin-depleted or in insulin-supplemented medium. Since insulin at physiological concentrations is known to accelerate significantly the adipose conversion of ob17 cells, its role is most likely, and merely, to modulate the expression of the different enzyme markers, as shown below.

Lipogenic and mitogenic effects of insulin during adipose conversion

Chronic exposure of confluent cells to insulin favors their development into lipid-filled cells and the appearance of fat cell clusters (Fig.2). Fifty to eighty percent of the cells accumulate triglycerides. Experiments described in

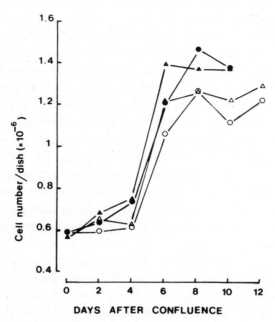

Fig. 3. Changes in the cell number of post-confluent cells maintained in insulin-depleted or in insulin-supplemented medium.
Ob17 cells were grown in insulin-depleted medium (○,●) or in control serum (△,▲), with 17 nM insulin added (●,▲) or not (○,△) at confluence. Cell enumeration was carried out with a Couter counter.

Fig.4 were carried out on confluent cells as a function of insulin or proinsulin concentrations added at confluence.

The specific activity of GPDH determined at day 13 indicates an EC_{50} value and a concentration of insulin maximally effective of 0.6 and 10 nM, respectively, whereas the corresponding figures for proinsulin are 9 and 300 nM (Fig.4A). Similar values are obtained when determining the cellular triacylglycerol content (Fig.4B). In contrast to these lipogenic responses, the mitogenic response of ob17 cells determined by cell enumeration shows that insulin and proinsulin have similar potencies (Fig.4C). The growth-promoting ability of both hormones is low at physiological concentrations and becomes significant only at supraphysiological concentrations. The mitogenic effect of insulin and analogues was also investigated by determining the reinitiation rates of DNA synthesis as a function of hormone concentration. The EC_{50} values are 300 and 500 nM for porcine insulin and proinsulin, respectively. In contrast to both hormones, IGF-I behaves as a potent mitogen at low concentrations ; it is approximately 500 times more potent than insulin and proinsulin (Fig.4D).

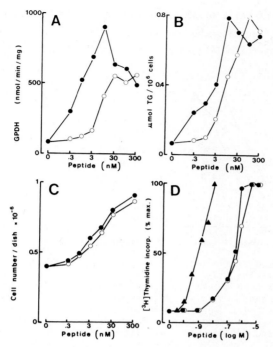

Fig. 4. Lipogenic and mitogenic effects as a function of insulin and proinsulin concentration.

Confluent ob17 cells in 35-mm diameter dishes were maintained for 13 days under standard conditions (8), in the presence of 1.5 nM triiodothyronine and varying concentrations of insulin (●) or proinsulin (○). The cell number and the triglyceride content at day 1 were 294,000 cells/dish and 0.035 mol/10^6 cells, respectively. (A) Glycerol-3-phosphate dehydrogenase (GPDH) activity. (B) Triglyceride content. (C) Cell number. (D) Reinitiation of DNA synthesis carried out as previously described (14).

In summary, the mitogenic response of ob17 cells to insulin appears at concentrations which are near-maximally effective for the lipogenic responses, and becomes important only at supraphysiological concentrations (31). The lipogenic responses are half-maximal within the nanomolar range of insulin concentrations and thus are most likely mediated through binding to insulin receptors , whereas the mitogenic response might be mediated through binding to specific receptors of insulin-like growth factors, which bind also insulin weakly and which are likely to be present in ob17 cells as they are in rat adipocytes (32).

A more detailed examination of the lipogenic responses between day 9 and day 13 after confluence, that is a time when post-confluent mitoses have ceased and when the total number of cells susceptible and insusceptible remains constant (Fig.3), reveals an interesting phenomenon. The sensitivity to insulin and to proinsulin increases approximately 10-fold between day 9 and day 13 in insulin-supplemented medium (Fig.5).

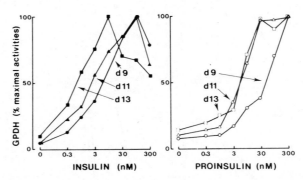

Fig. 5. Increased responsiveness to insulin and to proinsulin during adipose conversion.
Confluent ob17 cells were maintained after confluence for 9, 11 and 13 days as described in the legend of Fig.4. Determinations of glycerol-3-phosphate dehydrogenase (GPDH) activity were carried out at times indicated. The specific activities expressed in nmol/min/mg were found to be 626, 1030 and 903 for insulin and 396, 365 and 575 for proinsulin at day 9, 11 and 13, respectively.

As already reported, this increased responsivenes is dependent upon the "age" of confluent cells and upon the exposure time to insulin (31). For instance when insulin-pretreated cells are deprived of insulin for a couple of days, this deprivation leads to activity levels of GPDH (and other markers) similar to those determined in control cells of the same age and which have *never* been exposed to insulin. Nevertheless, insulin re-addition shows that insulin-pretreated cells do keep a "memory" and are able to respond much faster than control cells by increasing the activity levels of enzyme markers of adipose conversion. Although the molecular basis for this increased responsiveness to insulin is presently unknown, it is due neither to changes in the rate of insulin degradation nor to changes in the affinity/or in the number of insulin receptors (not shown).

Modulation of lipid-synthesizing enzymes by insulin in differentiated ob17 cells

After adipose conversion, insulin removal from differentiated cells gives within 24-48h a large decrease (50-80%) in fatty acid synthetase, glycerol-3-phosphate dehydrogenase and acid:CoA ligase activities, as well as in the rate of fatty acid synthesis determined by [^{14}C]acetate incorporation into lipids. All parameters are restored by insulin re-addition to initial values within

24-48h (Table 3). Dose-response curves (not shown) of insulin on the restoration of fatty acid synthesis give half-maximally effective concentrations close to 1 nM, in agreement with the "high-affinity" binding sites for insulin present in ob17 cells ($K_d \simeq 1$ nM). The continuous presence of insulin is necessary, and ob17 differentiated cells adjust their steady-state activities of triglycerol-pathway enzymes as a function of physiological concentrations of insulin. Additional experiments of immunotitration have demonstrated that changes in fatty acid synthetase activity reflect parallel changes in the enzyme content (not shown). Thus it is clear that the effect of insulin on differentiated ob17 cells is direct and takes place most likely through binding to insulin receptors. The co-ordinate nature of the responses suggests that the insulin effect involves a common intermediary step after insulin binding.

DISCUSSION

The role of insulin in the control of lipogenesis and triglyceride synthesis has long been known. The availability of pre-adipocyte cell lines has allowed us to gain some insights in the different roles possibly played by this hormone during adipose cell differentiation. The use of insulin-depleted medium shows that insulin is not involved in i) the committment of stem cells (adipoblasts) to preadipocytes ii) the onset of the differentiation program iii) the control of post-confluent mitoses. The possibility remains that growth and maintainance in insulin-depleted medium lead to unmasking of "super-receptors" able to bind insulin with an extremely high affinity and thus allow the cells to respond to insulin within a picomolar range of concentrations. This possibility is not very likely, since binding studies carried out on cells exposed to insulin-depleted serum or to control serum give superimposable Scatchard plots (not shown). Insulin behaves significantly as a mitogen only at concentrations above those observed *in vivo* during hyperinsulinemia in rodents (33), and one could hypothesize that its growth-promoting effect on ob17 cells is mediated through its binding to receptors of insulin-like growth factors. Therefore, the effects of insulin on these adipose precursor cells are limited to its ability to modulate reversibly the expression of the different enzyme markers of adipose conversion. These effects occur within a physiological range of concentrations and are most likely mediated through its binding to insulin receptors. Changes in enzyme activity reflect corresponding changes in enzyme content, as shown for lipoprotein lipase (6) and fatty acid synthetase (7). In conclusion, established adipocyte-like cells such as the ob17 clonal line represent useful cellular models to delineate the long-term effects of insulin on the process of adipose cell differentiation and on the hypertrophy of adipose cells.

TABLE 3

REVERSIBLE MODULATION AFTER INSULIN REMOVAL OR RE-ADDITION OF THE ACTIVITIES OF ENZYME MARKERS OF ADIPOSE CONVERSION

Experiment a : Confluent ob17 cells were maintained for 15 days in standard medium (8) supplemented with 17 nM insulin and 1.5 nM triiodothyronine. Cells were then exposed (zero time) to standard medium supplemented with 1.5 nM triiodothyronine. Enzyme assays were carried out according to published procedures (14). The 100% initial activities correspond to 960, 8.2, 12 and 2,800 mUnits/mg of protein for glycerol-3-phosphate dehydrogenase, fatty acid synthetase, acid:CoA ligase and lactate dehydrogenase, respectively. The incorporation of [^{14}C]acetate into lipids at zero time, and determined as previously described (14), was 27,500 dpm/2h per mg of protein. Experiment b : Confluent ob17 cells (a series different from that used in experiment a) were maintained for 14 days in standard medium (8) supplemented with 17 nM insulin and 1.5 nM triiodothyronine. Cells were then maintained in the same medium but in the absence of insulin. After 48h the medium was removed and replaced (zero time) by the standard medium supplemented with 17 nM insulin and 1.5 nM triiodothyronine. Enzyme activities and [^{14}C]acetate incorporation into lipids were determined as a function of time after insulin re-addition. The specific activities at day 14 before insulin removal, and taken as 100%, were 1045, 10.15, 11.3 and 2625 mUnits/mg of protein for glycerol-3-phosphate dehydrogenase, fatty acid synthetase, acid:CoA ligase and lactate dehydrogenase, respectively. The corresponding value for the [^{14}C]acetate incorporation into lipids was 35,000 dpm/2h per mg of protein.

	% initial activity remaining after insulin removal at time (hours)[a]				% activity recovered after insulin re-addition at time (hours)[b]				
	0	7	24	48	0	8	15	24	48
[^{14}C]acetate incorporation into lipids	100	47	28	23	22	54	100	-	-
Fatty acid synthetase	100	49	36	25	29	-	-	100	-
Glycerol-3-phosphate dehydrogenase	100	51	30	23	26.5	45	90	100	-
Acid:CoA ligase	100	95	70	52	42	44.5	-	80	100
Lactate dehydrogenase	100	-	103	98	100	-	-	100	-

ACKNOWLEDGEMENTS

We are indebted to Mrs. B. Barhanin and G. Oillaux for technical and secretarial assistance, respectively.

This work was supported by grants from the "Centre National de la Recherche Scientifique" (ATP 7300-01, grants n° 4162 and 4639), from the "Commissariat à l'Energie Atomique" and from the "Fondation pour la Recherche Médicale". The authors are very grateful to Dr. R. Humbel (Zürich-Switzerland) for the kind gift of IGF-I.

REFERENCES

1. Saggerson, E.D. and Greenbaum, A.L. (1970) Biochem.J. 119, 221-242.
2. Bruckdorfer, K.R., Khan, I.H. and Yudkin, J. (1972) Biochem.J. 129, 439-446.
3. Volpe, J.J. and Vagelos, P.R. (1976) Physiol.Rev. 56, 339-417.
4. Jansen, H., Garfinkel, A.S., Twu, J.S., Nikazy, J. and Schotz, M.C. (1978) Biochim.Biophys.Acta 531, 109-114.
5. Cryer, A. (1981) Intern.J.Biochem. 13, 525-541.
6. Vannier, C., Jansen, H., Négrel, R. and Ailhaud, G. (1982) J.Biol.Chem. 257, 12387-12393.
7. Grimaldi, P., Forest, C., Poli, P., Négrel, R. and Ailhaud, G. (1983) Biochem.J., in press.
8. Négrel, R., Grimaldi, P. and Ailhaud, G. (1978) Proc.Natl.Acad.Sci.USA 75, 6054-6059.
9. Forest, C., Grimaldi, P., Czerucka, D., Négrel, R. and Ailhaud, G. (1983) In Vitro 19, 344-354.
10. Ailhaud, G., Amri, E., Cermolacce, C., Djian, P., Forest, C., Gaillard, D., Grimaldi, P., Khoo, J., Négrel, R., Serrero-Davé, G. and Vannier, C. (1983) Diab. Métabol. in press.
11. Forest, C., Négrel, R. and Ailhaud, G. (1981) Biochem.Biophys.Res.Commun. 102, 577-587.
12. Grimaldi, P., Négrel, R., Vincent, J.P. and Ailhaud, G. (1979) J.Biol.Chem. 254, 6849-6852.
13. Ailhaud, G., Grimaldi, P., Négrel, R., Serrero, G. and Verrando, P. (1979) in "Obesity : molecular and cellular aspects" 87, 51-64 (Editions Scientifiques de l'INSERM).
14. Djian, P., Grimaldi, P., Négrel, R. and Ailhaud, G. (1982) Exp.Cell Res. 142, 273-281.
15. Ailhaud, G. (1982) Molec.Cell.Biochem. 49, 17-31.
16. Serrero, G., McClure, D. and Sato, G. (1979) in "Cold Spring Harbor Conferences on Cell Proliferation" 6, 523-530 (Ross, R. and Sato, G., eds.).
17. Gaillard, D., Négrel, R., Serrero-Davé, G., Cermolacce, C. and Ailhaud, G. (1983) submitted.
18. Grimaldi, P., Djian, P., Négrel, R. and Ailhaud, G. (1982) The EMBO J. 1, 687-692.
19. Morikawa, M., Nixon, T. and Green, H. (1982) Cell 29, 783-789.
20. Green, H. and Kehinde, O. (1975) Cell 5, 19-27.
21. Darmon, M., Serrero, G., Rizzino, A. and Sato, G. (1981) Exp.Cell Res. 132, 313-327.

22. Serrero, G. and Khoo, J. (1982) Anal.Biochem. 120, 351-359.
23. Serrero, G. and Sato, G. (1982) in "Cold Spring Harbor Conferences on Cell Proliferation" 9, 943-955 (Sato, G., Pardee, A. and Subasku, D., eds.).
24. Hiragun, A., Sato, M. and Mitsui, H. (1980) In Vitro 16, 685-693.
25. Greenberger, J.S. (1979) In Vitro 15, 823-828.
26. Kodama, H.A., Amagai, Y., Koyama, H. and Kasai, S. (1982) J.Cell.Physiol. 112, 83-88.
27. Lanotte, M., Scott, D., Dexter, T.M. and Allen, T.D. (1982) J.Cell.Physiol. 111, 177-186.
28. Sager, R. and Kovak, P. (1982) Proc.Natl.Acad.Sci.USA 79, 480-484.
29. Murphy, M.G., Négrel, R. and Ailhaud, G. (1981) Biochim.Biophys.Acta 664, 240-248.
30. Gharbi-Chihi, J., Grimaldi, P., Torresani, J. and Ailhaud, G. (1981) J. Recept.Res. 2, 153-173.
31. Grimaldi, P., Djian, P., Forest, C., Poli, P., Négrel, R. and Ailhaud, G. (1983) Mol.Cell.Endocrin. 29, 271-285.
32. King, G.L., Kahn, R., Rechler, M.M. and Nissley, S.P. (1980) J.Clin.Invest. 66, 130-140.
33. Bray, G.A. and York, D.A. (1979) Physiol.Rev. 59, 719-809.

RESUME

Les cellules de la lignée préadipocytaire ob17 se différencient en culture en présence de concentrations physiologiques en insuline et en triiodothyronine. Les cellules adipeuses ainsi formées possèdent les propriétés morphologiques et biochimiques des adipocytes isolés à partir du tissu adipeux.

L'exposition chronique à l'insuline des cellules ob17 confluentes conduisant à une accélération de leur développement en cellules adipeuses, le rôle de l'insuline a été analysé à l'aide de sérum appauvri en insuline après chromatographie d'immunoaffinité.

Les résultats démontrent clairement que l'insuline ne joue aucun rôle dans i) l'engagement des cellules souches (adipoblastes) à devenir des cellules susceptibles (préadipocytes) ii) la phase proliférative post-confluente spécifique des préadipocytes iii) la mise en route du programme de différenciation. L'insuline joue par contre, à des concentrations physiologiques, un rôle amplificateur sur l'expression des phénotypes caractéristiques du programme de différenciation adipocytaire. Ces effets de l'insuline peuvent être dissociés de l'effet mitogénique de l'hormone excercé à des concentrations supraphysiologiques. La réponse mitogénique des cellules ob17 paraît dépendre de la fixation de l'insuline aux récepteurs des facteurs insulino-mimétiques, alors que les réponses lipogéniques sont vraisemblablement liées à la fixation de l'hormone aux récepteurs insuliniques.

INTRACELLULAR SIGNALS
SIGNAUX INTRACELLULAIRES

KINETICAL AND PHYSICOCHEMICAL PROPERTIES OF V1 AND V2 VASOPRESSIN RECEPTORS : RELATION TO CYCLIC AMP DEPENDENT AND CALCIUM DEPENDENT ACTIVATION PROCESSES.

GILLES GUILLON* and DANIEL BUTLEN**
* Centre de Pharmacologie - Endocrinologie, Rue de la Cardonille, 34033 Montpellier Cedex, (FRANCE) and
** Laboratoire de Physiologie Cellulaire, Collège de France, 11, Place Marcelin Berthelot, 75231 Paris Cedex, (FRANCE).

INTRODUCTION

Vasopressin belongs to the class of polypeptidic hormones. It is composed of nine amino acids and is produced by the post-hypophyseal gland.

As seen in Table I, according to the tissues considered, this hormone exerts different actions in mammals. Vasopressin was first recognized as a pressoric agent (13) its antidiuretic action and its role in the regulation of body fluid osmolarity was then described (14). Recently, various other biological effects of vasopressin have been discovered on other tissues such as the liver and the brain.

Tritiated vasopressin binding experiments and pharmacological studies with vasopressin structural analogues have clearly demonstrated that all the biological effects of this hormone are mediated by specific vasopressin receptors (for review, see 15). Moreover, these effects are ellicited by vasopressin doses which are in the range or at the upper limits of physiological blood levels. Recent data clearly show that all these specific vasopressin receptors can be distinguished from each other on the basis of recognition patterns of vasopressin analogues (for review, see 16). Moreover, according to the target tissue considered, the molecular mechanisms implied in the biological action of vasopressin are different. Vasopressin exerts its antidiuretic action on kidney through cyclic AMP-dependent processes (17) when its glycogenolytic activity on liver is mediated through calcium dependent processes (18). This raises the problem of the existence of the vasopressin isoreceptor. Mitchell has already proposed the following classification : V1 is referred to vasopressin receptors acting through calcium-dependent processes and V2 to vasopressin receptors acting through cyclic AMP-dependent processes.

The purpose of this report is to summarize and compare the data relating to a V1 and a V2 vasopressin receptors from the same animal (rat liver and rat kidney vasopressin receptors respectively).

MATERIALS AND METHODS

(^{3}H)Tyr$_2$Lys$_8$ vasopressin (8-11 Ci/mmol) was prepared as previously described (20, 21). Its biological activity was identical to that of the unlabelled peptides. Purified liver plasma membranes were prepared according to Neville up to step 11, (22). Membrane

Table 1 : Effects of vasopressin in mammals [a]

Target tissue	Biological response	Reference
Kidney	- Increased water reabsorption by collecting ducts	(1)
	- Increased solute transport by the ascending limb of Henle's loop	(2)
	- Contraction of glomerular mesangial cells	(3)
	- Increased prostaglandin synthesis by medullary intestinal cells	(4)
	- Inhibition of isoproterenol-induced renin release	(5)
Liver	- Increased glycogenolysis and neoglucogenesis	(6)
Blood Vessels	- Contraction	(7)
Adenohypophysis	- Increased corticotropin secretion	(8)
Platelets	- Aggregation	(9)
Blood-Brain Barrier	- Increased fluid secretion	(10)
Central Nervous System	- Affects animal behaviour (in particular memory)	(11)
	- Localized changes in catecholamine turn-over	(12)

[a] Modified from JARD, S. (16)

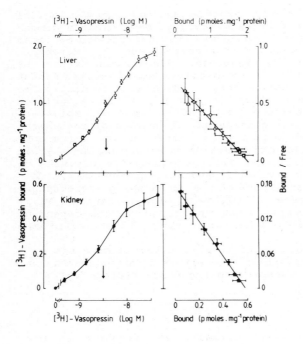

Figure N° 1 : Dose dependency for vasopressin binding to rat kidney and rat liver membranes.

Left panel : Specific (3H)-vasopressin binding curves.
Right panel : Scatchard representation corresponding to the dose binding curves (arrows indicate the Kbind).

fractions from the medullo-papillary portions of wistar rat kidney were prepared as earlier described (23). Adenylate cyclase was measured according to and Salomon (25, 26). Binding of (^{3}H)-vasopressin was realized by filtration through Millipore filters as already described (25, 27). Adenylate cyclase and vasopressin-receptor complexes were solubilized using a non ionic detergent (Triton X-100) (28, 29). The physicochemical parameters of the solubilized molecules studied were determined by sucrose gradient ultra centrifugation and gel filtration experiments (28, 29).

RESULTS

I <u>Properties of vasopressin membranous receptors.</u>

A) <u>Characterization of membranous vasopressin specific sites.</u>

Vasopressin binding experiments were realized by incubating liver or kidney membranes at 30°C with increasing amounts of (^{3}H)-vasopressin. Steady state binding measurements were performed after an incubation period of 10 to 15 min for kidney and liver membranes respectively. As depicted in Fig. 1, specific binding was saturable and non cooperative in both cases (Hill coefficient = 1.03 ± 0.03. These results were adequately described by a reversible binding of (^{3}H)-vasopressin to a homogenous population of binding sites, as indicated by the linear Scatchard plot of the binding curves (Fig. 1). The apparent dissociation constants (concentration of peptides leading to half maximum specific binding) were very similar for the two tissues tested, Kbind = 5.0 nM ± 0.8 and 4.3 nM ± 0.6 for liver and kidney membranes respectively. Nevertheless, the maximal specific binding capacities were different : 2.9 ± 0.60 and 0.59 ± 0.07 pmoles of vasopressin bound per mg. of protein for liver and kidney membranes respectively.

B) <u>Recognition pattern of vasopressin receptors.</u>

To compare the structure of vasopressin receptors from kidney and liver membranes, we have analysed the structure activity relationships of a series of vasopressin analogues on these two systems. The determination of the apparent dissociation constant of the unlabelled analogues (Kbind) was realized by binding competition experiments (27). Results obtained with the entire series of analogues are given in Table II. For both liver and kidney vasopressin receptors, iodination of the Tyrosyl residue in position 2, drastically reduced the affinity of the peptide for its receptor. However, for the other structural analogues, modifications brought on the molecule did not involve a large variation of the apparent affinities of these peptides. For the kidney vasopressin receptor extreme Kbind values are 14.5 nM for dPAVP and 0.70 nM for d(CH$_2$)$_5$Tyr(et)-VAVP. On the other hand for the liver vasopressin receptor, structural modification induced higher variations in affinities than those observed for the renal receptor (extreme Kbind values determined are 0.81 nM for HO-AVP and 0.45 μM for HO-VDAVP). These results indicate that the renal vasopressin receptor discriminates vaso-

TABLE II : RECOGNITION PATTERNS OF VASOPRESSIN ISORECEPTORS

COMPOUND	REFERENCE	KEY FOR SYMBOL	ABBREVIATIONS	PK BIND VALUE FOR LIVER MEMBRANES	PK BIND VALUE FOR KIDNEY MEMBRANES
(8-LYSINE)-VASOPRESSIN	27	1	LVP	8.3	8.3
(2-2'-MONOIODOTYROSINE),8-LYSINE)VASOPRESSIN	‡	2	I-LVP	6.1	5.7
(2-(2'4'-DIIODOTYROSINE),8-LYSINE)VASOPRESSIN	‡	3	I2-LVP	5.4	5.0
(1-DEAMINO, 7-L^3-PROLINE, 8,-ARGININE)VASOPRESSIN	‡‡	4	D,^{3}PRO^7 AVP	8.7	9.0
(1-L-2-HYDROXY-3-MERCAPTO PROPANOIC ACID, 8-ARGININE)-VASOPRESSIN	27	5	HO-AVP	9.2	9.1
(1-L-2-HYDROXY-3-MERCAPTO PROPANOIC ACID,4-VALINE,8-D-ARGININE)-VASOPRESSIN	‡‡	6	HO-VDAVP	6.4	9.1
(1-DEAMINO, 2-O-METHYL TYROSINE,4-VALINE, 8-ARGININE)-VASOPRESSIN	24	7	O-ME-TYR^2 DVAVP	6.9	9.0
(1-DEAMINOPENICILLAMINE, 8-ARGININE)-VASOPRESSIN	24	8	D-PAVP	8.4	7.8
(1-(8-MERCAPTO-8,8, CYCLOPENTAMETHYLENE PROPIONIC ACID),4 VALINE,8-D-ARGININE)-VASOPRESSIN	27, 25	9	D(CH$_2$)5-VDAVP	7.4	7.9
(1-(8-MERCAPTO-8,8,CYCLOPENTAMETHYLENE PROPIONIC ACID),2-O-ETHYL,TYROSINE, 4-VALINE, 8-ARGININE)-VASOPRESSIN	24	10	D(CH$_2$)5-TYR(ET)VAVP	8.5	9.2
(8-D-ARGININE)-VASOPRESSIN	27, 25	11	DAVP	7.0	9.3
(1-DEAMINO,8-D-ARGININE)-VASOPRESSIN	27, 25	12	D-DAVP	7.0	9.6
(1-DEAMINO,4-THREONINE,8-D-ARGININE)-VASOPRESSIN	27, 25	13	D-TDAVP	7.0	9.6
(4-VALINE,8-ARGININE)-VASOPRESSIN	27, 25	14	VAVP	8.0	9.5
(4-VALINE,8-D-ARGININE)-VASOPRESSIN	27, 25	15	VDAVP	6.3	9.7
(1-DEAMINO,4-VALINE,8-D-ARGININE)VASOPRESSIN	27, 25	16	D-VDAVP	6.5	9.6
OXYTOCIN	27	17	OT	7.1	7.5
(8-ARGININE-VASOPRESSIN)	27, 25	18	AVP	8.5	9.4
(1-DEAMINOPENICILLAMINE,4-VALINE,8-D-ARGININE)-VASOPRESSIN	29	19	D-PVDAVP	7.6	8.9

‡ BUTLEN, D. ET AL. ACCEPTED FOR PUBLICATION IN COLL. CZECH. CHEM. COMMUN.
‡‡ BUTLEN, D. UNPUBLISHED RESULTS

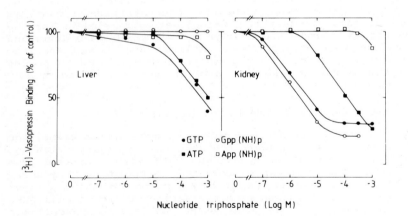

Figure N° 2 : Effect of nucleotides on the vasopressin binding.
Rat kidney or rat liver membranes were incubated with 5 nM of (3H)-vasopressin in the absence (control) or in the presence of increasing amounts of nucleotides. Specific binding values were determined in each condition.

pressin analogues much less efficiently than compared to the liver vasopressin receptor. If we assume that the observed differences between the recognition pattern of the two vasopressin receptors very likely reflects a difference in the structure of their respective binding sites, it is clear that the liver and the kidney receptors are quite different.

C) Effects of triphospho-nucleotides on vasopressin binding.

As seen in Fig. 2, adenyl and guanyl triphospho-nucleotides affect the specific (^{3}H)-vasopressin binding in different ways depending on the tissues tested. In the case of kidney receptors, GTP is more efficient in comparison to the ATP. The concentration of GTP leading to 50 % of maximal binding inhibition are around 1 μM whatever the guanyl triphospho-nucleotide used : GTP or Gpp (NH)p (a non hydrolysable analogue of GTP). The results are very different for the liver receptor. In this case, Gpp(NH)p is completely inactive even at higher concentrations (1 mM). Only GTP or ATP inhibit the (^{3}H)-vasopressin binding but at concentrations higher than that used for experiments with kidney vasopressin receptors. Both for the liver and the kidney vasopressin receptors, this nucleotide effect is relatively slow (increase of the Kbind by a factor 3 to 6 in the presence of the triphosphonucleotide). It only affects the apparent dissociation constant of the receptor with no modification of the maximal specific vasopressin binding capacity. Moreover, this effet is agonist specific and was not observed for vasopressin antagonists of the adenylate cyclase response and vasopressin antagonists of the glycogenolytic response on kidney and liver membranes respectively. These results may signify that both the kidney receptor and the liver vasopressin receptor could be influenced by the same modulators. However, the nucleotide induced change in the affinity of the liver vasopressin receptor may be explained by a phosphorylation reaction since, 1) the non hydrolysable analogues of GTP or ATP are not efficient enough to produce an inhibition of (^{3}H)-vasopressin binding, 2) the concentrations of GTP and ATP needed to produce such an inhibition are in the range of nucleotide concentrations needed for phosphorylation reactions. No such interpretation could be invoked for the kidney vasopressin receptor since the non hydrolysable analogue of GTP is the most efficient effector. The agonist specific character of the nucleotide effect might indicate that ATP or GTP plays an important role in the functional coupling of liver vasopressin receptor to their still unknown primary effector. In the case of the kidney vasopressin receptor these results confirm the crucial role of the GTP binding proteins implicated in the coupling of hormonal receptor and adenylate cyclase (30).

D) Effects of magnesium on vasopressin binding.

When kidney or liver membranes were depleted in endogenous magnesium by extensive washings with EDTA, only few specific sites of vasopressin binding were detected. Normal binding could be restored by the addition of magnesium. The apparent Km for magnesium effect on vasopressin binding were very different depending on the type of vasopressin receptor tested (Km = 1 mM and 10 μM for liver and kidney receptor respectively). As seen in Fig. 3, the reduction of the magnesium concentration to inframaximal values, introduces an apparent heterogeneity in the population of vasopressin binding sites. Under these conditions, some of the receptors were found to have a low affinity state (Kbind was about 50 to 100 nM).

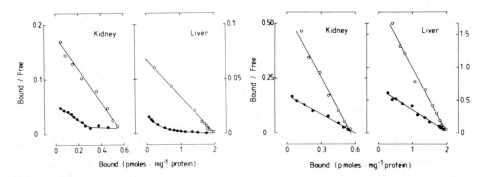

Figure N°3 : Effects of magnesium on the vasopressin binding (left panel).
Rat kidney membranes were incubated in the presence of increasing amounts of (3H)-vasopressin with 750 nM of magnesium (open symbols) or with 5 µM of magnesium (closed symbols). Rat liver membranes were incubated in the same conditions with 1 mM of magnesium (open symbols) or without magnesium (closed symbols). Specific binding values were determined and used to calculate the corresponding Scatchard plots.

Figure N° 4 : Effects of the duration of the incubation on the vasopressin binding (right panel).
Rat kidney membranes were incubated at 30 °C in the presence of increasing amounts of (3H)-vasopressin either 10 min (closed symbols) or 60 min (open symbols). Specific binding values were determined and used to calculate the corresponding Scatchard plots.

E) Effects of the duration of the incubation time between hormones and membranes on vasopressin binding.

As depicted in Fig. 4, the time during which the membranes and (^{3}H)-vasopressin were incubated modify the apparent affinity of the renal or hepatic receptor with no modification of the maximal specific binding capacity. This effect was relatively slow (decrease of Kbind by a factor 3 to 8 with increasing the duration of the incubation time). In the case of both liver and kidney systems, this effect is agonist specific. The affinity of glycogenolytic or antidiuretic antagonists of vasopressin on liver and kidney, remains unchanged with the incubation time. For the renal vasopressin receptor, this effect seems to correlate with its sensibility towards guanyl nucleotides. Addition of Gpp(NH)p does not modifiy the affinity of vasopressin towards its receptors for short periods of incubation (Kd 10 min = 3.5 nM ; Kd 10 min + Gpp(NH)p = 3.7 nM) since it increases its apparent affinity by a factor about 3 for long periods in incubation (Kd 60 min = 4.1 nM ; Kd 60 min + Gpp(NH)p = 11.0 nM). Such a phenomena was not observed in the case of the liver vasopressin receptor for which the effect of GTP is independent of the duration of incubation.

This time-dependent modification of vasopressin affinity may reflect the first coupling steps of hormone-receptor complexes with endogenous effectors implied in the biological response of vasopressin.

II <u>Biological mechanisms triggered by vasopressin receptors.</u>

A) <u>The kidney vasopressin receptors</u>.

Earlier studies have clearly established that the effects of vasopressin on the osmotic permeability to water of mammalian collecting ducts (1) and on the solute transport by the ascending limb of Henle's Loop (2) are mediated via the stimulation of the adenylate cyclase which raises the concentration of intracellular cyclic AMP (31, 32).

Numerous biochemical experiments of vasopressin binding and adenylate cyclase activity measurements on partially purified membranes of rat kidney have clearly shown that the detected vasopressin binding sites are the specific receptors involved in adenylate cyclase activation (for review, see 15, 33). These conclusions are based on the following data : 1) The regions in which the specific vasopressin binding sites were detected (cortical and medullary portions of kidney) correspond to those in which a vasopressin-sensitive adenylate cyclase activity is present (34). 2) As illustrated in Fig. 5, (left panel), there is a good correlation between the Kbind of (^{3}H)-vasopressin and the concentration of the same hormone which activates the membranous adenylate cyclase at 50 % of its maximal value (Kact). These results have been confirmed for a series of vasopressin and oxytocin analogues (25). A good correlation exists between their Kbind and their corresponding Kact (Fig. 5, right panel). 3) Demonstration of parallel reduction of vasopressin receptors and vasopressin-stimulated adenylate cyclase activity in the different experimental or physiopathological situations (i.e. hereditary diabetes insipidus or vasopressin-induced desensitisation (35)) and 4) Existence of a parallel ontogenic development of vasopressin-sensitive adenylate cyclase activity and vasopressin receptors (36).

B) <u>The liver vasopressin receptors</u>.

In comparison to kidney receptors, a great deal is known about the coupling mechanisms between the liver vasopressin receptor and endogenous factors implicated in the hydrolysis of hepatic glycogene. As seen in Fig. 6, the binding of vasopressin to their specific receptors leads to the activation of phosphorylase a. This was verified by pharmacological studies realized on intact hepatocytes (27) where a close correlation between the concentration of the analogue leading to half maximal binding and to half maximal stimul-

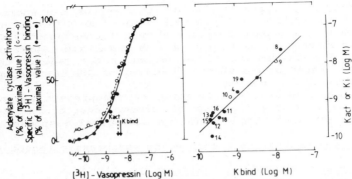

Figure N° 5 : Correlation between K binding and K activation or K inhibition of adenylate cyclase for lysine-vasopressin and analogues.
Left panel : Rat kidney membranes were incubated with increasing amounts of (3H)-vasopressin. In each case, the specific vasopressin binding and the activity of adenylate cyclase were measured.
Right panel : The K binding for each vasopressin analogues was plotted as function of its corresponding K activation or K inhibition value (Key to Symbols, see Table II).

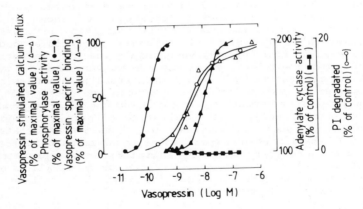

Figure N° 6 : The concentration dependence of lysine-vasopressin stimulated events in isolated hepatocytes.
Isolated hepatocytes were incubated with increasing amounts of lysine-vasopressin. For each concentration, hormonal binding, calcium influx, phosphorylase a activity were measured. Sources of other data are phosphatidyl inositol degradation, Kirk et al. (1981) Biochem. J. <u>194</u>, 155-165. With the permission of Biochem. J.

ation of phosphorylase a were established. As demonstrated by several authors (33, 37, 38), this hormonal activation of phosphorylase a was not mediated by a cyclic AMP process.

One of the primary effects which induces the glycogenolytic response to vasopressin is a rapid increase of the concentration of free cytosolic calcium (18). The absolute requirement of calcium for vasopressin-induced phosphorylase activation was based on the following observations : 1) No phosphorylase activation was observed when calcium was omitted in the incubation medium (39), 2) The calcium ionophore A23187 imitated the effect of vasopressin on phosphorylase a provided that calcium was present in the incubation medium (38) and 3) The increase of cytosolic calcium induced by vasopressin treatment is sufficient to produce the activation of phosphorylase b kinase (enzyme which converts phosphorylase b to a (18). Two distinct hypotheses have been proposed to explain this increase of cytosolic calcium in liver, induced by a vasopressin treatment : A mobilisation of the intracellular stores of calcium or a stimulation of the calcium influx from the external medium. The first mechanism is based on the observed mitochondrial release of calcium after vasopressin treatment (40). The second on 1) a rapid increase (less than 15 sec) of vasopressin stimulated calcium influx through hepatocyte membranes (38 and Mauger, unpublished results.) 2) the existence of a correlation between the concentration range of vasopressin analogues leading to hormonal binding and hormonal activation of calcium influx (Fig. 6), 3) an agonist specific increase of liver plasma membrane permeability to calcium after vasopressin treatment (41) and 4) a partial blockage of the increase of the hepatocyte cytoplasmic concentration of calcium induced by vasopressin by a depletion of external calcium (Claret submitted for publication). Probably these two mechanisms were implicated in the glycogenolytic response of vasopressin on liver. The first event should be a rapid stimulation of calcium influx. Then, the modification of the calcium cytosolic content may provoke calcium movements from other intracellular sources (endoplasmic reticulum, mitochondria, plasma membranes) by calcium-induced calcium-released mechanisms as earlier described (42).

As proposed by Mitchell and collaborators (43) the mechanisms which control the hormone-induced calcium increase in rat hepatocyte cytosol might be linked to the modification of the phosphatidylinositol metabolism. The following arguments favor this hypothesis : 1) Phosphatidylinositol breakdown was activated by vasopressin (27, 44 and Fig. 6), 2) Existence of a close correlation between the concentration of vasopressin analogues leading to half maximum activation of both phosphorylase a activity and phosphatidylinositol breakdown (46, 47), 3) Existence of a close correlation for a series of vasopressin analogues between their Kbind and their concentration leading to half maximal effect on phospatidylinositol metabolism (27, 45), 4) Antagonist analogues of the vasopressor response induced by vasopressin on the rat are ineffective to stimulate phosphatidylinositol breakdown (45), 5) The hormonal modification of the phosphatidylinositol metabolism is calcium independent (47). All these data are compatible with the following sequence of events : hormone binding to its specific receptor, modification in phosphatidylinositol metabolism, increase in cytosolic calcium concentration, activation of a calcium dependent kinase (phosphorylase b kinase), activation of the liver phosphorylase a, hydrolysis of hepatic glycogene and liberation of glucose in the blood. The mechanisms by which phosphotidylinositol breakdown may increase the cytosolic calcium concentration are unknown. Different hypotheses have been proposed : 1) a release of calcium bound to phosphatidylinositol during its hydrolysis and 2) the liberation of an endogenous calcium ionophore (such as phosphatidic acid) during phosphatidylinositol breakdown (48). However, it must be stressed that the relationship between modification in the metabolism of phosphatidylinositol and modification in calcium repartition is still unclear. Both events could arise separately from a primary response to vasopressin activation.

III Solubilization and characterization of molecules implicated in antidiuretic and glycogenolytic action of vasopressin.

As already described, (49) non ionic detergents were often used for the solubilization of membranous proteins. Previous studies (28, 29, 50) have revealed that Triton X-100 was one of the most efficient non ionic detergents. It allows an almost complete solubilization (70 - 90%) of the molecules studied under an active catalytic form.

A) Solubilization of the rat kidney system.

Determination of the hydrodynamic parameters of solubilized vasopressin-sensitive adenylate cyclase and vasopressin receptor complexes.

As seen in Fig. 7 (right panel) two peaks of adenylate cyclase activity were detected after elution through an ultrogel column, whereas three peaks of radioactivity corresponding to vasopressin-receptor complexes were objectified. The first peaks of both adenylate cyclase activity and radioactivity correspond to partially solubilized molecules since they were eluted in the void volume of the column. The remaining peaks correspond to complete solubilized forms of adenylate cyclase and hormone-receptor complexes. Their corresponding Stokes Radii are summarized in Table III. The sucrose gradient sedimentation profiles of these molecules (Fig. 7, left panel) allow the determination of their respective apparent sedimentation constants. Similar experiments were carried out with D_2O gradients in order to determine their partial specific volumes. The data, summarized in Table III, clearly show that 1) All these solubilized molecules are asymetrical detergent-protein components, 2) Two forms of solubilized vasopressin-receptor complexes are detected and 3) The vasopressin sensitive adenylate cyclase is a molecule distinct for the vasopressin receptors.

Effects of guanyl triphospho-nucleotides on solubilized adenylate cyclase and (^3H)-vasopressin-receptor complexes.

Guanyl triphospho-nucleotides are known to stimulate membranous adenylate cyclase (51). Even after treatment with detergents, the

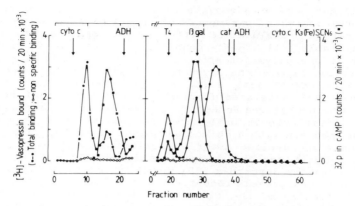

Figure N° 7 : Gel filtration and ultracentrifugation profiles of solubilized rat kidney vasopressin-receptor complexes and rat kidney adenylate cyclase.
Rat kidney membranes were incubated in the presence of (3H)-vasopressin and solubilized with Triton X-100. The soluble extracts were layered either on an Ultrogel ACA 34 column (left panel) or on 3-10 % linear sucrose gradient (right panel). Gel filtration and ultracentrifugation were performed. In each fraction radioactivity and adenylate cyclase activity were measured. Arrows indicate the elution peaks of the markers used to calibrate the column and the gradients.

TABLE III : MOLECULAR PARAMETERS OF SOLUBILIZED VASOPRESSIN RECEPTORS AND ADENYLATE CYCLASE

Parameters [a]	Rat Kidney Adenylate Cyclase		Rat Kidney Vasopressin Receptor		Rat Liver Vasopressin Receptor	
	Basal State	Gpp(NH)p activated state	Light form	Heavy form	Light form	Heavy form
Stokes Radius (nm) [b]	70 ± 1 (3)	66 ± 2 (3)	56 ± 1 (8)	65 ± 1 (6)	55 ± 1 (5)	65 ± 2 (5)
Apparent sedimentation coeff. (Svedberg unit) [c]	6.4 ± 0.1 (3)	7.4 ± 0.1 (3)	3.6 ± 0.1 (10)	6.1 ± 0.2 (10)	3.7 ± 0.1 (7)	5.8 ± 0.2 (7)
Standard sedimentation coeff. (Svedberg unit) [d]	6.5 (h)	7.6 (h)	3.8	6.2	3.8	5.7
Partial specific volume (ml.mg^{-1}) [e]	-	-	0.78 ± 0.01 (10)	0.74 ± 0.01 (30)	0.75 ± 0.01 (12)	0.72 ± 0.01 (9)
Detergent protein complex molecular weight (Dalton) [f]	200,000 (h)	230,000 (h)	108,000	180,000	90,000	150,000
Protein molecular weight (Dalton) [g]	-	-	85,000	171,000	81,000	150,000

[a] Values given are means ± SD for the number of determinations in parentheses
[b] Determined according to Laurent and Killander (J. of Chromatography 1964, 14, 317-330)
[c] Determined according to Martin and Ames (J. Biol. Chem. 1961, 236, 1372-1379)
[d] Calculated according to Bon et al. (Eur. J. Biochem. 1973, 35, 572-579)
[e] Calculated according to Meunier et al. (FEBS Letters 1972, 24, 63-68)
[f] Calculated according to Siegel and Monty (Biochim. Biophys. Acta. 1966, 112, 346-362)
[g] Calculated according to Smigel and Fleischer (J. Biol. Chem. 1977, 252, 3689-3696)
[h] Calculated on the basis of a partial specific volume = 0.75 ml.g^{-1} (value determined on solubilized pig kidney adenylate cyclase which exhibits the same other molecular parameters)

solubilized form of adenylate cyclase remains activated (28, 52). The physicochemical parameters of this activated state of solubilized adenylate cyclase are different from those of the unactivated enzyme (Table III). This accounts for an increase of the molecular weight of the enzyme of about 30,000. Such results may signify an interaction between the light form of adenylate cyclase (enzyme in the basal state) with a GTP binding protein. This complex may represent the activated form of adenylate cyclase (heavy form). These interpretations agree with the results of Gilman and collaborators. These authors have shown the existence of a specific GTP binding protein with a molecular weight of 40,000 interacting with the adenylate cyclase after an activation with either triphospho guanyl nucleotides or a neurotransmitter (53).

As previously described (30) Gpp(NH)p reduce the affinity of the membranous vasopressin receptor of rat kidney by a factor 3 to 5 with no modification of the maximal binding capacity of the membrane. Unlike the adenylate cyclase, solubilization experiments of vasopressin-receptors formed in the presence of Gpp(NH)p failed to reveal any differences in the physicochemical parameters of the two forms of receptors. Nevertheless, guanyl nucleotide treatment led to a decrease of about 60 % of the heavy form of the receptor. Experiments realized with partially purified forms of solubilized vasopressin-receptor complexes indicate that these two forms have different dissociation constant rates (k-1 = 0.066 and 0.036 min^{-1} for the light and the heavy forms respectively). Moreover, only the dissociation rate constant of the heavy form is affected by the presence of Gpp(NH)p.

Effects of the duration of the incubation time between vasopressin and kidney membranes.

When the duration of the incubation time between hormone and membranes was increased from 5 to 30 min, the relative amount of the heavy form of the receptor was increased by a factor 2 approximately. No significant modification of the specific vasopressin binding to membranes was observed during this period. This could probably signify a conversion of the light form to a heavy form occuring during incubation.

Molecular interpretation.

The determination of the properties of the two solubilized-vasopressin receptor complexes raises the question of the nature of these two forms. On the basis of their molecular weight, the heavy form of the receptor may represent a dimer of the light form (see Table III). However, it is not likely that the association of two insensitive forms to guanyl nucleotide may produce a sensitive one. The heavy form of the receptor probably represents a complex between the light form of the receptor and another component sensitive to guanyl nucleotides. The molecular weight of this component can be roughly estimated by the difference of the molecular weight of the two vasopressin receptors. The value we obtained (86,000) is in line with those determined directly by Gilman's group for the guanyl-nucleotide binding proteins (called G/F) known to interact with the β-adrenergic receptor (52, 53).

The experiments concerning the properties of the basal form and the Gpp(NH)p activated form of adenylate cyclase have also demonstrated the existence of a GTP binding protein of 30,000 molecular weight associated with the adenylate cyclase catalytic moiety after activation by guanyl nucleotides.

These results could signify that two distinct GTP binding proteins are engaged in the molecular mechanisms leading to the activation of rat kidney adenylate cyclase. On the basis of our results, such a hypothesis cannot be excluded. Nevertheless, as a great number of analogies exist between hormonal and neurotransmitter activation of adenylate cyclase, it seems reasonable to consider, in line with the recent works on the β-adrenergic receptor coupled to adenylate cyclase, that the GTP binding protein of 30,000 molecular weight liberated during the activation processes would represent a subunit of the 86,000 molecular weight molecule (52, 54, 55, 56).

According to this hypothesis, the scheme of hormonal activation of adenylate cyclase became clear. The formation of a ternary complex between the hormone, its specific receptor and a GTP binding protein of 86,000 molecular weight represents the first step of this mechanism. The interaction of GTP with this ternary complex then lead to the liberation of the 30,000 molecular weight GTP binding protein. By interactions with the catalytic moiety of the adenylate cyclase this protein activates the enzyme.

B) <u>Solubilization of the rat liver vasopressin-receptor complexes</u>.

Determination of the hydrodynamic parameters of vasopressin receptors complexes.

The ultrogel column elution profiles and sucrose gradient sedimentation profiles of (^{3}H)-vasopressin receptor complexes solubilized from liver membranes by the use of Triton X-100 are very similar to those obtained with solubilized kidney vasopressin receptors. Two major peaks of radioactivity corresponding to solubilized hormone-receptor complexes formed on intact membranes are detected. Their physicochemical parameters were summarized in Table III. These data clearly indicate that few differences can be exempt from the two vasopressin systems studied, despite the existence of marked differences observed in their functional properties (see Section I).

Properties of the vasopressin-receptor complexes solubilized from rat liver membranes.

As previously shown, (Fig. 2) GTP at high concentrations, increases the dissociation rate of membranous vasopressin-receptor complexes. Experiments realized on solubilized vasopressin-receptor complexes seem to indicate that this effect is reduced on crude supernatant and suppressed on partially purified extracts. Moreover, the two forms of solubilized vasopressin-receptor complexes have approximately the same dissociation rate constant (k_{-1}=0.11 and 0.12 min_{-1} for the light form and the heavy form respectively).

As described in the case of the kidney vasopressin-receptor the duration of the incubation time between membrane and hormone increases the relative amount of the heavy form of the solubilized vasopressin-receptor complexes. This effect is somewhat weaker.

These preliminary results are not sufficient to propose, as in the case of kidney vasopressin-receptor, a general model of interpretation.

SUMMARY AND CONCLUSIONS

Specific vasopressin binding sites have been characterized on rat kidney and rat liver plasma membranes. In these two systems, it has been clearly established that these binding sites are the receptors involved in the biological effects of vasopressin on these tissues.

Comparative studies of these two vasopressin receptors have revealed the existence of similar properties : Same Kbind for vasopressin, same molecular parameters, same requirement for magnesium and same effect of the duration of the incubation time between hormones and membranes. Marked differences were also noted : different recognition patterns for a series of vasopressin structural analogues and different molecular mechanisms triggered by vasopressin action.

The molecular mechanisms involved in vasopressin activation of renal adenylate cyclase are now well established. Still, little is known about the molecular events leading to the activation of hepatic glycogenolysis after vasopressin stimulation. Moreover, the analogies between the V1 and the V2 vasopressin receptor previously summarized lead us to propose a hypothesis concerning the molecular mechanisms. The vasopressin-receptor complex may interact with an unknown molecule. The formation of this ternary complex (heavy form of the receptor) is necessary for the increase of the cytosolic calcium level of the hepatocytes. This complex may represent : an activated form of an enzyme implicated in the degradation of phosphatidylinositides, a calcium channel or a molecule which generated an intracellular messenger implicated in calcium mobilisation. A greater understanding of these molecular mechanisms could be probably obtained by knowing more about the nature of this molecule.

REFERENCES

1. Grantham, J.J. and Burg, M.B. (1966) Am. J. Physiol., 211, 255-259.
2. Hall, D.A., and Varney, D.M. (1980) J. Clin. Invest. 66, 792-802.
3. Ausiello, D.A., Kriesberg, J.I., Roy, C. and Karnovsky, H.J. (1980) J. Clin. Invest. 65, 754-760.
4. Zusman, R.M. and Kreiser, H.R. (1977) J. Clin. Invest. 60, 215-223.
5. Konrads, A., Hofbaver, K.G., Werner, U. and Gross, F. (1978) Pflugers Arch. 377, 81-85.
6. Hems, D.A. and Whitton, P.D. (1973) Biochem. J. 136, 705-709.

7. Saameli, K. (1968) in Handbook of Experimental Pharmacology (B. Berde eds.) 23, 545-612, Springer Verlag, Berlin and New York.
8. Doepfner, N. In Neurophyphophysial hormones and similar peptides (B. Berde eds.) Handbook of Experimental Pharmacology, Vol. 23 Springer Verlag, Berlin and New York.
9. Haslam, R.J. and Rosson, G.H. (1972) Am J. Physiol. 233, 958-967.
10. Nagasubramanian, S. (1977) Trans. Opthalmol. Soc. U.K. 97, 686-701.
11. De Wied, D. and Bohus, B. (1978) Prog. Brain. Res. 48, 327-336.
12. Tanaka, M., De Kloet, E.R., De Wied, D. and Versteeg, D.H.G. (1977) Life Sci. 20, 1799-1803.
13. Olivier, G. and Shafer, E.A. (1895) J. Physiol. London 18, 277-279.
14. Verney, E.B. (1947) Proc. R. Soc. Sci. B. 135, 27-106.
15. Jard, S., Roy, C., Barth, T., Rajerison, R. and Bockaert, J. (1975) in Advances in Cyclic Nucleotides Research, Vol. 5, 31-52. Drummond, G.I., Greengard, P. and Robinson, G.A. Eds. Raven Press, New York.
16. Jard, S. (1983) Current Topics in Membranes and Transport 18, 255-285.
17. Orloff, J. and Handler, J.S. (1967) Amer. J. Med. 42, 757-768.
18. De Wulf, H., Keppens, S., Vandenheede, J.R., Haustraete, F., Proost, C. and Carton, H. (1980) In hormone and cell regulation. Nunez, J. and Dummont, J. Eds. North Holland Publ. Amsterdam.
19. Mitchell, R.H., Kirk, C.J. and Billah, M.H. (1979) Biochem. Soc. Trans. 7, 861-865.
20. Pradelles, P., Morgat, J.L., Fromageot, P., Camier, M., Bonne, D., Cohen, P., Bockaert, J. and Jard, S. (1972) FEBS Letters 26, 189-192.
21. Sturmer, E. (1968) Handbook of Experimental Pharmacology, (Eds. B. Berde) Springer Verlag, Berlin and New York Vol. 23, 130-189.
22. Neville, D.M. (1968) Biochim. Biophys. Acta 154, 540-552.
23. Bockaert, J., Roy, C., Rajerison, R. and Jard, S. (1973) J. Biol. Chem. 248, 5922-5931.
24. Guillon, G., Butlen, D., Cantau, B., Barth, T. and Jard, S. (1982) Eur. J. Pharmacol. 85 291-304.
25. Butlen, D., Guillon, G., Rajerison, R., Jard, S., Sawyer, W.H. and Manning, M. (1978) Molecular Pharmacol. 14 1006-1017.
26. Salomon, Y., Londos, C. and Rodbell, M. (1974) Anal. Biochem. 58, 541-548.
27. Cantau, B., Keppens, S., De Wulf, H. and Jard, S. (1980) J. of Receptor Research 1, 137-168.
28. Guillon, G., Couraud, P-O. and Roy, C. (1979) Biochim. Biophys. Research. Commun. 87, 855-861.
29. Guillon, G., Couraud, P-O., Butlen, D., Cantau, B. and Jard, S. (1980) Eur. J. Biochem. 111, 287-294.
30. Rajerison, R. (1979) Thèse de Doctorat. Université de Pierre et Marie Curie, Paris.
31. Imbert, M., Chabardes, D., Montaigu, M., Clique, A. and Morel, F. (1975) Pflugers Arch. 357, 173-186.

32. Imbert, M., Chabardes, D., Montaigu, M., Clique, A. and Morel, F. (1975) C.R. Acad. Sci. Paris, 280, 21329-2132.
33. Jard, S. (1981) J. Physiol. Paris 77, 621-628
34. Rajerison, R., Marchetti, J., Roy, C., Bockaert, J. and Jard, S. (1974) J. Biol. Chem. 249, 6390-6400.
35. Rajerison, R., Butlen, D. and Jard, S. (1977) Endocrinology 161, 1-12.
36. Rajerison, R., Butlen, D. and Jard, S. (1976) Mol. Cell. Endocrinol. 4, 271-285.
37. Kirk, C.J. and Hems, D.A. (1974) FEBS Letters 47, 128-137.
38. Keppens, S., Vandenheed, J.R. and De Wulf, H. (1977) Biochim. Biophys. Acta. 496, 448-457.
39. Khoo, J.C. and Steinberg, D. (1975) FEBS Letters 57, 68-72.
40. Taylor, W.H., Prpic, V., Exton, J.H. and Bygrave, F.L. (1980) Biochem. J. 188, 443-450.
41. Guillon, G., Pogglioli, J. and Claret, M. Abstract 5th International Conference on Cyclic Nucleotides and Protein Phosphorylation.
42. Endo, M. (1977) Physiological Review, 57, 71-108.
43. Mitchell, R.H., Kirk, C.J. and Billah, M.H. (1979) Biochem. Soc. Trans. 7, 861-865.
44. Kirk, J.C., Verrinder, T.R. and Hems, D.A. (1977) FEBS Letters 83, 267-271.
45. Kirk, C.J., Mitchell, R.H. and Hems, D.A. (1981) Biochem. J. 194, 155-165.
46. Kirk, C.J., Creba, J.A., Downes, C.P. and Mitchell, R.H. (1981) Biochem. Soc. Trans, 9, 135.
47. Billah, M.H. and Mitchell, R.H. (1979) Biochem. J. 182, 661-668.
48. Barritt, G.J., Dalton, K.A. and Whiting, J.A. (1981) FEBS Letters 125, 137-140.
49. Helenius, A. and Simons, K. (1975) Biochim. Biophys. Acta. 415, 29-79.
50. Guillon, G., Roy, C. and Jard, S. (1978) Eur. J. Biochem. 92, 341-348.
51. Rodbell, M. (1980) Nature 284, 17-22.
52. Howlett, A.C. and Gilman, A.G. (1980) J. Biol. Chem. 255, 2861-2866.
53. Sternweiss, P.C., Northup, J.K., Smigel, M.D. and Gilman, A.G. (1981) J. Biol. Chem. 256, 11517-11526.
54. Limbird, L.E., Gill, D.M. and Lefkowitz, R.J. (1980) Proc. Natl. Acad. Sci. USA 77, 775-779.
55. Hanski, E., Sternweiss, P.C., Northup, J.K., Dromerick, A.W. and Gilman, A.G. (1981) J. Biol. Chem. 256, 12911-12919.
56. Stadel, J.M., Shorr, R.G., Limbird, L.E. and Lefkowitz, R.J. (1981) J. Biol. Chem. 256, 8718-8723.

Résumé

La vasopressine, hormone polypeptidique de neuf acides aminés produite par l'hypophyse régule chez les mamifères les mouvements d'eau au niveau du rein. Plus récemment, son activité glycogénolytique a été démontré au niveau du foie. L'existence chez un même animal de récepteur à la même hormone localisé sur des organes differents et impliqués dans des régulations biologiques distinctes pose le problème de l'existence d'isorécepteur à la vasopressine.

L'utilisation de lysine-vasopressine tritiée a permis de démontrer l'existence de récepteurs spécifiques situés sur les membranes plasmiques des tissus cibles de l'hormone.

Des études pharmacologiques et biochimiques de ces 2 types de récepteur ont révelées des similarités de propriétés, 1) les 2 récepteurs ont la même affinité pour la vasopressine, 2) la liaison de l'hormone est regulée par des agents identiques (magnesium et nucleotides triphosphate), 3) la taille du récepteur hépatique solubilisé est voisine de celle du rein. Cependant, de grandes différences ont aussi été observées : a) selon le tissu considéré, les mécanismes moléculaires impliqués dans l'action de la vasopressine sont distincts, (élévation de l'AMP cyclique cellulaire dans le cas du rein, élévation du calcium cytosolique dans le cas du foie.) b) l'ordre d'affinité d'une série d'analogue de la vasopressine est différent selon le récepteur considéré.

L'ensemble de ces différences justifie donc la classificaiton préconisée par Mitchell et collaborateurs ou V1 correspond au récepteur vasopressique de type hépatique et V2 au récepteur de type rénal.

THE MECHANISM OF CONTROL OF cGMP PHOSPHODIESTERASE BY PHOTOEXCITED RHODOPSIN IN RETINAL CELLS. ANALOGIES WITH HORMONE CONTROLLED SYSTEMS.

Marc CHABRE[1], Claude PFISTER[1], Philippe DETERRE[1] and Hermann KÜHN[2]

[1]Laboratoire de Biologie Moléculaire et Cellulaire (ER 199 CNRS), DRF, Centre d'Etudes Nucléaires, 85 X, F 38041 Grenoble (France) and [2]Institute für Neurobiologie, K.F.A. Jülich, D-5170 (West Germany).

INTRODUCTION

It was first observed in 1971 that illumination of retinal rod cells depressed the cyclic nucleotide level in the photosensitive outer segment (1). The effect was later characterized as resulting from the activation of a cGMP selective phosphodiesterase (2) (3). GTP was found to be an important cofactor in the process (4,5,6), and analogies were suspected between this system and that involved in stimulation of adenylate cyclase in some hormone sensitive cells. By 1980 the main components of the photosensitive reaction chain controlled by cyclic nucleotides had been characterised and purified (7,8,9). The major components are : the receptor molecule, rhodopsin (R), the GTP binding protein and the cGMP phosphodiesterase (PDE). Progress has been very fast in the last few years in the elucidation of the molecular mechanism involved in the light induced control of cGMP. These now seem to be understood in more details than for the hormone sensitive systems, due to two major advantages presented by retinal rod outer segments : 1) - the proteins involved in the light induced control of cGMP constitute more than 95 % of the total protein content of the organelle and 2) light is a particularly convenient stimulating agent. This system may then, besides its implications in the visual transduction process, become a very powerful and general model of modulation of the intracellular cyclic nucleotide level. However the very special morphology of this organelle may have a significant influence on the fast rate and high amplification of the light induced reactions controlling the cGMP level. The regularly stacked discs represent a considerable area of membrane. The intrinsic or peripheral proteins are present in or on these membranes in quantities equivalent to that of millimolar concentrations over the cytoplasmic volume of the cell. The cytoplasm is split by the disc in extremely thin (120 Å) and extended layers in which soluble components have a very short free path, the diffusion being almost restricted to two dimensions. We shall not discuss here the electrophysiology except for the fact that a diffusible transmitter (cGMP and/or Ca++), controlled by light, has to interact with the Na+ channels or transporters in the cell membrane. But we shall review the main characteristics of the major proteins involved in the light induced modulation of cGMP.

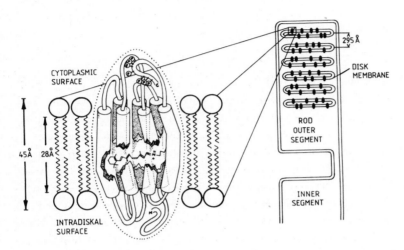

Fig. 1 - Schema of a retinal rod and of rhodopsin. In a frog rods there are ≃ 1500 stacked discs of 6μ diameter, with 10^6 rhodopsin molecules per disc. Transducin and cGMP phosphodiesterase, not shown here, are peripherally bound to the cytoplasmic surface. The kinase is soluble in the cytoplasm. The cytoplasm layer between two discs is only about 120 Å thick (reproduced from (11) with permission).

THE COMPONENTS OF THE REACTION CASCADE

The receptor : Rhodopsin. Rhodopsin, the "photon receptor" may be considered as the equivalent of an hormone receptor in a cell membrane. It is an intrinsic membrane protein. It spans the thickness of the membrane, and bears carbohydrate residues on the intradiscal surface : this is originally the extracellular surface in the growing discs which originate from infoldings of the cell membrane. The intradiscal volume is topologically equivalent to an extracellular medium, with probably high concentrations of Na+ and Ca++ ions. Rhodopsin is extraordinarly concentrated in the disc membrane, of which it constitutes about 50 % of the dry weight, and the near

totality of the intrinsic protein components. Its equivalent molar
concentration, referred to the total volume of the rod outer segment is 2.5
millimolar.

Bovine rhodopsine has been fully sequenced. Biophysical and biochemical
studies have now converged toward a well defined model for its structure : a
hydrophobic core formed of seven transmembrane α helices, with two
hydrophilic domains of about equivalent size (10 KD) on both sides of the
membrane (11) (see fig 1). The retinal chromophore is in the median plane of
the molecule, buried in the hydrophobic interior.

The chromophore may be visualised as the equivalent of a small hormone,
already bound, but in an inactive state, to its membrane receptor. The
photoisomerisation of the retinal may be envisionned as an activation of a
silent hormone, which in turns activates its receptor protein : It has been
demonstrated (12) that the photoisomerisation step, completed within
picoseconds, is very endoenergetic for the chromophore in its proteins site :
it requires 35Kcal/mol out of the 50 Kcal/mol of photon energy. For a
retinal free in solution the same cis-trans isomerisation would be nearly
isoenergetic (E = 0.5 Kcal/mol). Constraint energy is thus stored in the
protein upon the isomerisation, and forces the protein to change conformation. It converts R into an activated R* state, reached within 10^{-3} sec.
This long delay, on the time scale of molecular motions, is indicative of
significant conformation changes. R*,defined as the state which interacts
with the GTP dependant protein, has been identified with the state Meta II
rhodopsin,reached within a millisecond and stable for many seconds(13,14). At
this stage the chromophore is still bound into its site. The detachment of
the chromophore from the protein is a much slower process which, in vivo has
probably no relevance for the regulation of the cyclic nucleotides.

The very high concentration of rhodopsin in a rod cell is accounted for by
the low efficiency of photon capture, as compared for example with the high
efficiency of hormone receptor molecules : photons do not wander around the
cell until they bind by high affinity to a receptor, as an hormone molecule
does. The photon crosses the cell only once on a straight line. The only
possibility to increase the light catching efficiency is to accumulate the
maximum number of chromophores on their path. But the full photoactivation of
the rod cell requires only a very minor proportion of the rhodopsin molecules
to be photoexcited : in a dark adapted rod, containing a few 10^7 rhodopsin
molecules, the photoresponse is detectable upon photoexcitation of only a few

rhodopsins, and saturation is reached when less than 10^3 rhodopsin are photoexcited : this last number is in the same range as that of activable membrane receptors in an hormone sensitive cell.

Rhodopsin is highly mobile in the disc membrane : its lateral diffusion rate is indeed the highest ever measured for a membrane protein (15). This is probably of importance for the fast interaction of R* with many GTP binding proteins.

<u>The GTP dependant protein</u> : Transducin (T). Various other names are used in the litterature for this protein : GTPase, GTP binding protein, G protein. We use here Transducin specifically for the GTP dependant protein found in retinal rod. It consists of three subunits T_α (39KD) T_β (37KD) and T_γ (6KD) (7,10,17,18). Transducin is present in the rod at a stochiometry of 1 to 10 with respect to rhodopsin and represents about 20 % of the protein content of the cell. In the dark T is bound to the disc membrane if the medium has an ionic composition comparable to that of the cytoplasm. But it is extractable by washing at low ionic strength. This demonstrates a peripheral and non specific binding to the membrane surface. H. Kühn (10) made the capital observation that illumination and addition of GTP modify dramatically the interaction of T with the membrane. This led to a very elegant method of purification, and gave much insight into the mechanism of coupling of T with R*. The basic finding was that upon illumination of fragmented rods, washed free of any GTP, T becomes strongly bound to the membrane, through a specific, ionic strength independant binding to R*. Extensive washing may then remove all other peripheral proteins from the sedimentable membrane fraction. Subsequent addition of GTP or non hydrolysable analogs release specifically and quantitatively pure Transducine. Upon washing at low ionic strength one gets the three subunits, but at moderate ionic strength only T_α is released, most of $T_{\beta\gamma}$ remains bound to the membrane fraction (16,17). This method, which requires only a centrifuge and some light, amount to a purification by affinity binding of T on its natural receptor protein. Of the three subunits, only T_γ has yet been sequenced (19). Amino acid composition and peptide maps of T_α and T_β have been recently been compared (20) to that of GTP binding proteins involved in other cyclic nucleotide systems. The analogies are striking, especialy for the T_β subunit; we shall discuss that later.

The Effector : cGMP Phosphodiesterase. The PDE is also a peripheral membrane protein. It is extractable from dark adapted or illuminated disc membranes by washing with a low ionic strength, magnesium free medium. Its stochiometry is approximatly 1 per 100 rhodopsins. It is constituted of three subunits of respectively 88 KD, 86 KD and 10 KD. The lighter subunit is a heat stable, trypsin sensitive inhibitor (21,22). A short trypsinisation of the native holoenzyme leads to a considerable increase of its activity and recombination with native inhibitor hinders this activity.

Other proteins interacting with R*. The first protein ever demonstrated to interact with R* was an ATP dependant kinase which phosphorylates specifically photoexcited rhodopsin(8). Inactive on R, it can insert up to 9 phosphates on R*, most of them near the C terminal end (23). Upon illumination in the absence of ATP it remains bound to R*. Its relative stochiometry in situ is below 1 per 1000 R and its molecular weight is 65 KD.

Another protein, highly soluble, quite abundant in situ (3 per 100 R) and of 48 KD molecular weight is also observed to bind to rhodopsin after illumination (16). This binding is however slow and requires the presence of millimolar quantities of nucleotide triphosphate. The time delay and nucleotide dependance suggests that the binding is on a late state of photoexcited rhodopsin, different from the state R* responsable for the interaction with T (24).

THE AMPLIFYING CASCADE REACTION

By using as monitor the proton released upon the hydrolysis of cGMP, Liebman and Pugh measured a high speed of hydrolysis of this nucleotide upon weak flash illumination (25).This implied that one R* could control the activation of many P.D.E.. Fung, Hurley and Stryer (18,26) later elucidated the role of T, demonstrating that the interaction of R* with T catalysed a GTP/GDP exchange on T_α and that the coupling of T_{GTP} with the PDE occured after the dissociation of the R*-T complex : one R* is then able to interact sequentially with many T. The kinetics were however not accessible to the biochemical approaches. Kühn et al. (27) then demonstrated that light scattering transients observed in the near I.R. (which does not photoexcite rhodopsin), upon flash photoexcitation of rods, are related to the onset of the R*-T-PDE cascade. This provides very convenient signals to monitor the

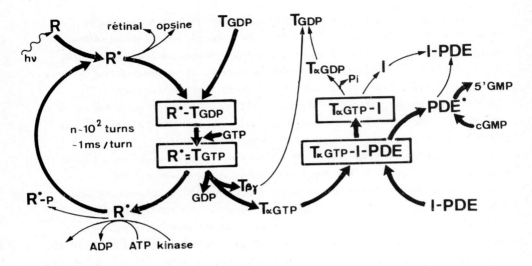

Fig. 2 - The rhodopsine-transducine-phosphodiesterase amplifying cascade. The thick arrows represent fast processes triggered within the lifetime of one R*.

kinetics and stochiometry of differentes steps of this cascade reaction. Further recent work (14,24,28,29) combining biochemical approaches, reconstitution with purified components, near I.R. light scattering and other optical approaches, have led to the amplification scheme of fig. (3). The main steps are :

1) The R→R* transconformation. A binding site for T_{GDP} is created on one or more of the loops connecting the transmembrane helices; this site does not include the free C terminal end. The kinase phosphorylation sites are precisely located on this C terminal end, which has been observed to become also more sensitive to proteolytic attack upon the R R* transconformation (30), as if the transconformation released in the cytoplasm a previously buried C terminal end.

2) T_{GDP} binds to R*. If GTP has been suppressed from the medium, this binding remains permanent. It induces a change of conformation of T, which will allow the GTP/GDP exchange. That a change of conformation preceedes the exchange is indicated for example by the fact that T bound to R* is ADP ribosylable by choleratoxine which is inactive on free T, in the dark. The

binding of T to R* reacts also on R* : R* bound to T undergoes a chromophore spectral decay different from that of free R* (13,24). The R*-T binding gives rise in the near I.R. to a characteristic signal which has been used to demonstrate the one to one stochiometry of the complex (27).

3) <u>The GDP/GRP exchange on T</u>. With the high concentration of GTP (10^{-3}M) normally present in the cell, the bound T_α exchanges its GDP for a GTP. The exchange, catalysed by the R*-T interaction, is very fast, and the R*-T complexe dissociates. R*, as a catalyst, has remained unmodified, and is therefore capable of interacting with a new T_{GDP}. A cycle of binding, exchange and released is completed in situ within about a millisecond. One R* is therefore able to activate a few hundred T in the few hundred millisecond lag time of the visual response at very low level of illumination. This speed is probably controled by the diffusion rates of R* and T respectively in and on the membrane.

4) <u>T_α activates the PDE</u>. The T_{GTP} formed dissociates not only from R*, but also, in part from the membrane : $T_{\alpha GTP}$ separates from $T_{\beta\gamma}$, the former subunit being markedly more soluble at physiological ionic strength than the later. $T_{\alpha GTP}$ seems not to require the presence of $T_{\beta\gamma}$ to interact with the PDE inhibitor. There are indications that $T_{\alpha GTP}$ suppress the inhibition of the PDE by physically releasing the inhibitor subunit from the catalytic ones (31). This point needs further confirmation.

It may seem strange that $T_{\alpha GTP}$ becomes soluble to shuttle between two membrane bound proteins R* and PDE. But it makes sense when one considers the special morphology of the organelle : the thinness of the cytoplasmic layer between two discs (~120Å) ensures a high rate of collision of the soluble protein with the enzymes bound on both the membrane surfaces which limit the cytoplasmic layer.

5) <u>GTP hydrolysis</u>. Upon the hydrolysis of the GTP on T_α, its interaction with the PDE inhibitor vanishes, $T_{\alpha GDP}$ is released and the PDE inhibition is restored. $T_{\beta\gamma}$ seems not required for this hydrolytic activity on GTP. But $T_{\alpha GDP}$ has to bind back with $T_{\beta\gamma}$ in order to be able to interact again with R*. This "regeneration" cycle of T is however too slow to be significant for the regulation of the amplification process : In vitro, the half life of $T_{\alpha GTP}$ is of the order of 30 sec.

THE REQUIREMENT OF GTP HYDROLYSIS FOR AMPLIFICATION.

Non hydrolysable analogs of GTP lead to an irreversible activation of the PDE. The hydrolysis of GTP appears as a turn off signal. This leads sometime to a misunderstanding on the necessity of this hydrolysis, considered as a wasteful consumption of GTP. The confusion arises from the fact that the T_{GTP} (or T_{GNP}) state of transducin, being that which interacts with the effector is considered as the "active" state and the $T_{GTP} \rightarrow T_{GDP}$ transition as a "deactivation" step. Indeed the free energy (or part of) liberated by the GTP hydrolysis must be kept by the protein in its T_{GDP} state. T_{GDP} must be able to interact spontaneously with R* : the conformational change of T_{GDP} bound to R*, which preceeds the GDP/GTP exchange, is only facilitated, not driven by R*. As a catalyst, R* does not provide any energy in the process. R* must be left unchanged upon completions of the process otherwise it would be unable to interact in sequence with a large number of T_{GDP}. T_{GDP} is therefore driving the reaction with the energy stored upon the hydrolysis of GTP in the preceeding turn of the cyclic process. Wathewer the exact stage of its injection into the cycle, the energy of the hydrolysable phosphate bond is required to complete a cycle of information transfer from R* to PDE. No cyclic reaction can occur without an energy input. With the non hydrolysable analogs of GTP the cycle is blocked. It just happens that the blocking occurs at a stage where T_α remains irreversibly bound to the PDE inhibitor since the energy required to release T_α comes from the hydrolysis of the nucleoside triphosphate.

PHOSPHORYLATION OF R* AND ATP DEPENDANT REGULATIONS.

Photoexcited rhodopsin, in the R* state is highly phosphorylable on its C terminal end by the ATP dependant kinase. Inhibitory competition, probably of steric origin, can be observed between T binding and kinase action on R* (24): In the absence of GTP and with T in excess with respect to R* (e.g. 1 % rhodopsin photoexcited) all R* remain blocked in R*-T complexes, and the kinase action is strongly hindered. Upon addition of micromolar amounts of GTP or non hydrolysable analog (not a substrat for the kinase), the R*-T complexes dissociates very fast, all the T_{GDP} pool is processed and the kinase becomes active. On the other hand, if R* is in excess over T (e.g. > 10 % rhodopsin photoexcited) the kinase is always active on this excess free R*, independantly of the presence of GTP.

The phosphorylation of R* is probably a regulatory process. The regulation would depend on the relative rates of kinase action on R* and T "activation" by R*. It has been demonstrated that ATP, the kinase substrate, quenches the light induced PDE activation within a few seconds after a small flash (32). The detailed mechanism by which phosphorylated R* becomes unable to further interact with T is however not yet understood and may involve more than rhodopsin phosphorylation.

Other regulations must also exist, independant of R* : in vitro, the rate of GTP hydrolysis by T_{GTP} is very slow, T_{GTP} has an half life of the order of 30 seconds. This would maintain a high level of PDE activity for minutes, even after all R* has been blocked. Other processes might exist in vivo, to switch off faster the interaction of T_{GTP} with the PDE inhibitor.

ANALOGIES WITH OTHER SYSTEMS OF CONTROL OF CYCLIC NUCLEOTIDES

Both biochemical and functionnal evidences now exist which demonstrate strong analogies between the various GTP dependant proteins involved in cyclic nucleotide regulation. These analogies extend to excitatory (e.g. cyclase activity increase) as will as inhibitory system and to cGMP control as well as cAMP.

On the biochemical side, the recent finding by Manning and Gilman (20), that the β subunit of the various systems are even better conserved than the α subunits comes as a surprise: the T_α subunit carries the GDP/GTP binding site and seems, in the retinal rod at least, to be sole responsible for the release of the PDE inhibition; the role of $T_{\beta\gamma}$ is not clear, it does not seem to be directly involved in the effector activation, but seems critical for the receptor recognition and R*-T binding. The strong conservation of this subunit might imply that the various receptor molecules in the different systems have a similar recognition site for T_β on their cytoplasmic surface.

Reconstitutions experiments with mixed system, where for example visual transducin complement cyclase of other origin have been performed (33). More needs to be done on this line, with purified subunits, to better define their specific roles.

Another line of comparison is provided by the study of toxine effects. T is ADP ribosylable by choleratoxin, with a maximum sensitivity after illumination in the absence of GTP, that is when T_{GDP} is bound to R*, or when it is blocked with a non hydrolisable analog of GTP, such as GNP (34). This is analogous, to the effect of choleratoxin on hormone sensitive adenylate cyclase systems.

A more recent, and more puzzling finding is that transducin is also sensitive to ADP ribosylation by Pertussis toxin (from Bordetella per tussis), but this time preferentially in its T_{GDP} state (L. Stryer and H. Bourne, private communication). In hormone sensitive systems, pertussis toxin is thought to act through the ADP ribosylation of an inhibitory protein postulated to be different from the "excitatory" GTP dependant protein sensitive to choleratoxin. In the retinal system, both toxins act on different sites of the same protein and block it in different states: an "excitatory" state for choleratoxin active after illumination, and an "inhibitory" state for pertussis toxin active in the dark. This is a clear case where the retinal system, with its simple biochemistry and its easily controlled excitation by light, may provide significant clues for the comprehension of other, apparently more complex, hormonal systems.

REFERENCES

1. Bitenski, M., Gorman, R. and Miller, W. (1971) Proc. Natl. Acad. Sci. U.S. 68 561-562.

2. Goridis, C. and Virmaux, N. (1974) Nature 248 57-58.

3. Chader, G.J., Herz, L.R. and Fletcher, R.T. (1974) Biochim. Biophys. Acta 347 491-493.

4. Wheeler, G.L., Matuo, Y. and Bitenski, M.W. (1977) Nature 269 822-823.

5. Bignetti, E., Cavaggioni, A. and Sorbi, R.T. (1978) J. Physiol., Lond. 279 55-69.

6. Robinson, W.E. and Hagins, W.A. (1979) Nature 280 398-400

7. Godchaux, W. and Zimmerman, W.F. (1979) J. Biol. Chem. 254 7874-7884

8. Kühn, H. (1978) Biochemistry 17 4389-4395

9. Baehr, W., Delvin, M.J. and Applebury, M.L. (1979) J. Biol. Chem. 254 11669-11677.

10. Kühn, H. (1980) Nature 283 587-589.

11. Dratz, E.A. and Hargrave, P.A. (1983) Trends in Bioch. Sci. 8 128-131.

12. Cooper, A. (1979) Nature 531-533.

13. Emeis, D., Kühn, H., Reichert, J. and Hofman, K.P. (1982) F.E.B.S. Letters 143 29-34.

14. Bennett, N., Michel-Villaz, M. and Kühn, H. (1982) Eur. J. Biochem. 127 97-103.

15. Poo, M. and Cone, R.A. (1974) Nature 247 438-440.

16. Kühn, H. (1981) in current topics in membrane and transport 15 171-201.

17. Kühn, H. (1982) Methods in Enzymology (Academic Press New York) 81 556-564.

18. Fung, B.K., Hurley, J.B. and Stryer, L. (1980) Proc. Natl. Acad. Sc. US 77 2500-2504.

19. Hurley, J. Private communication.

20. Manning, D.R. and Gilman, A.G. (1983) J. Biol. Chem. 258 7059-7063.

21. Hurley, J.B., Barry, B and Ebrey, T.G. (1981) Biochim. Biophys. Acta 675 359-365.

22. Hurley, J.B. and Stryer, L. (1982) J. Biol. Chem. 257 11094-11099.

23. Wilden, V. and Kühn, H. (1982) Biochemistry 21 3014-3016.

24. Pfister, C., Kühn, H. and Chabre, M. (1983) Eur. J. Biochem. Under the press.

25. Yee, R. and Liebman, P.A. (1978) J. Biol. Chem. 253 8902-8909.

26. Fung, B.K. and Stryer, L. (1980) Proc. Natl. Acad. Sci. U.S. 77 2500-2504.

27. Kühn, H., Bennett, N., Michel-Villaz, M. and Chabre, M. (1981) Proc. Natl. Acad. Sci. U.S. 78 6873-6877.

28. Kühn, H. and Hargrave, P.A. (1981) Biochemistry 20 2410-2417.

29. Bennett, N. (1982) Eur. J. Bioch. 123 133-140.

30. Kühn, H., Mommertz, O. and Hargrave, P.A. (1982) Biochim. Biophys. Acta 679 95-100.

31. Yamazaki, A., Stein, P.J., Chernoff, N. and Bitenski, M. (1983) J. Biol. Chem. 258 8188-8194.

32. Liebman, P.A. and Pugh, E.N. (1980) Nature 287 734-736.

33. Bitenski, M.W., Wheeler, M.A., Rasenick, M.M., Yamazaki, A. Stein, P.J., Halliday, K.R. and Wheeler, G.L. (1982) Proc. Natl. Acad. Sci. U.S.A. 79 3408-3412.

34. Abood, M.A., Hurley, J., Pappone, M.C., Bourne, H.R. and Stryer, L. (1982) J. Biol. Chem. 257 10540-10543.

Résumé

La photoexcitation de la rhodopsine dans les bâtonnets rétiniens entraîne une baisse rapide du niveau de GMP cyclique dans cette cellule. Cet effet est transmis par une protéine GTP dépendante, et le mécanisme paraît analogue à celui par lequel des récepteurs hormonaux activent les AMP cyclases. Les composants de la cascade amplificatrice entre la rhodopsine photoexcitée (R*), équivalente du récepteur hormonal, et la phosphodiestérase à GMP cyclique (PDE) sont très bien caractérisés dans les bâtonnets : la rhodopsine a été séquencée et sa structure commence à être bien décrite. La protéine GTP dépendante, Transducine T est facile à purifier, c'est une protéine périphérique, à trois sous-unités dont l'une, T_α devient soluble lorsqu'elle a chargé un GTP. La phosphodiestérase est aussi une protéine périphérique, dont l'activité est contrôlée par un inhibiteur. L'interaction catalytique R*-T permet à une R* de catalyser l'échange GDP/GTP sur des centaines de T, d'où l'amplification. $T_{\alpha GTP}$ se lie à l'inhibiteur de la PDE. Des analogies biochimiques et fonctionnelles importantes ont été démontrées entre T et l'unité G/N des systèmes hormonaux.

CYCLIC NUCLEOTIDES AND CALCIUM IN PARAMECIUM:
A NEUROBIOLOGICAL MODEL ORGANISM.

JOACHIM E. SCHULTZ, G. BOHEIM*, DORIS GIERLICH, W. HANKE*, REGINHARD VON HIRSCHHAUSEN, GERTRUD KLEEFELD, SUSANNE KLUMPP, MANFRED K. OTTO, and ULRICH SCHÖNEFELD
Pharmazeutisches Institut der Universität Tübingen, Morgenstelle 8, 7400 Tübingen, and *Lehrbereich Zellphysiologie, Ruhr-Universität Bochum, 4630 Bochum. Federal Republic of Germany (FRG)

INTRODUCTION

The concentration of extra- and intracellular Ca in eucaryotic cells differs by about 4 orders of magnitude, inside low. Therefore, Ca-ions may have two functions in cellular homeostasis: (i) Ca-ions can be used as signal carriers across the cell membrane for transmission of externally received information, and (ii) careful regulation of intracellular Ca-levels by a variety of Ca-transport systems and Ca-binding proteins allows the cell to orchestrate the regulation of several Ca-dependent enzymes. A cellular system where Ca-ions are used for both purposes is the ciliated protozoon Paramecium. It can respond to many stimuli, e.g. of thermal, chemical or mechanical nature, with an alteration of its swimming speed and direction. The well known backward swimming of this unicellular organism in response to adverse stimuli ('avoiding response') is controlled by a regenerative and graded Ca/K action potential resembling the Na/K action potential in higher systems as nerve and muscle cells (1-3). Ca-ions carrying the depolarizing inward current enter the cell via voltage-sensitive Ca-channels localized predominantly in the membrane covering the cilia (3,4). The rise in the intraciliary Ca-concentration is responsible for the reversal of the ciliary beat by a yet unknown mechanism (5). While Ca-ions are absolutely essential for this process, it is not known whether the Ca-ions which carry the action potential are sufficient by themselves for this purpose. Ca-ions also control a slow K^+-outward current (6) and may be related to adaptation of Paramecium to ionic stimuli.

The cilia of Paramecium seem to be a suitable model to integrate electrical and biochemical events at a molecular level since many behavioral mutants with defined defects in the electrical properties of the excitable ciliary membrane are available (7,8). These mutants should be of considerable help in elucidating the biochemistry of signal reception, -transmission and -response in Paramecium.

MATERIALS AND METHODS

Cell culture and preparation of membrane vesicles - Paramecium tetraurelia wild type 51s and the 'pawn' mutants pwA (d4-94), leaky pwB (d4-96) and pwA/pwB, (a constructed double mutant, genotype pwA/pwA pwB/pwB; provided by Dr.C. Kung, Univ. Wisconsin, Madison) were grown axenically as described (9). Cilia from stationary phase cultures were used in all experiments. Cells were deciliated by a calcium shock. Cilia were purified by differential centrifugation and were found 'pure' by electron microscopy (10). For Ca-flux measurements ciliary membrane vesicles were prepared at 4°C with a French Press in the presence of 15 mM arsenazo III. Vesicles were collected by centrifugation and washed several times with buffer containing 50 μM EGTA to completely remove free arsenazo III and Ca^{2+}. The vesicles were purified by a Percoll density gradient (11). The low density fraction containing ciliary membrane vesicles (fraction I, for details see (11)) was used in Ca-flux experiments. Also, adenylate cyclase and guanylate cyclase distribution was examined in this gradient (see Fig. 4).

Calcium flux studies - Kinetics of Ca-influx into membrane vesicles were studied in a stopped-flow apparatus, mixing time was 40 ms (12). Total Ca-permeability was analyzed in a Zeiss PM6 spectrophotometer (10). To monitor possible changes in turbidity, we measured at three wavelengths: 572 nm (an isosbestic point of arsenazo III), 652 nm (maximal difference between free dye and its calcium complex), and 740 nm (neither dye nor its Ca-complex absorb). The absorbance time course at 652 nm in the absence of Ca was subtracted from the trace at 652 nm in the presence of Ca and the percentage saturation of arsenazo III with Ca^{2+} was computed assuming a 1:1 complex. All values were corrected for vesicle lysis (usually <5%) (11).

For measurement of adenylate cyclase and guanylate cyclase, ciliary membrane vesicles were purified by a sucrose gradient (13). Removal of axonemal tubulin and purity of ciliary membranes was monitored by electron microscopy and SDS-PAGE (11). Enzyme activities were determined as reported (14,15).

RESULTS

So far, ion channels of excitable membranes have been characterized mainly by electrophysiological and pharmacological methods. Direct biochemical measurements of ion fluxes with membrane vesicles were limited by the difficulty to discriminate between specific and leakage fluxes. The excitable ciliary membrane from Paramecium contains voltage dependent Ca-channels which have been studied in detail by electrophysiological methods. Due to the brilliant work of Kung et alii (for review see (7,8)), a number of behaviorally and electrophysiologically well characterized mutants ('pawns') is available which lack this Ca-conductance. Using these mutants similarly to specific drugs used in pharmacological studies, one is able to unequivocally discriminate between specific Ca-fluxes and leakage currents.

When Ca^{2+} was added to ciliary membrane vesicles prepared from freshly harvested Paramecium (50 µM free Ca^{2+}), a rapid influx occurred (Fig. 1A). About 80% of the arsenazo III in the vesicles from wildtype cells was complexed by Ca^{2+}. However, in vesicles prepared from Ca-channel deficient mutants (pwA, d4-94 and pwA/pwB) saturation of arsenazo III by Ca^{2+} never exceeded 50%. The observed saturation of arsenazo III by calcium was not due to dye leakage, but to calcium influx into the vesicles. On addition of 0.1% Triton X-100, >95% of arsenazo III was immediately complexed by Ca^{2+} indicating that (i) sufficient calcium had been added initially and, (ii) in pawns, a large fraction of the dye enclosed by the vesicles was not accessible to the external calcium. Vesicles from cilia which had been stored at -180°C exhibited a reduced calcium influx for both wildtype and pawn mutants. However, the relative differences in calcium permeability between wildtype and mutant vesicles remained the same, indicating identical calcium permeation sites (Fig. 1B).

Apparently, influx of Ca^{2+} into the vesicles proceeded until the concentration of free Ca^{2+} inside the vesicles equalled the external concentration of Ca^{2+}. This was evidenced by the finding that the calcium fluxes were reversible. Consequently, at equilibrium, the dye saturation was a measure for the vesicular volume accessible to externally added calcium. Therefore, the large difference in Ca-influx between wildtype and pawn ciliary vesicles can only be ascribed to the mutation in calcium conductance in the mutant membrane, since this is the only phenotypic character of the ciliary membrane known to be affected by the mutation (7,16).

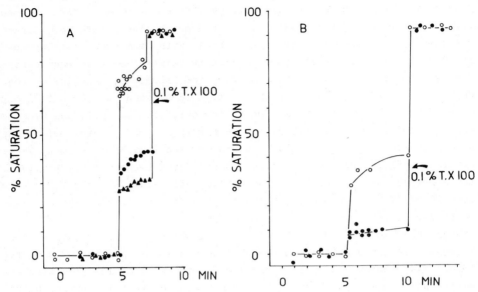

Fig. 1. Calcium influx into membrane vesicles of fresh cilia (A) and cilia stored at -180°C (B). O, Paramecium tetraurelia wild type 51 s; ▼, mutant pwA (d4-94); ●, mutant pwA/pwB. At 5 min, CaCl$_2$ was added to a final concentration of 50 μM free Ca . The absorbances at 740, 652, and 572 nm were recorded at 15 s intervals and the % saturation of arsenazo III by calcium was calculated; 100% saturation corresponds to an influx of 15 nmol of Ca^{2+}/mg of protein. Means of 5 experiments are shown; SEM < 5%. Arrows: addition of Triton X-100 to 0.1%. Note that a sizeable fraction of the membrane vesicles is remarkably tight toward external calcium. (from Thiele & Schultz, 1981 (10)).

The influx/efflux sequence could be repeated several times indicating that under the isopotential experimental conditions used most of the calcium gates remained open at 50 μM free Ca^{2+} within the vesicles.

To further characterize the nature of the ciliary Ca-gate, a stopped-flow spectrophotometric technique was used to study the kinetics of Ca-flux into ciliary membrane vesicles from Paramecium tetraurelia wild type and several 'pawn' mutants with defective Ca-conductances (12). As observed in stationary experiments, the absolute amount of Ca-permeable vesicles was significantly reduced in preparations from the 'pawn' mutants compared to wild type. However, influx kinetics were identical for vesicles from wild type and 'pawn' mutant Paramecia when the fraction of Ca-permeable vesicles was taken into account. The influx rate was about 500 nmol Ca/mg protein x min^{-1}

for vesicles from wildtype and pawn mutants. Ca-influx was rapid with a time constant of about 1.5 s and an initial saturation rate of arsenazo III of about 50%/single vesicle x s^{-1}. Ca-influx rates were half maximal at approximately 20 μM Ca. Comparisons of Ba toxicity tested with a behavioral assay, Ca-inward conductances under voltage clamp conditions and Ca-influx kinetics between wild type and the 'pawn' mutants pwA (d4-94), leaky pwB (d4-96) and the double mutant pwA/pwB indicated that Ca-transport in all types of ciliary membrane vesicles occurred through similar Ca-gates (12).

The biochemical identification of the Ca-channel rests entirely on the differences in Ca-permeability between wildtype and pawn vesicles. However, differences in gross membrane properties such as phospholipid content and composition of membrane proteins could result in membrane vesicles of different size and structure and could possibly cause these differences. This can be ruled out since: (i) no difference in phospholipid content and composition has been found between pawns and wildtype (17,18), (ii) no difference in the pattern of the major membrane proteins between several pawn mutants and wildtype has been detected (13); and (iii) inspection of the vesicles by electronmicroscopy showed the same size distribution for preparations from wildtype or pawn mutants (average diameter = 300 nm (11)).

One may further argue that only the difference in Ca-permeability between wild type and pawn vesicles can be ascribed to the 'pawn' mutation, i.e. the voltage-dependent Ca-conductance. However, the kinetic properties of only these differences of Ca-influx were similar to fluxes observed for wildtype or pawn vesicles (12). Therefore, the data support the idea of a qualitatively similar Ca-gate in wild type and 'pawn' vesicles.

The reduced calcium influx into vesicles from pawn mutants can then be conceived as a large reduction of the number of functional calcium channels in the mutants. This raises the possibility that the influx occurred only through the voltage-sensitive Ca-channels in all types of vesicle preparations and that the 'pawn' mutation may represent a regulatory defect of the channel biosynthesis rather than a structural one. This would also explain the lack of clear differences in protein patterns of wild type and 'pawn' cilia (13). Experiments with artificial vesicle preparations are currently in progress to further elaborate on the kinetic properties of the Ca-influx.

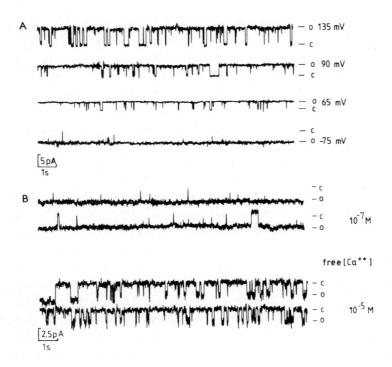

Fig. 2 Reconstitution of an ion conducting channel from ciliary membrane fragments into planar lipid bilayers. In the presence of cholesterol, single channel current fluctuations are observed which are voltage and Ca-dependent.
Virtually solvent-free planar bilayers were formed from a mixture of lipid (1-stearoyl-3-myristoyl-glycero-2-phosphatidylcholine/soybean -phosphatidylethanolamine/cholesterol, 80/10/10, mol%) in n-hexane using sandwich teflon septa (19). Salt solutions on both sides: 100 mM KCl, 5 mM HEPES, pH 7.2. A: 0.1 µM free Ca^{2+}; B: free Ca as indicated. Applied voltages: A: as indicated; B: -80 mV. Temperature: 22°C. Ciliary membrane vesicles were added to the cis-side of the membrane system to which also voltage signs refer. This means, trans-side is virtual ground.
Note that the channel is open (o) at low Ca-concentrations and less positive values. Higher Ca-concentrations and/or higher voltages lead to an increase in the probability to adopt the closed state (c).

Another approach to study the ciliary ion conductancies is by single channel analysis. Yet, at present a direct access to individual membrane channels in Paramecium by electrophysiological methods is impossible because of the small size of the cilia. One way out of this difficulty is reconstitution in planar lipid bilayer membranes. By this method, highly purified ciliary membrane vesicles were incorporated into a lipid matrix by fusion (19). Two different types of channels were observed in 100 mM KCl: (i) a slow cation selective channel of 15 pS conductance (19,20), and (ii) a faster bursting channel of 45 pS conductance which is voltage and Ca-dependent in its gating kinetics (Fig. 2). The Ca-dependence of this ciliary channel is invers to that of the Ca-dependent K-channel from mammalian skeletal muscle (21), i.e. Ca^{2+} closes the ciliary channel. Whereas the latter ion channel can only be detected in the presence of cholesterol in the lipid matrix and with low Ca-concentrations, the muscle channel is active in the absence of the sterol and only in the presence of 50 mM $CaCl_2$. Using ciliary membrane fragments from a pawn mutant it was shown that this ciliary channel type exhibits a much shorter mean lifetime of the open state compared to wild type (20). Further experiments of this type will certainly be valuable in order to characterize the ciliary ion-conductancies in more detail.

How the biochemical processing of the ionic signal in the cilia is accomplished to finally result in a reversal of the ciliary power stroke, remains to be established. In the central nervous system of higher animals cyclic nucleotide formation and protein phosphorylation/dephosphorylation certainly participate in regulation of nervous activity, although a final role has not yet been found. Therefore, we looked for the presence of components of the cyclic nucleotide system in the excitable organelle of Paramecium, the cilium. We have identified and partially characterized an adenylate cyclase (15), guanylate cyclase (22), several protein kinases (23) and their endogenous substrate proteins (24), cyclic nucleotide phosphodiesterase, calmodulin (26), and, by immunocytochemical localization, calcineurin, a phosphoprotein phosphatase (27).

Characterization of guanylate cyclase - Using a sucrose step gradient known to separate ciliary membranes, incompletely demembranated cilia and the mechanical machinery, the axonemes (13), guanylate cyclase with an activity up to 6 nmol/min x mg^{-1} was found in the ciliary membrane. Both, Mn^{2+} and Mg^{2+} were used as metal cofactor. ATP was only a poor substrate (2-5% of that with GTP). Under normal assay conditions, EGTA at micromolar concentrations proved to be a powerful inhibitor of guanylate cyclase activity (ID_{50} = 20 μM). Ciliary membrane vesicles washed free of remaining Ca-ions with Chelex treated reagents exhibited only 20% of control activity. Activity could be completely restored by addition of Ca (ED_{50} = 8 μM, Fig. 3.). The maximal effect was obtained with 60 μM Ca^{2+}. Calcium increased V_{max} while Km was not affected.

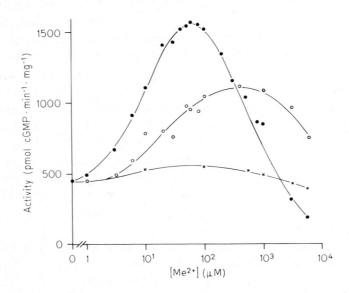

Fig. 3. Effect of divalent cations on guanylate cyclase activity. Membranes were washed with 60 μM EGTA followed by two washing steps with Ca-free buffer prior to the assay. All other reagents had been purified by Chelex treatment and Ca-contamination was < 1 μM. Activity was determined in the presence of various concentrations of Ca^{2+} (●), Sr^{2+} (O) or Ba^{2+} (X) with 21 μg protein/tube. (from Klumpp and Schultz, 1982 (14)).

The stimulatory effect was specific for Ca-ions. Only Sr was capable to partially replace Ca^{2+}, Ba was inactive (Fig. 3.). Many effects of intracellular calcium are mediated by calmodulin. It was found that antibodies against calmodulin from Tetrahymena and soybean inhibited guanylate cyclase by 70% indicating that a membrane bound calmodulin participated in the regulation of guanylate cyclase (28). Surprisingly, EGTA washing (up to 500 μM) did not release calmodulin, however, low concentrations of lanthanum (13 μM), often used as calcium analogue, inhibited the ciliary guanylate cyclase by dissociation of endogenous calmodulin (29).

Reconstitution was achieved with calmodulin from Paramecium, Tetrahymena, pig brain, soybean, spinach, and zucchini (30). Ca-binding proteins lacking trimethyllysine like calmodulin from Dictyostelium, parvalbumin and troponin C, failed to restore enzyme activity. Reassociation of the homologous and other calmodulins with the enzyme was comparably weak. Even in the presence of 60 µM Ca^{2+}, calmodulin remained in the supernatant upon a single centrifugation (48,000 x g for 60 min) while the totally inactive catalytic subunit sedimented with the membrane fraction. The regulation of the enzyme by cations was also altered. Whereas Ca was the most potent and specific activator of the native enzyme, in the reconstituted system Sr was far more effective and even Ba^{2+}-ions showed considerable activity (28).

Due to its regulation via Ca/calmodulin, the guanylate cyclase is a likely candidate for a Ca-sensor which may translate the ionic signal during an action potential into a biochemical reaction sequence via cyclic GMP stimulated protein phosporylation.

Identification of adenylate cyclase - In addition to the guanylate cyclase a particulate adenylate cyclase was identified in the cilia (15). When the sucrose step gradient mentioned above was used to pinpoint the subciliary localization of adenylate cyclase, about 90% was found in the membrane fraction. This was remarkably similar to the localization of the guanylate cyclase. Although both ciliary cyclases are completely membrane bound, quite a number of differences in enzyme properties exist: (i) Adenylate cyclase was not affected by the calcium chelator EGTA in the assay while guanylate cyclase was strongly inhibited. (ii) Guanylate cyclase is powerfully inhibited by La^{3+} which dissociates its tightly bound calmodulin. Yet, La-treated membrane preparations had the same adenylate cyclase activity as untreated membranes. Even in the presence of 100 µM La^{3+}, a concentration which completely inhibits guanylate cyclase, adenylate cyclase activity was not impaired. (iii) Adenylate cyclase was most sensitive to preincubation at 37° C while guanylate cyclase activity was not. (iv) Guanylate cyclase had a pH optimum around pH 8 while adenylate cyclase activity increased with rising pH values.

Differential distribution of calcium channels, adenylate cyclase and guanylate cyclase in the ciliary membrane - Purification of ciliary membrane vesicles for Ca-flux measurements was carried out using a 25% Percoll gradient of low osmolarity upon fragmentation of cilia by French Press treatment (11). Two distinct fractions were obtained (Fig. 4). As determined by electron microscopy (11), the upper band consisted of low density membrane vesicles contaminated with axonemal fragments. The second band of slightly higher bouyant density contained mainly intact 9+2 axonemes and a small amount of membrane vesicles. Three markers were used to monitor the content of ciliary membrane vesicles in the Percoll gradient fractions: (i) the entrapped dye arsenazo III as marker for the intravesicular volume, (ii) guanylate cyclase activity, and (iii) adenylate cyclase activity (Fig. 4). To our great surprise, the membrane markers were clearly separated by the gradient. The content of arsenazo III was highest in the vesicle fraction of low buoyant density which was used for Ca-flux studies. Guanylate cyclase activity was found in the region of higher density. About 50% of the total adenylate cyclase activity was localized in membrane vesicles of low buoyant density which carry the bulk of the voltage-sensitive Ca-channels. The remainder of adenylate cyclase activity was associated with membrane fragments of higher density. Strikingly, the second peak of adenylate cyclase reproducibly preceeded the peak guanylate cyclase activity (4 separate experiments (15)). This differential distribution of adenylate cyclase, guanylate cyclase and the calcium conductance indicates an interesting heterogeneity in the enzymic pattern of the excitable ciliary membrane.

In order to exclude any possibility of an inactive guanylate cyclase activity in the first part of the Percoll gradient (e.g. guanylate cyclase without calmodulin), the initial fractions were tested for guanylate cyclase activity with optimal concentrations of calmodulin purified from Paramecium at various concentrations of Ca. We did not obtain any indication that the low density vesicles contained masked guanylate cyclase activity.

The possibility had to be excluded that the apparent separation of the ciliary membrane into two functionally different components by the Percoll gradient was the result of adsorption of membrane fragments onto axonemal components. Therefore, sucrose step gradients were used to analyze both fractions from the Percoll gradient individually. About 90% of the arsenazo III entrapped in membrane vesicles, 90% of guanylate cyclase activity and of adenylate cyclase activity appeared on top of the respective 45% sucrose layers containing exclusively ciliary membrane vesicles. The bands of incompletely demembranated cilia (45-55% sucrose) and axonemes (55-66% sucrose) contained virtually no arsenazo III, adenylate cyclase or guanylate cyclase activity. Thus, it can be concluded that (i) the observed membrane heterogeneity is no experimental artifact, and (ii) the adenylate cyclase and guanylate cyclase activities are due to separate protein moieties which, most likely, will be subject to individual regulation.

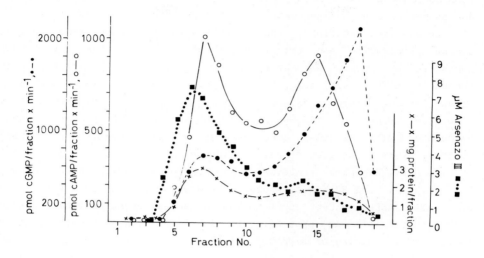

Fig. 4. Distribution of arsenazo III (■), adenylate cyclase (O) and guanylate cyclase (●) activities in a Percoll gradient (11,15).

The first conclusion is further supported by the observation that the ciliary membrane contains two domains of different membrane fluidity as indicated by two distinct fluorescence lifetimes of diphenylhexatriene of 7.9 and 12.4 ns, respectively (31). The possible biological relevance of the spatial separation of the calcium-sensor enzyme guanylate cyclase, adenylate cyclase and the calcium channels remains to be seen.

Cyclic AMP and cyclic GMP can stimulate protein kinases which have been identified and partially purified from the cilia (23). Total ciliary protein kinase activity consists of two soluble cyclic AMP stimulated protein kinases corresponding to type I and type II enzymes characterized in higher systems, and a cyclic GMP-dependent protein kinase. About 50% of the total protein kinase activity is membrane bound (23).

Several endogenous substrate proteins of the cilia were phosphorylated in vitro by these kinases (24,25). Labeling was strongly stimulated by cyclic AMP and to a lesser extent by cyclic GMP. Since ATP breakdown was most rapid in cilia and in subciliary fractions we used multiple substrate additions during incubations to show that phosphorylation was almost completed within 30 s. Only very little dephosphorylation by phosphoprotein phosphatases occurred during 5 min of incubation. Possibly, phosphoprotein phosphatases were removed during preparation and fractionation of cilia or the activity was inhibited by unknown regulatory factors. Alternatively, the persistent phosphorylation could suggest that this effect is somehow linked to production of longer lasting changes in excitability. Proteins of molecular weight of 103,000 and 46,000 were shown to be particularly associated with axonemal structures of the cilia. Cyclic AMP receptor proteins were identified by use of the photoaffinity label 8-azido-[^{32}P] cyclic AMP. Receptor proteins with apparent molecular weights of 43, 39, 37, 31, and 30 K were probably related to the regulatory subunits of cyclic AMP-dependent protein kinases as evidenced by inhibition of incorporation of the photoaffinity label by low concentrations of cyclic AMP. Tagging of a protein of 85 K was specifically inhibited by cyclic GMP, thus in all likelihood represented a cyclic GMP-dependent protein kinase. Corresponding autophosphorylated protein bands were observed with [α-^{32}P]ATP.

We have also found a reasonable phosphodiesterase activity in the cilia. In purified cilia, most of the activity is of particular nature. Its kinetic properties are currently under investigation.

Immunocytochemical studies - The localization of cyclic GMP, cyclic GMP-dependent protein kinase, calmodulin and the calmodulin-binding protein calcineurin in Paramecium tetraurelia cells has been examined with immunocytochemical methods (27). These molecules appeared to be localized to a large extent in the cilia. To ascertain that antibodies had access to all cellular compartments we have used three different preparations for immunocytochemistry: (i) with 'whole cell' preparations immunofluorescent staining for all four molecules was mainly visible in the cilia; (ii) in 'deciliated' Paramecium, staining for cyclic GMP and calmodulin was found in regular patterns on the cell surface most likely representing kinetosomes; (iii) using 'sectioned cells', additional cytoplasmic calmodulin appeared to be associated with glycogen particles as evidenced by the disappearance of the granular staining pattern after preincubation with α-amylase. In contrast, cyclic GMP, cyclic GMP-dependent protein kinase and calcineurin fluorescence was only very weak and diffuse in cell bodies. The physical presence of calcineurin has, however, not been confirmed by its isolation. Such experiments are in progress.

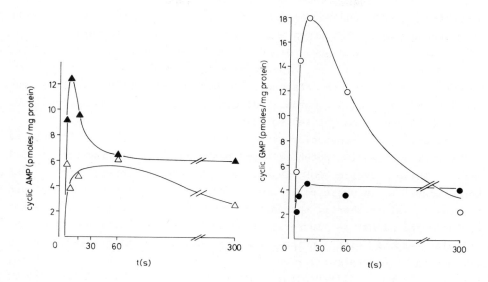

Fig. 5. Alterations of cyclic AMP (left) and cyclic GMP (right) levels in whole Paramecium after a sudden increase (at 0 min) of Ca-ions (filled symbols) or Ba-ions (open symbols).

CONCLUSION

As evident from the data summarized above, the cilia from Paramecium constitute a unique organelle in that they not only possess proteins necessary for regulation of ion fluxes, i.e. voltage-dependent Ca-channels and ATPases (32), and a complete set of enzymes necessary to amplify the ionic signal, but they are also the organelle responsible for the immediate behavioral response of Paramecium to external stimuli, i.e. the alteration of swimming speed and direction, by virtue of their mechanical apparatus. Using Paramecium as a neurobiological model, many meaningful suggestions can be made about the functional role of the cyclic nucleotide cascades in regulation of excitation and ion fluxes. However, we have not yet tapped the full potential of this neurobiological model organism. First, we have not really used the rich supply of mutants available with defective electrogenesis or mechanics which will help substantially in the assignments of specific functions to the molecular components identified so far. Second, we are only beginning to carry out 'in vivo' studies to learn about the physiological regulation of cyclic nucleotide levels and to compare and integrate it to the wealth of data on membrane potential changes and concomitant behavioral responses.

Initial data would be compatible with the general idea that cyclic AMP levels increase on hyperpolarizing stimuli while cyclic GMP concentrations rise in response to depolarizing stimuli. Also, very preliminary data with a mutant indicate that cyclic GMP may actually be involved in processes related to adaptation.

Stimulation of Paramecium by a sudden increase in Na^+ or K^+ results in a series of action potentials and backward jerks lasting for up to 1 min. While cyclic AMP levels are barely affected, cyclic GMP concentrations in whole cells seem do double within 1 min and to return to basal values shortly thereafter. When the concentration of divalent earth metal ions is suddenly changed in the incubation buffer, the animals respond with an immediate increase in swimming speed, usually reflecting a hyperpolarization of the membrane. Under these conditions, dramatic and differential alterations of intracellular cyclic nucleotide levels are seen (Fig. 5). In response to a 20-fold increase in Mg^{2+}, Ca^{2+}, and Sr^{2+} concentrations, cyclic AMP levels increase about 3-fold within 3 s from a baseline of 8 pmol/mg protein. The fall is rapid and starts after 15 s. Cyclic GMP levels on the other hand were most dramatically elevated by addition of Ba^{2+} (5-fold) and to a lesser extent by Ca^{2+} and Sr^{2+}. The time course of the changes was very similar (Fig. 5).

While drawing a detailed model from these preliminary data would certainly overextend the realm of permissible speculation, it may be fair to conclude that the blend of molecular ingredients available at this moment will make Paramecium a promising neurobiological model to study in the future.

ACKNOWLEDGEMENTS

This work was supported by the Deutsche Forschungsgemeinschaft (SFB 76 and 114) and by the Fonds der Chemischen Industrie (J.E.S.).

REFERENCES

1. Naitoh, Y. and Eckert, R. (1968) Z.Vgl.Physiol. 61,427-452.
2. Meech, R.W. (1978) Annu.Rev.Biophys.Bioeng. 7,1-18.
3. Eckert, R. and Brehm, P. (1979) Annu.Rev.Biophys.Bioeng. 8,352-382.
4. Dunlap, K. (1977) J.Physiol.(London) 271,119-133.
5. Naitoh, Y. and Kaneko, H. (1972) Science 176,523-524.
6. Satow, Y. and Kung, C. J.Exp.Biol. 88,293-303.
7. Oertel, D., Schein, S.J. and Kung, C. (1977) Nature(London) 29,49-60.
8. Kung, C., Chang, S.Y., Satow, Y., Houten, J. van, and Hansma, H. (1975) Science 188,898-904.
9. Thiele, J., Honer-Schmid, O., Wahl, J., Kleefeld, G. and Schultz, J.E. (1980) J.Protozool. 27,118-121.
10. Thiele, J. and Schultz, J.E. (1981) Proc.Natl.Acad.Sci.USA 78,3688-3691.
11. Thiele, J., Klumpp, S., Schultz, J.E. and Bardele, C.F. (1982) Eur.J.Cell Biol. 28,3-11.
12. Thiele, J., Otto, M.K., Deitmer, J.W. and Schultz, J.E. (1983) J.Membr.Biol.in press.
13. Adoutte, A., Ramanathan, R., Lewis, R.M., Dute, R.R., Ling, K.-Y., Kung, C. and Nelson, D.L. (1980) J.Cell Biol. 84,717-738.
14. Klumpp, S. and Schultz, J.E. (1982) Eur.J.Biochem. 124,317-324.
15. Schultz, J.E. and Klumpp, S. (1983) FEBS Letters 154,347-350.
16. Satow, Y. and Kung C. (1980) J.Exp.Biol. 84,57-71.
17. Andrews, D. and Nelson, D.L. (1979) Biochim.Biophys.Acta 550,174-187.
18. Kaneshiro, E.S., Beischel, L.S., Merkel, S.J. and Rhoads, D.E. (1979) J.Protozool. 26,147-158.
19. Hanke, W., Eibl, H. and Boheim G. (1981) Biophys. Struct. Mech. 7,131-137
20. Boheim, G., Hanke, W., Methfessel, C., Eibl, H. Kaupp, U.B., Maelicke, A. and Schultz, J.E. (1982) in Transport in Biomembranes: Model Systems and Reconstitution. (R. Antoli et al. eds.) Raven Press, New York. pp.87-97.
21. Methfessel, C. and Boheim, G. (1982) Biophys. Struct. Mech. 9,35-60.
22. Schultz, J.E. and Klumpp, S. (1980) FEBS Letters 122,64-66.
23. Schultz, J.E. and Jantzen, H.M. (1980) FEBS Letters 116, 75-78.

24. Eistetter, H., Seckler, B., Bryniok, D. and Schultz, J.E. (1983) Eur.J.Cell Biol.in press.

25. Lewis, R.M. and Nelson, D.L. (1981) J.Cell Biol. 91, 167-174.

26. Walter, M.F. and Schultz, J.E. (1981) Eur.J.Cell Biol. 24,97-100.

27. Klumpp, S., Steiner, A.L. and Schultz, J.E. (1983) Eur.J.Cell Biol.in press.

28. Klumpp, S., Kleefeld, G. and Schultz, J.E. (1983) J.Biol.Chem.in press.

29. Schultz, J.E. and Klumpp, S. (1982) FEMS Microbiol.Letters 13, 303-306.

30. Schultz, J.E. and Klumpp, S. (1984) Adv.Cycl.Nucl.Res.in press.

31. Otto, M.K., Krahbichler, G., Thiele, J., Ölkrug, D. and Schultz, J.E. submitted.

32. Doughty, M.J. (1978) Comp.Biochem.Physiol. 60B,339-345.

Résumé

Dans la membrane qui entoure l'axonème, machinerie mécanique responsable du battement ciliaire du Paramecium tetraurelia, on trouve des canaux Ca^{++}-voltage dependants, impliqués dans un potentiel d'action Ca^{++}/K^{+}. Des études biochimiques de la membrane ciliaire excitable ont permis de localiser les canaux Ca^{++} dans une fraction particulière de la membrane, distincte d'une guanylate cyclase Ca^{++}/calmoduline dépendante, liée à la membrane. Une adénylate cyclase est apparemment distribuée de manière assez uniforme dans l'entièreté de la membrane ciliaire. Les protéines kinases cyclic nucléotide-dépendantes, les substrats endogènes qui y correspondent et une phosphodiestérase ont été identifiés du point de vue biochimique. Une tentative de localisation d'une phosphoprotéine phosphatase, la calcineurine, a été effectuée dans les cils par immunocytochimie. In vivo, les taux de nucléotide cyclique augmentent en réponse à des changements de concentration des cations mono et bivalents. Vu que de nombreux mutants du Paramecium, bien caractérisés du point de vue de l'électrophysiologie et du comportement, sont disponibles, ce système représente un modèle neurobiologique excellent pour déterminer la fonction biochimique des nucléotides cycliques dans une organelle excitable.

CONTROL OF INTERMEDIARY METABOLISM

CONTRÔLE DU MÉTABOLISME INTERMÉDIAIRE

REGULATION OF FATTY ACID SYNTHESIS BY INSULIN, GLUCAGON AND CATECHOLAMINES.

D. GRAHAME HARDIE, ROSS HOLLAND AND MICHAEL R. MUNDAY
Department of Biochemistry, Dundee University, Dundee, DD1 4HN, Scotland.

INTRODUCTION

Since most mammals can satisfy their requirements for membrane fatty acids from the diet, the pathway of fatty acid synthesis in these organisms is concerned mainly with the conversion of excess dietary carbohydrate into triacylglycerol, which provides a more concentrated store of energy for the long term. Given that this process is fairly costly of energy, it is clearly necessary that the pathway be stringently regulated. Fatty acids are synthesised only at times when the carbohydrate available is surplus to requirements, and are not synthesised during periods of starvation or stress. This regulation occurs at two levels. Firstly, the amounts of enzyme protein of lipogenic enzymes (i.e. ATP-citrate lyase, acetyl-CoA carboxylase, fatty acid synthase and others discussed in reference [1]) are subject to considerable variation according to the diet. The amounts of these enzymes are probably regulated at the level of transcription [2-4] but beyond this little is known of the mechanisms involved. In any case since the half-lives of the lipogenic enzymes are generally greater than 24 hours, it is clear that a regulatory system with a more rapid response is also required. The main function of this review is to discuss the second level of regulation of fatty acid synthesis, i.e. the acute modulation of enzyme activity. Depending on the precursor used, there are several steps during the conversion of a carbohydrate precursor to a fatty acid where regulation may potentially occur [5]. However, the most important site of regulation under most circumstances appears to be the step catalyzed by the enzyme acetyl-CoA carboxylase, which is the first step that is committed to fatty acid synthesis. For the major part of this review, we will discuss the regulation of acetyl-CoA carboxylase activity, considering first the evidence from studies with purified enzyme preparations and then discussing studies conducted either _in vivo_ or in intact cell preparations.

STUDIES WITH PURIFIED ACETYL-CoA CARBOXYLASE

1. Allosteric Regulation

The activation of fatty acid synthesis in vitro by citrate was first noted by Brady and Gurin in 1952 [6]. The activation of purified acetyl-CoA carboxylase by citrate was discovered ten years later [7]. Acetyl-CoA carboxylase purified under conditions which prevent proteolysis is found to consist of a single subunit of molecular weight 240,000 [8, 9]. In the absence of citrate the purified enzyme is completely inactive and is found in the "small" or "protomer" form ($\simeq$ 13S) which is believed to be a dimer [10]. After incubation with high concentrations of citrate, it is converted into an active "polymer" ($\simeq$ 50S) which consists of up to 30 subunits aggregated in a linear array [11].

These results have led to the hypothesis that the active form of the enzyme is exclusively in the polymerised state. While this hypothesis has been accepted almost as dogma by many workers in the field, it is still perfectly feasible that the polymerisation process is a secondary event and that the active conformation of the enzyme is merely susceptible to this aggregation. The only direct method that has been used to monitor the polymerisation is centrifugation, so it is impossible to say whether it correlates with activation by citrate, which in the case of the enzyme purified from lactating rat mammary gland is complete within one minute (D.G.H, unpublished). Whatever the mechanism of activation, the effect of citrate can be easily rationalised on physiological grounds. In most species (excluding ruminants) the major precursor of fatty acid synthesis is mitochondrial acetyl-CoA, which is believed to be exported from that organelle in the form of citrate, followed by reconversion to acetyl-CoA by the enzyme ATP-citrate lyase [5]. Citrate can be regarded therefore as a classical feed-forward activator of fatty acid synthesis.

The other well-characterised allosteric effector is long chain acyl-CoA which antagonises the effects of citrate and favours conversion of the active form to the inactive "protomer" form [12]. While many enzymes are inhibited non-specifically by the detergent properties of long chain aceyl-CoA esters [13], the effects on acetyl-CoA carboxylase appear to be more specific. Thus the inhibition occurs using concentrations of Co-A ester equimolar with the enzyme and considerably less than the critical micelle concentration [14] with Ki values of 1 to 10 nM for saturated fatty acid esters of chain length 16-20 [15]. Fatty acyl-CoA would therefore represent a classical feedback inhibitor of fatty acid synthesis, and elevation of fatty acyl-CoA is a plausible explanation for the inhibition of fatty acid synthesis in perfused liver by fatty acid [16].

Two other potential allosteric effectors of acetyl-CoA carboxylase have been described recently. Kim and co-workers [17] have reported that CoA activates the enzyme by reducing the apparent Km for acetyl-CoA, while Witters et al [18] report an activation by guanine nucleotides. However, in neither case is the physiological relevance of these effects clear. The CoA effect is only observed with purified acetyl-CoA carboxylase if it is first treated with Dowex 1-X8 resin, and is completely abolished in the presence of free Mg^{2+} [17], so that it is not clear that CoA would have any effect *in vivo*. The guanine nucleotide effect is not observed with purified enzyme, and in any case the effect is maximal at about 100μM GMP, GDP or GTP. The total guanine nucleotide concentration in the cell would be relatively constant and would be likely to be greater than 100μm.

2. Protein Phosphorylation

In 1973 Carlson and Kim [19] reported that if a crude preparation of rat liver acetyl-CoA carboxylase was incubated with Mg-ATP, a time-dependent inactivation of acetyl-CoA carboxylase occurred. Using antiserum against acetyl-CoA carboxylase, they subsequently showed that this inactivation was associated with the incorporation of radioactivity from $[\gamma-^{32}P]$ATP into immunoprecipitable protein. They also showed that both the inactivation and the incorporation of ^{32}P-radioactivity could be reversed by treatment with a preparation containing protein phosphatase activity [20]. These workers concluded that acetyl-CoA carboxylase was inactivated by a phosphorylation mechanism.

This interpretation was criticised on the grounds that incubation with Mg.ATP would lead to the formation of carboxylated enzyme due to the presence of bicarbonate in the buffers, and it had already been shown that the carboxylated form of acetyl-CoA carboxylase purified from chicken liver was unstable [21]. Unfortunately since nearly all of the early experiments of Kim and coworkers were carried out using endogenous protein kinase, control experiments in the absence of protein kinase were not possible.

Since these earlier phosphorylation studies were all carried out with crude preparations of acetyl-CoA carboxylase, we elected to study the phenomenon using homogeneous enzyme. We used lactating mammary gland as our starting material, since this tissue is the richest source of acetyl-CoA carboxylase, while all evidence to date suggests that mammary gland, liver and adipose tissue acetyl-CoA carboxylase proteins are identical. We purified and characterised acetyl-CoA carboxylase from lactating rabbit [22], and rat [23] mammary glands by a new procedure involving polyethylene glycol precipitation. We also showed that the purified enzyme could be phosphorylated by at least two protein kinases, these being cyclic AMP-dependent kinase and a cyclic nucleotide - independent protein kinase which we termed acetyl-CoA carboxylase kinase-2 [22] (ACCK-2). ACCK-2 was initially described as an endogenous contaminant of the rabbit acetyl-CoA carboxylase [22], but we have recently partially purified this kinase from lactating rat mammary gland (Munday and Hardie, manuscript in preparation). Acetyl-CoA carboxylase kinase-2 is not the free catalytic subunit of cyclic AMP-dependent protein kinase, since: (1) it is completely insensitive to the heat-stable protein inhibitor of the cyclic AMP-dependent kinase; (2) it has approximately twice the molecular weight of the free catalytic subunit as judged by gel filtration on TSK-3000 columns; and (3) it has a substrate specificity which is quantitatively different from cyclic AMP-dependent protein kinase, ACCK-2 being much more specific for acetyl-CoA carboxylase.

Acetyl-CoA carboxylase is now prepared from rat mammary gland in our laboratory using the avidin-Sepharose procedure [24], which has two advantages: (1) it minimises proteolytic degradation of the enzyme, and (2) contamination of the final product by endogenous protein kinase is negligible. Both cyclic AMP-dependent protein kinase and ACCK-2 phosphorylate homogeneous acetyl-CoA carboxylase stoichiometrically, and phosphorylation is associated with a 60-70% decrease in enzyme activity measured at low citrate concentration (Fig.1). The inactivations produced by the two protein kinases are not additive, suggesting that the same phosphorylation site may be involved in both cases. The inactivations are all reversed rapidly by addition of homogeneous protein phosphatase-2A, which also removes the bulk of the radioactive phosphate incorporated into the enzyme (Fig.1).

The stoichiometry of phosphorylation by cAMP-dependent protein kinase clearly indicated that multiple sites were being phosphorylated. We have examined this using reversed-phase high performance liquid chromatography (HPLC) of chymotryptic peptides derived from ^{32}P-labelled enzyme. The results (Fig.2) show that both cAMP-dependent protein kinase and ACCK-2 yield the same, major ^{32}P-labelled peptide (peptide 1). Peptide 1 produced using cAMP-dependent protein kinase or ACCK-2 comigrates both by reversed phase HPLC and thin layer isoelectric focussing (not illustrated). Several of the minor peptides (numbered 2 to 7) also comigrate, but are much less prominent, especially in the case of ACCK-2 (Fig 2).

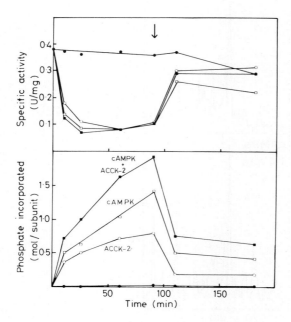

Figure 1. Reversible phosphorylation and inactivation of acetyl-CoA carboxylase by the catalytic subunit of cAMP-dependent protein kinase (CAMPK) and ACCK-2. Highly purified rat mammary acetyl-CoA carboxylase was incubated with Mg[γ-^{32}P]ATP without additions [●], with CAMPK [□], with ACCK-2 [○], or with CAMPK and ACCK-2 [■]. At the point shown by the arrow, excess EDTA was added to stop further phosphorylation, and catalytic subunit of protein phosphatase 2A was added. At various times aliquots were removed for assay of acetyl-CoA carboxylase activity at 0.5mM citrate (upper panel) or for determination of incorporation of ^{32}P into protein lower panel.

Since the site corresponding to peptide 1 is the only site that produces a peptide in anything approaching stoichiometric amounts, it was likely that phosphorylation at this site caused the enzyme inactivation. To confirm this, we analysed the distribution of radioactivity in peptides 1 to 7 during a time course of inactivation by cAMP-dependent protein kinase or ACCK-2. The results showed that peptide 1 is the only peptide phosphorylated in reasonable amount with a time-course which correlates with enzyme inactivation.

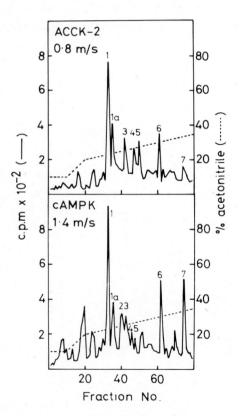

Figure 2. [^{32}P] peptides derived from purified rat mammary acetyl-CoA carboxylase phosphorylated by ACCK-2 (0.8 mol/subunit; upper panel) or cAMP-dependent protein kinase (1.4 mol/subunit; lower panel). Limit chymotryptic peptides were separated by reversed phase HPLC on a C_{18} column using 0.1% trifluoroacetic acid in H_2O/acetonitrile as eluant. The continuous line shows [^{32}P] radioactivity in fractions of the eluate; the dashed line shows the % acetonitrile in the eluant.

We have characterised the effect of phosphorylation by these two
protein kinases on the kinetic properties of acetyl-CoA carboxylase in
some detail (Table 1). As expected from the similarity of the phosphorylation sites, their effects are almost identical. They both produce a
decrease in Vmax, but more significant is the large increase
in the Ka for the allosteric activator, citrate. The physiological
concentration of citrate in e.g. liver cytosol is probably of the
order of 0.1 to 1.0 mM [25] so that at physiological citrate concentrations the effect of phosphorylation will be relatively large.

TABLE 1

EFFECT OF PHOSPHORYLATION ON KINETIC PARAMETERS OF ACETYL-COA CARBOXYLASE.

Acetyl-CoA carboxylase was phosphorylated to the stoichiometry indicated
(measured in a parallel incubation with [γ-^{32}P]ATP). The enzyme
activity was measured at different citrate concentrations using the
[^{14}C] bicarbonate fixation assay and kinetic parameters were determined
by computer fitting to the equation $v = \dfrac{V_{max} \cdot [citrate]^n}{K_a + [citrate]^n}$

Treatment	Phosphate incorporated mol/subunit	Vmax (Units/mg)	Ka citrate (mM)	Hill coefficient(n)
+MgATP	0.01	1.71±0.05	4.3±0.3	1.12
+MgATP + CAMPK	2.47	0.83±0.04	9.9±0.9	1.27
+MgATP + ACCK-2	1.44	0.88±0.04	8.6±0.8	1.32

The results summarised above provide conclusive evidence that
cyclic AMP-dependent protein kinase and ACCK-2 inactivate purified acetyl-
CoA carboxylase by phosphorylation in vitro. Evidence that the
phosphorylation by cyclic AMP-dependent protein kinase is involved in
the action of hormones which activate adenylate cyclase will be presented
in a later section. Kim and coworkers [26] have provided evidence for
an alternative mechanism of inactivation by cyclic AMP, in which cyclic
AMP binds to acetyl-CoA carboxylase itself and promotes phosphorylation
by a distinct, cyclic AMP-independent protein kinase. However, this
effect requires supraphysiological concentrations of cyclic AMP, and
5'-AMP is just as effective [27].

Several other protein kinases apparently affecting acetyl-CoA carboxylase activity have been partially characterised. Lent and Kim [28] and Shiao et al [29] have purified from rat liver protein kinases which inactivate acetyl-CoA carboxylase from the same source. In both cases the inactivation could be reversed by treatment with preparations of protein phosphatase. Both authors claimed that their preparations contained a single, major Coomassie Blue staining band after polyacrylamide gel electrophoresis in the presence of sodium dodecyl sulphate. However in neither case was any evidence presented that this band corresponded to the protein kinase itself. In view of this the conclusions of Lent and Kim [28] that their activity was different from that of Shiao et al are premature. The activity described by Shiao et al causes almost complete inactivation of acetyl-CoA carboxylase [30] even when the enzyme is assayed at 20mM citrate. This is clearly different to the effects of cyclic AMP-dependent protein kinase or ACCK-2, which are minimal at these saturating citrate concentrations (Table 1).

The protein kinase activity purified by Lent and Kim is not sensitive to the heat stable protein inhibitor of cyclic AMP-dependent protein kinase, showing that it is distinct from the latter enzyme. This cyclic AMP-independent protein kinase activity appears to partially associate with acetyl-CoA carboxylase during ammonium sulphate fractionation and gel filtration, and was therefore assumed to be the same activity responsible for the endogenous MgATP-dependent inactivation of crude rat liver acetyl-CoA carboxylase studied earlier in the same laboratory. These workers have previously described an interesting effect of adenine nucleotides using this latter system [27]. AMP promotes the inactivation of acetyl-CoA carboxylase by the endogenous, cyclic AMP-independent protein kinase, and this effect is antagonised by ATP. The authors suggest that this represents a plausible mechanism whereby a low energy charge could limit fatty acid synthesis. However it must be said that such a low energy charge is unlikely to occur in the liver except under pathological conditions of anoxia. The physiological roles of the protein kinase activities purified by Lent and Kim and Shiao et al are therefore still not clear.

Brownsey and Denton [31] have observed a MgATP-dependent activation of partially purified rat adipocyte acetyl-CoA carboxylase by plasma membrane-enriched fractions of the same tissue. This activation was associated with an increased phosphorylation of acetyl-CoA carboxylase. However, no direct evidence was presented that the phosphorylation and activation were casually related. In fact, the authors admitted that the activation occurred more slowly than the increased phosphorylation [31], but argued that this may be because the conformational change caused by the phosphorylation may occur slowly. If the hypothesis of these authors is correct, acetyl-CoA carboxylase would be the first enzyme which was both activated and inactivated by phosphorylation. This is certainly a novel concept but further experiments (such as reversal of the activation by treatment with a protein phosphatase) are necessary to substantiate the hypothesis.

3. Characterisation of Protein Phosphatases acting on Acetyl-CoA Carboxylase.

Several laboratories have described activation of acetyl-CoA carboxylase activity by treatment with partially purified protein phosphatases [20, 23, 24, 29, 32, 33]. Recently we have characterised the protein phosphatase activities in rat liver cytosol which dephosphorylate purified acetyl-CoA carboxylase labelled using $[\gamma-^{32}P]ATP$ and cyclic AMP-dependent protein kinase [34]. All of the acetyl-CoA carboxylase phosphatase activity in rat liver cytosol active against this substrate can be accounted for by three enzymes, protein phosphatases 1, 2A and 2C, with 2A being quantitatively the most important. These three protein phosphatases have been previously purified and well characterised using other ^{32}P-labelled proteins, e.g. phosphorylase or glycogen synthase, as substrate [35]. We could find no evidence for protein phosphatase activities specific for acetyl-CoA carboxylase.

Krakower and Kim [36] have described the purification from rat adipose tissue of an acetyl-CoA carboxylase phosphatase. The purified product only yielded one polypeptide detectable by iodination. However, no evidence was presented that this polypeptide was associated with the phosphatase activity, and this possibility is rendered unlikely by the observation that the molecular weight of the iodinated polypeptide, determined by gel filtration in 6M urea, was 71,000, whereas the apparent molecular weight of the native phosphatase activity determined by gel filtration, was 26,000. The molecular weight of the purified phosphatase, its lack of dependence on divalent metal ions, its resistance to heat-stable inhibitor proteins, and its stability to ethanol treatment suggest that it may represent the free catalytic subunit of protein phosphatase-2A. However further studies will be required to substantiate this.

EFFECT OF HORMONES ON ACETYL-COA CARBOXYLASE ACTIVITY

Several hormones have been shown to acutely regulate the rate of fatty acid synthesis in liver and adipose tissue, and in every case the same hormone has been shown to produce an effect on the activity of acetyl-CoA carboxylase which survives homogenisation and dilution for assay [37]. These hormones can be divided into three classes according to their probable mechanism of action, and we will now summarise the studies that have been carried out with each type of hormone.

1. Hormones which elevate cyclic AMP levels

In the case of fatty acid synthesis this class of hormones is represented by glucagon, and catecholamines acting through β-receptors. Glucagon has been shown to inhibit fatty acid synthesis in liver _in vivo_, in the perfused organ and in isolated hepatocytes [37]. This inhibition has previously been explained in terms of the allosteric effects of citrate on acetyl-CoA carboxylase. It was reported that if glucagon was added to isolated chicken hepatocytes, fatty acid synthesis was inhibited by more than 95%, and this inhibition correlated with an 85% decrease in cytosolic citrate content [38]. However in these experiments the glucagon treatment was commenced immediately after isolation of the cells. From our experience with rat hepatocytes, freshly isolated cells are in a distinctly non-physiological state, having low ATP/ADP ratios, low citrate levels, and low rates of fatty acid synthesis. The very large effect of glucagon addition at this time may be ascribed to the fact that it prevents the recovery of citrate to physiological levels due to inhibition of glycolysis. In our experiments with isolated rat hepatocytes, cells are preincubated for 90 minutes before addition of hormone to allow recovery from the trauma of isolation. During this preincubation the ATP/ADP ratio, the citrate content, and the rate of fatty acid synthesis return to levels similar to those observed _in vivo_.

If isolated hepatocytes are incubated with increasing concentrations of lactate in the presence of 5mM glucose, the rate of fatty acid synthesis is progressively stimulated [Fig.(3)]. This stimulation does correlate with an increased cellular citrate content. However, addition

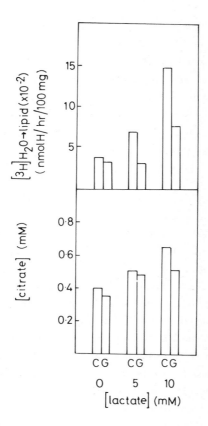

Figure 3. Effects of lactate and glucagon on fatty acid synthesis and citrate content of isolated rat hepatocytes. Cells were incubated with 5mM glucose plus the indicated concentration of lactate + pyruvate (molar ratio 10:1) for 60 minutes. Glucagon (10^{-7}M) was added to some flasks at this time and fatty acid synthesis (upper panel) and cellular citrate concentration (lower panel) was estimated between 15 and 30 minutes later. Fatty acid synthesis was estimated by the incorporation of ^{3}H from ^{3}H$_2$O into saponifiable lipid. Citrate was estimated enzymically, using ATP-citrate lyase, in perchloric acid extracts made by centrifuging cells through an oil layer into perchloric acid. Cell water content was estimated by the ^{3}H radioactivity in the perchloric acid layer.

of glucagon to preincubated cells causes marked (up to 50%) inhibition of fatty acid synthesis without significant decreases in citrate content. Thus while there is good evidence that the concentration of citrate does affect the rate of fatty acid synthesis in isolated hepatocytes, decreases in citrate concentration caused by depression of glycolysis do not account for inhibition of fatty acid synthesis by glucagon in cells which have been preincubated to allow a return to a more physiological state.

We would argue that inhibition of glycolysis is also unlikely to be a major factor in the effects of glucagon on hepatic fatty acid synthesis in vivo, since the major precursor for the pathway is probably extrahepatic lactate. It is now known that a substantial proportion of the glucose absorbed by the intestinal mucosa is converted to lactate in that tissue [39], so that lactate concentrations in the portal vein can be as high as 4mM. Even in starved animals the concentration of lactate in peripheral blood is 1-2mM [40].

We will now present the evidence that phosphorylation of acetyl-CoA carboxylase by cyclic AMP-dependent protein kinase does account for the effects of glucagon. Acetyl-CoA carboxylase was first identified as a phosphoprotein in liver cells in experiments on ^{32}P-labelled isolated hepatocytes by Witters et al [41]. These authors showed that acetyl-CoA carboxylase became labelled, using immunoprecipitation to identify the protein, and also demonstrated that glucagon elevated the ^{32}P-labelling of the enzyme as judged by densitometry of autoradiograms.

In collaboration with Dr. Lee Witters we have now analysed in some detail the effects of glucagon on phosphorylation of acetyl-CoA carboxylase in isolated [^{32}P]labelled hepatocytes. Cells are preincubated in normal medium for 30 minutes, in medium containing [^{32}P] phosphate for a further 60 minutes, and then with or without hormone for a further 15 minutes. Acetyl-CoA carboxylase is purified to homogeneity by avidin-Sepharose chromatography in the presence of EDTA and fluoride, which prevent phosphorylation and/or dephosphorylation during the isolation. The specific radioactivity and kinetic parameters of the isolated enzyme are then determined. We also isolate AMP, ADP and ATP by anion exchange HPLC from an aliquot of cells stopped in perchloric acid. From the specific radioactivity of the γ phosphate of the ATP we can then calculate the total phosphate content of the isolated acetyl-CoA carboxylase. This calculation assumes that all of the phosphate in the enzyme equilibrates with the γ phosphate of intracellular ATP during the 60 minute preincubation. To confirm this we isolated a large batch of enzyme from cells at the end of the preincubation, and measured its phosphate content directly using a spectrophotometric assay. The results agreed very well with the value obtained from the radioactivity measurements. They are also in excellent agreement with values obtained for enzyme isolated from intact liver by the same method, showing that the isolated cells are in this respect a good model of the liver in vivo. The results are summarised in Table 2:

TABLE 2
EFFECT OF GLUCAGON (10^{-7}M) TREATMENT OF ISOLATED HEPATOCYTES ON PROPERTIES OF ACETYL-CoA CARBOXYLASE PURIFIED TO HOMOGENEITY FROM THE CELLS.

Parameter	Control	Glucagon
Phosphate content (mol/subunit)	4.4 ± 0.2*	5.2 ± 0.3*
Vmax (μmoles/min/mg)	0.51 ± 0.01	0.36 ± 0.01
Ka citrate (mM)	3.8 ± 0.2	5.9 ± 0.7

* mean ± standard error of the mean for 3 experiments.

It is clear from these data that the effect of glucagon on enzyme activity survives purification to homogeneity (making it very unlikely that an allosteric effector is responsible) and also that the effect of glucagon is very similar to the effect of phosphorylation of the purified enzyme by cyclic AMP-dependent protein Kinase (cF Table 1). It may also be seen by comparison of Tables 1 and 2 that the Vmax for the enzyme purified from control hepatocytes is very low compared to that of the enzyme purified from lactating mammary gland. This difference correlates with a marked difference in phosphate content(5.2 ± 0.1 (4) for enzyme from control hepatocytes as opposed to 2.1 ± 0.2 (4) for the mammary enzyme:alkali-labile phosphate in mol/subunit expressed as mean ± standard error for the number of preparations shown in brackets). If the enzyme purified from control hepatocytes is treated with protein phosphatase 2A, a very large increase in Vmax (up to 10 fold) is observed. This demonstrates that the enzyme in the intact liver cells is present in a highly phosphorylated, low activity state, even before stimulation by glucagon. The protein kinase(s) responsible for this basal phosphorylation is not yet identified, but it is not cyclic AMP-dependent kinase since the latter enzyme cannot explain the large effect on Vmax. It may correspond to the activity originally described by Carlson and Kim [19], which appeared to exert a large Vmax effect [42].

It can also be seen from Table 2 that although the percentage stimulation of phosphorylation by glucagon is small (about 15%) it is significant because it represents an increase of almost one molecule of phosphate per subunit. Figure 4 shows that this increase is almost entirely accounted for by a single chymotryptic peptide. This peptide comigrates with peptide 1, the peptide most heavily phosphorylated in purified acetyl CoA carboxylase using cyclic AMP-dependent kinase, both on reversed phase HPLC (cf fig.2) and on thin layer isoelectric focussing (not illustrated). The stimulation by glucagon of phosphorylation of this peptide is $92\pm14\%$ (mean $\pm$ standard error for 6 experiments) while no other peptide increased by more than 27%.

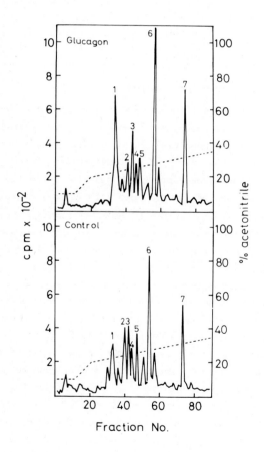

Figure 4. Effect of glucagon on ^{32}P-labelling of chymotryptic peptides of acetyl-CoA carboxylase. The separation of peptides is as described for Fig. 2.

Figure 4 also shows that several peptides other than peptide 1 are heavily labelled in control hepatocytes. One or more of these phosphorylations is presumed to be responsible for the low Vmax of the hepatocyte enzyme.

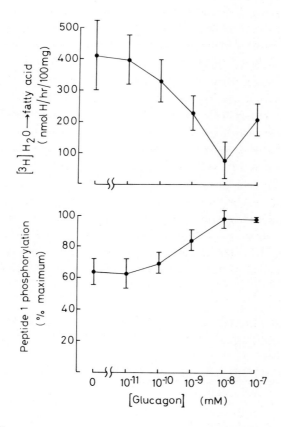

Figure 5. Dose-response curves for the effects of glucagon on fatty acid synthesis (upper panel) and ^{32}P-labelling of peptide 1 (lower panel). Fatty acid synthesis was estimated by the incorporation of ^3H from ^3H_2O into saponifiable lipid; ^{32}P-labelling of peptide 1 was estimated as in Fig. 4.

Figure 5 shows that dose-response curves for the effects of glucagon on fatty acid synthesis and labelling of peptide 1 are inversely related, and that both effects occur at physiological glucagon concentrations (half-maximal effects at about $10^{-9}M$ glucagon).

These results clearly show that the effects of glucagon on fatty acid synthesis in liver cells are accounted for by phosphorylation of acetyl-CoA carboxylase by cyclic AMP-dependent protein kinase. We have obtained very similar results for the effects of adrenaline and glucagon in white adipose cells. An earlier study of the effects of adrenaline was carried out by isolating acetyl-CoA carboxylase using immunoprecipitation and analysing ^{32}P-labelled peptides by isoelectric focussing [43]. We have confirmed and extended these studies, and also looked at effects of glucagon, using the more quantitative avidin-Sepharose/HPLC technique described above (Holland, Hardie, Clegg and Zammit, unpublished). Table 3 shows parameters for acetyl CoA carboxylase isolated from adipocytes and the effects of adrenaline and glucagon.

The basal level of phosphorylation in adipocytes is lower than that in hepatocytes, and this correlates with a higher Vmax. Analysis of chymotryptic peptides by HPLC reveals a pattern that is qualitatively similar to that in hepatocytes (not illustrated). Treatment with glucagon leads to a selective increase in peptide 1 exactly as in isolated hepatocytes, while treatment with adrenaline significantly increases peptides 1 and 5-7 correlating both with a small decrease in Ka and a large decrease in Vmax as shown in Table 3. The reason for the difference in the effects of adrenaline and glucagon is not yet clear. One possible explanation is that adrenaline is acting through both β- and α-adrenergic mechanisms.

TABLE 3
EFFECTS OF TREATMENT OF ISOLATED ADIPOCYTES WITH ADRENALINE OR GLUCAGON ON PHOSPHORYLATION AND KINETIC PARAMETERS OF ACETYL-CoA CARBOXYLASE ISOLATED USING AVIDIN SEPHAROSE.

Parameter	Control	Adrenaline	Glucagon
Phosphate content (mol/subunit)	3.3	4.7	3.7
Vmax (μmole/min/mg)	1.70 ± 0.11	0.33 ± 0.01	1.00 ± 0.06
Ka citrate (mM)	6.1 ± 0.8	5.0 ± 0.5	7.4 ± 0.9
Hill coefficient	0.93	1.49	1.14

2. Hormones which elevate cytosolic Ca^{2+} levels

In the case of fatty acid synthesis, this class of hormones is represented by catcholamines acting at α_1 receptors, vasopressin and angiotensin II. All of these hormones cause two coupled effects on rat liver cells [44]: (1) they very rapidly cause breakdown of polyphosphoinositides in the plasma membrane, releasing inositol phosphates and diacylglycerol; and (2) they cause release of bound Ca^{2+} which transiently elevates cytosolic Ca^{2+} and brings about Ca^{2+} efflux from the cells. The three hormones, while possessing independent receptors, appear to act via a common mechanism, since desensitisation of hepatocytes by one of these hormones also desensitises the cells against the other two [45].

Hems and coworkers were the first laboratory to study effects of these hormones on fatty acid synthesis. They showed that adrenaline, vasopressin and angiotensin II all inhibited fatty acid synthesis in perfused mouse liver. They also determined an inhibition by all three hormones of acetyl-CoA carboxylase activity measured in cell extracts [46-48]. The effects of these hormones on rat liver are somewhat confusing. Ly et al [49] have reported an inhibition of acetyl-CoA carboxylase measured in extracts of cells treated with the α-agonist phenylephrine. They also determined that the agonist increased the ^{32}P-radioactivity of acetyl-CoA carboxylase in labelled cells [49]. However Assimacopoulos-Jeannet et al [50] have reported an activation of both fatty acid synthesis and acetyl-CoA carboxylase activity by vasopressin in isolated rat hepatocytes. This latter finding represents something of an anomaly at present. Since vasopressin stimulates accumulation of pyruvate by the cells [50], it will be important to rule out the possibility [51] that in these experiments pyruvate carboxylase activity was interfering with the acetyl-CoA carboxylase assays.

An appealing candidate to mediate the intracellular effects of the Ca^{2+} linked hormones is the Ca^{2+} and phospholipid-dependent protein kinase described by Nishizuka et al [52]. At physiological Ca^{2+} levels this enzyme is dramatically activated by diacylglycerol, which is known to be formed within seconds of hormonal stimulation [44]. We have shown that a preparation of this protein kinase from rat brain will phosphorylate acetyl-CoA carboxylase in a Ca^{2+}- amd phospholipid-dependent manner (Naylor and Hardie, unpublished). Recently Castagna et al [54] have shown that the phorbol ester 12-0-tetradecanoyl phorbol 13-acetate (TPA) will substitute for diacylglycerol in the activation of the protein kinase. This compound is potentially a powerful tool since it now appears that it may represent a specific activator of this protein kinase which is effective in intact cells [54]. Preliminary experiments in our laboratory show that treatment of isolated rat hepatocytes with vasopressin or TPA, but not the Ca^{2+} ionophore A23187, stimulate the phosphorylation of acetyl

-CoA carboxylase. This suggests that the increased phosphorylation of the enzyme after phenylephrine treatment [49] results from release of diacylglycerol and consequent activation of the Ca^{2+}- and phospholipid-dependent protein kinase. This hypothesis is now eminently testable.

3. Insulin

Insulin stimulates fatty acid synthesis in isolated hepatocytes and, more markedly, in isolated adipocytes. In both cases elevations of acetyl-CoA carboxylase activity measured in cell extracts have also been reported [37]. From studies with purified acetyl-CoA carboxylase (see earlier section) one might expect that increased activity would result from decreased phosphorylation of the enzyme. However, insulin increases radioactivity in acetyl-CoA carboxylase in ^{32}P-labelled adipocytes [55] or hepatocytes [56]. Recently two groups have reported that the increased ^{32}P-labelling of the enzyme is due to an increase at a site which is distinct from that phosphorylated by cyclic AMP-dependent protein kinase. Brownsey and Denton [55] proposed that phosphorylation at this site activates the enzyme, although direct evidence for this is lacking. Indeed Witters et al [57] showed that the effect on enzyme activity did not survive avidin-Sepharose isolation, while the effect on phosphorylation was preserved. This suggests either that the increased phosphorylation is unrelated to enzyme activation or that some dissociable factor(s) is required for the effect of insulin to be observed.

McNeillie and Zammit [58] have reported interesting experiments which suggest that in lactating mammary gland insulin can decrease the phosphorylation of acetyl-CoA carboxylase. They measured enzyme activity in mammary extracts before and after treatment with protein phosphatase, and used the ratio of these two measurements as a crude index of the enzyme's phosphorylation state. Studies revealed that starvation, or short-term streptozotocin induced diabetes, led to a marked increase in phosphorylation state, and that insulin treatment in vivo reversed the effects. Our laboratory has recently confirmed directly that 24 hour starvation increases the phosphate content of the isolated enzyme (Table 4). This is associated with a large decrease in Vmax, so that phosphorylation by cAMP-dependent protein kinase and/or ACCK-2 would not be sufficient to account for the effect (cf Table 1). Treatment of enzyme isolated from mammary glands with protein phosphatase 2A completely reverses the effects of starvation (Munday and Hardie, unpublished). We are currently testing whether infusion of insulin in vivo will also reverse the effect. Further experiments would be necessary to show whether such an effect of insulin was a direct effect on the mammary cells themselves.

TABLE 4
EFFECT OF 24 STARVATION ON ACETYL-CoA CARBOXYLASE ISOLATED FROM LACTATING MAMMARY GLAND. Results are expressed as means± standard error of the mean for the number of preparations shown in brackets.

Parameter	Fed	Starved
Phosphate content (mol/subunit)	2.1 ± 0.2 (4)	6.3 ± 1.1 (5)
Vmax (μmole/min/mg)	2.9 ± 0.8 (4)	0.57 ± 0.12 (5)
Ka citrate (mM)	1.4 ± 0.1 (4)	2.2 ± 0.4 (5)

CONCLUSIONS

It should be clear from this review that the evidence that hormones regulate fatty acid synthesis mainly by altering the phosphorylation state of acetyl-CoA carboxylase is now very strong. While the mechanisms of the hormonal effects are only well understood in the case of hormones which activate adenylate cyclase, hypotheses about the mechanisms of the other hormones are now becoming testable and it is confidently expected that rapid progress will be made in the next few years.

REFERENCES

1. Romsos, D.R. and Leveille, G.A. (1974) Advan. Lipid Res., 12, 97-146.
2. Flick P.K., Chen, J., Alberts, A.W. and Vagelos, P.R. (1978) Proc. Natl.Acad.Sci. USA, 75, 730-734.
3. Nepokroeff, C.M. and Porter, J.W. (1978) J.Biol.Chem., 253, 2279-2283
4. Morris, S.M., Nilson, J.H., Jenik, R.A., Winberry, L.K., McDevitt,M.A and Goodridge, A.G. (1982) J.Biol.Chem., 257, 3225-3229.
5. Hardie, D.G. (1981) Trends Biochem.Sci., 6, 75-77.
6. Brady, R.O. and Gurin, S. (1982) J.Biol.Chem., 199, 421-431.
7. Vagelos, P.R., Alberts, A.W. and Martin, D.B. (1962) Biochem.Biophys. Res.Commun., 8, 4-8.
8. Tanabe, T., Wada, K., Okazaki, T. and Numa, S. (1975) Eur.J.Biochem, 57, 15-24.
9. Hardie, D.G. and Cohen, P. (1978) Eur.J. Biochem., 92, 25-34.

10. Gregolin, C., Ryder, E., Warner, R.C., Kleinschmidt, A.K. and Lane, M.D. (1966) Proc.Natl.Acad.Sci. USA, 56, 1751-1758.
11. Gregolin, C., Ryder, E., Kleinschmidt, A.K., Warner, R.C. and Lane, M.D. (1966) Proc.Natl.Acad.Sci. USA, 56, 148-155
12. Numa, S., Ringelmann, E. and Lynen, F. (1965) Biochem.Z., 343, 243-257.
13. Taketa, K. and Pogell, B.M. (1966) J.Biol.Chem., 241, 720-726.
14. Ogiwara, H., Tanabe, T., Nikawa, J. and Numa, S. (1978) Eur.J. Biochem., 89, 33-41.
15. Nikawa, J., Tanabe, T., Ogiwara, H., Shiba, T. and Numa, S. (1979) FEBS Lett. 102, 223-226.
16. Topping, D.L. and Mayes, P.A. (1976) Biochem.Soc.Trans., 4, 717
17. Yeh, L., Song, C. and Kim, K. (1981) J.Biol.Chem., 256, 2289-2296.
18. Witters, L.A., Friedman, S.A., Tipper, J.P. and Bacon, G.W. (1981) J.Biol.Chem., 256, 8573-8578.
19. Carlson, C.A. and Kim, K.H. (1973) J.Biol.Chem., 248, 378-380.
20. Lee, K.H. and Kim, K.H. (1977) J.Biol.Chem., 252, 1748-1751.
21. Lane, M.D., Moss, J. and Polakis, S.E. (1974) Curr.Top.Cell.Regul. 8, 139-195.
22. Hardie, D.G. and Cohen, P. (1978) FEBS Lett., 91, 1-7
23. Hardie, D.G. and Guy, P.S. (1980) Eur.J.Biochem., 110, 167-177.
24. Tipper, J.P. and Witters, L.A. (1982) Biochem.Biophys.Acta, 715, 162-169.
25. Siess, E.A., Brocks, D.G. and Wieland, O.H. (1978) Hoppe-Seyler's Z.Physiol.Chem., 359, 785-798.
26. Lent, B.A., Lee, K.H. and Kim, K.H. (1978) J.Biol.Chem., 253, 8149-8156
27. Yeh, L., Lee, K.H. and Kim, K.H. (1980) J.BiolChem., 255, 2308-2314.
28. Lent, B. and Kim, K.H. (1982) J.Biol.Chem., 257, 1897-1901.
29. Shiao, M., Drong, R.F. and Porter, J.W. (1981) Biochem.Biophys.Res. Commun., 98, 80-87.
30. Abdel-Halim, M.N. and Porter, J.W. (1980) J.Biol.Chem., 255, 441-444.
31. Brownsey, R.W. Belsham, G.J. and Denton, R M. (1981) FEBS Lett., 145-150.
32. Hardie, D.G. and Cohen, P. (1979) FEBS Lett., 103, 333-338.
33. Krakower, G.R. and Kim, K.H. (1980) Biochem.Biophys.Res.Commun., 92, 389-395.
34. Ingebritsen, T.S., Blair, J., Guy, P., Witters, L. and Hardie, D.G. (1983) Eur.J.Biochem., 130, 185-193.
35. Ingebritsen, T.S. and Cohen, P. (1983) Eur.J.Biochem., 133, 255-261
36. Krakower, G.R. and Kim, K.H. (1981) J.Biol.Chem., 256, 2408-2413.
37. Hardie, D.G. (1980) in Molecular Aspects of Regulation, P. Cohen (Ed.) Elsevier/North Holland Biomedical Press, Amsterdam, Vol. 1, pp. 33-62.

38. Watkins, P.A., Tarlow, D.M. and Lane, M.D. (1977) Proc.Natl.Acad. Sci. USA, 149701501.
39. Nicholls, T.J., Leese, H.J. and Bronk, J.R. (1983) Biochem.J., 212, 183-187.
40. Hopkirk, T.J. and Bloxham, D. (1977) Biochem.Soc.Trans. 5, 1294-1297.
41. Witters, L.A., Kowaloff, E.M. and Avruch, J. (1979) J.Biol.Chem., 254, 245-248.
42. Carlson, C.A. and Kim, K.H. (1974) Arch.Biochem.Biophys. 164, 490-501
43. Brownsey, R.W. and Hardie, D.G. (1980) FEBS Lett., 120, 67-70.
44. Kirk, C.J., Creba, J.A., Downs, C.P. and Michell, P. (1981) Biochem. Soc.Trans., 7, 377-379.
45. Breant, B., Keppens, S. and de Wulf, H. (1981) Biochem.J., 200, 509-514.
46. Ma, G.Y. and Hems, D.A. (1975) Biochem.J., 152, 389-392.
47. Ma, G.Y., Gove, C.D. and Hems, D.A. (1977) Biochem.Soc.Trans., 5, 9860989.
48. Hems, D.A. (1977) FEBS Lett., 80, 237-245.
49. Ly, S. and Kim, K.H. (1981) J.Biol.Chem., 256, 11585-11590.
50. Assimacopoulos-Jeannet, F., Denton, R.M. and Jeaurenaud, B. (1981) Biochem. J., 198, 485-490.
51. Davies, D.R., van Schaftingen, E. and Hers, H.G. (1982) Biochem. J., 202, 559-560.
52. Kishimoto, A., Takai, Y. Mori, T., Kikkawa, U. and Nishizuka, Y. (1980) J.Biol.Chem., 255, 2273-2276.
53. Castagna, M., Takai, Y., Kaibuchi, K., Sano, K., Kikkawa, U. and Nishizuka, Y. (1982) J.Biol.Chem., 257, 7847-7851.
54. Ashendell, C.L., Staller, J.M. and Boutwell, R.K. (1983) Biochem. Biophys.Res.Commun., 111, 340-345.
55. Brownsey, R.W. and Denton, R.M. (1982) Biochem.J., 202, 77-86.
56. Witters, L.A. (1981) Biochem.Biophys.Res.Commun., 100, 872-878
57. Witters, L.A., Tipper, J.P. and Bacon, G.W. (1983) J. Biol.Chem., 258, 5643-5648.
58. McNeillie, E.M. and Zammit, V.A. (1982) Biochem.J., 204, 273-280.

Résumé

L'acétyl-CoA carboxylase est supposée catalyser l'étape limitante de la synthèse des acides gras dans la plupart des conditions. Des données récentes suggèrent que l'effet des hormones sur la synthèse des acides gras est médié dans une large mesure par des modifications de l'état de phosphorylation de cet enzyme. L'acétyl-CoA carboxylase purifiée jusqu'à homogénéité à partir de glande mammaire de rat en lactation est phosphorylée de manière réversible par la protéine kinase AMP cyclique dépendante. Cette phosphorylation est associée à l'inactivation de l'enzyme et est en corrélation avec la phosphorylation d'un site donnant un peptide chymotryptique distinct (peptide 1). Dans les hépatocytes ou les adipocytes intacts, l'enzyme est phosphorylé au niveau de sites multiples. Néanmoins, les hormones qui élèvent le taux d'AMP cyclique (ex. glucagon) augmentent la phosphorylation du peptide 1 et provoquent un changement dans les propriétés cinétiques de l'enzyme similaire à celui obtenu in vitro en utilisant la protéine kinase AMP cyclique dépendante. D'autres hormones, telles que l'insuline ou la vasopressine pourraient agir en modifiant la phosphorylation de l'enzyme au niveau de sites phosphorylés par une ou des kinases AMP cyclique indépendantes. Nous nous proposons d'analyser les études récentes concernant cette possibilité.

MOLECULAR MECHANISMS REGULATING GLUCOSE OXIDATION IN INSULIN DEFICIENT ANIMALS

PHILIP J. RANDLE, STEPHEN J. FULLER, ALAN L. KERBEY, GRAHAM J. SALE and THOMAS C. VARY

Nuffield Department of Clinical Biochemistry, Oxford University, John Radcliffe Hospital, Oxford OX3 9DU, U.K.

INTRODUCTION

PDH complex and glucose oxidation. Glucose oxidation is reduced markedly in 48h starved or diabetic animals and may be restored after several hours by carbohydrate refeeding or insulin treatment respectively. Glucose oxidation in vitro is also reduced in heart and skeletal muscle preparations from starved or diabetic rats but it is not restored by insulin in vitro. Therefore the action of insulin in vivo is either indirect or of long latency. The rate of glucose oxidation in animal tissues is regulated by the mitochondrial pyruvate dehydrogenase (PDH) complex which catalyses the irreversible reaction (K_{eq} approximately 10^7) shown in equation 1. The PDH complex is regulated by end

$$pyruvate + NAD^+ + CoASH \xrightarrow{TPP} acetylCoA + NADH + H^+ + CO_2 \qquad 1.$$

product inhibition and by reversible phosphorylation. The latter mediates effects of insulin deficiency on glucose oxidation and it is the mechanism of this regulation that is considered here. For a background review see [1].

The phosphorylated form of the PDH complex is inactive; the dephosphorylated form is active. Phosphorylation and inactivation are catalysed by PDH kinase which copurifies with the complex. Dephosphorylation and reactivation are catalysed by PDH phosphatase which separates from the complex during purification [2]. In rats the proportion of complex in the active form is diminished by 48h of starvation and by alloxan-diabetes in heart, skeletal muscle, liver, kidney and adipocytes [1,3]. The total concentration of complex (sum of active and inactive forms) is unchanged. The effects of starvation are reversed by refeeding and of diabetes by insulin treatment. It is concluded that effects of starvation and diabetes on glucose oxidation are due to inactivation of the PDH complex by phosphorylation i.e. to activation of the kinase reaction or to inhibition of the phosphatase reaction or to both.

THE PYRUVATE DEHYDROGENASE COMPLEX

Chemistry of the PDH complex. The PDH complex contains three enzymes;

E1 (pyruvate dehydrogenase or decarboxylase) forming hydroxyethyl-TPP-E_1 and CO_2; E_2 an acyltransferase forming acetyl CoA; and E_3 (lipoyl dehydrogenase) forming NADH. Lipoyl residues attached covalently to E_2 transfer acyl groups and H between component enzymes of the complex. E_1 has two dissimilar subunits [$M_r(\alpha)$ = 41000; $M_r(\beta)$ = 36000] and is a tetramer ($\alpha_2 \beta_2$); E_2 contains 60 copies of a subunit of M_r = 52000 arranged in a pentagonal dodecahedron; E_3 is a dimer of a subunit of M_r = 55000. As purified the E_2 core bears 30 or 60 E_1 tetramers and 12 E_3 dimers. M_r values are by SDS-PAGE except for E_2 which gives an anomolous M_r (76000) because of an extended configuration; the value of 52000 is based on sedimentation equilibrium [4]. PDH kinase is a dimer of two dissimilar subunits (M_r = 45000 and 48000) attached to the E_2 core and is an SH enzyme [5]. It is specific for ATP or its thiophosphoryl analogue. PDH phosphatase is a dimer of two dissimilar subunits (M_r = 50000 and 98000); it contains FAD, the function of which is unknown [6].

Reversible phosphorylation. Phosphorylation is confined to serine residues in an α-chain of E_1 and is half site in ox and pig complexes i.e. equivalent to only one α-chain of E_1. Phosphorylation results in more than 99% inactivation, and reactivation can be induced only by dephosphorylation. Three serine residues are phosphorylated; these are recovered in two tryptic phosphopeptides (TA and TB) the sequences of which (pig) are shown. Ox complex has asparagine in place of aspartate in TA [4,7]. Cleavage of the Asp-Pro bond

Tyr-His-Gly-His-Ser-Met-Ser-Asp-Pro-Gly-Val-Ser-Tyr-Arg   TA.
           SITE 1   SITE 2

Tyr-Gly-Met-Gly-Thr-Ser-Val-Glu-Arg   TB.
              SITE 3

in TA allows site occupancy to be determined [8]. The relative rates of phosphorylation by PDH kinase are sites 1> 2> 3. The relative rates of dephosphorylation by PDH phosphatase are sites 2> 1> 3. Phosphorylation and dephosphorylation reactions may be followed in purified complex, or in mitochondria, or in isolated tissues by assaying holocomplex activity; and by incorporation of ^{32}P from [γ-^{32}P]ATP (the latter may be generated in situ from $^{32}P_i$ in mitochondria or in isolated tissues). Close parallelism exists between the behaviour of purified complex and that of complex in mitochondria or in isolated tissues.

Phosphorylation of sites 1 or 2 is inactivating but that of site 3 is not

[9-11]. During phosphorylation or in the steady state in vivo more than 98% of inactivation is due to phosphorylation of site 1. The major function of sites 2 and 3 is therefore not inactivation. During dephosphorylation the contribution of site 2 to inactivation may increase from <2% to 8% suggesting that it may function to delay reactivation by the phosphatase [11] (discussed in more detail later).

Regulation of PDH kinase and phosphatase reactions. The kinase reactions are inhibited by ADP (competitive with ATP), by pyruvate (uncompetitive with ATP) and by dichloroacetate [2,12,13]. The kinase reactions are accelerated by increasing concentration ratios of ATP/ADP, NADH/NAD$^+$ and acetylCoA/CoA [14-16]. Activation by a further factor referred to as kinase/activator protein is discussed later. The phosphatase reactions require Mg^{2+} ($K_{0.5}$ approximately 1mM) and in its presence are activated by Ca^{2+} ($K_{0.5}$ approximately 1 μM)[17, 18]. Operation of these regulatory mechanisms for PDH kinase and phosphatase in mitochondria has been shown [11, 19-21].

MECHANISMS OPERATING IN INSULIN DEFICIENT ANIMALS

The effect of starvation and of alloxan diabetes on the proportion of PDH complex in the active form has already been described. Before discussing the mechanism involved brief reference will be made to other factors affecting the proportion of active complex in vivo. In heart and skeletal muscles, work increases the proportion of active complex [22, 23]. In muscles the oxidation of fatty acids and ketone bodies decreases the proportion of active complex [3, 24, 25]. Insulin in vitro increases the proportion of active complex in adipocytes but it does not increase the proportion of active complex in heart or skeletal muscle in vitro [26, 27]. In rat heart perfused for 40min with 11mM-glucose the proportion of active complex is actually reduced by insulin (by c. 60%); this is apparently due to stimulation of glucose transport by insulin ([27]; see also Table 2). Therefore the effect of insulin deficiency (in starved or diabetic rats) to decrease the proportion of active complex is not the result of a lack of a direct short term effect of the hormone.

The role of lipid fuels. In vivo, oxidation of fatty acids and ketone bodies is enhanced in starved or diabetic animals. In vitro, oxidation of fatty acids is enhanced in tissues of starved or diabetic animals because of enhanced degradation of intracellular triglycerides which are increased in concentration [28, 29]. Oxidation of these lipid fuels decreases the proportion of active complex presumably as a result of activation of the PDH kinase reaction by increased mitochondrial concentration ratios of acetylCoA/CoA and NADH/NAD$^+$ [30,31].

Use of 2-tetradecylglycidic acid (TDG) has shown that oxidation of fatty acids is essential for the effects of starvation and diabetes on the proportion of active complex. TDG inhibits fatty acid oxidation at the level of carnitine acyl transferase provided that the ratio of TDG to fatty acid is favourable. As shown in Fig.1a and [26] TDG reversed the effects of starvation

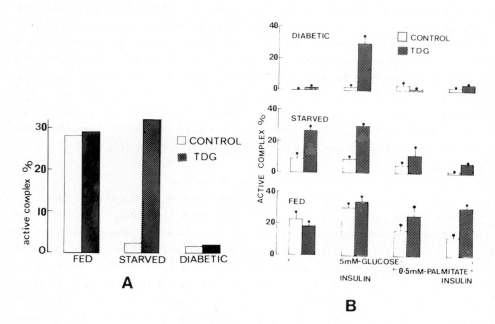

Fig. 1. The effect of 2-tetradecylglycidic acid (TDG) on the proportion of active complex in rat heart. In (a) sodium TDG complexed with albumin was injected intravenously (7 μmol/Kg) and hearts removed for analysis after 20 min. In (b) hearts were perfused for 10min with 0.1mM-sodium TDG/albumin (20g/l). Further details are in [26].

on the proportion of active complex in the heart in vivo, but not that of diabetes. In vitro TDG reversed the effect of starvation and of diabetes on the proportion of active complex in the heart; reversal in diabetes required insulin, reversal in starvation was independent of added insulin (Fig. 1b, first two vertical groups; [26]). The lack of an effect of TDG in diabetic animals in vivo and the requirement for insulin in vitro is attributed to an unfavourable ratio of TDG to fatty acids and to the likely effect of circulating ketone bodies in vivo (see [26] for details of the evidence). As shown in Fig 1b (vertical groups 3 and 4) the effect of TDG in starved and diabetic rats was reversed by palmitate. In hearts of fed rats (in which glucose oxidation accounts for 80% of O_2 consumption) TDG had no effect on the proportion of active complex in vivo (Fig. 1a) nor in vitro unless palmitate was added to the perfusion medium(Fig. 1b). TDG was also partially effective in kidney in vivo, but not in liver, skeletal muscle or adipocytes [26]. The reason for the lack of an effect of TDG in the latter tissues is not known.

Mechanisms other than oxidation of lipid fuels

General considerations. The proportion of active complex in perfused hearts from alloxan-diabetic rats (<1%) is much lower than that of hearts of normal rats perfused with fatty acids or ketone bodies (c. 7%). This suggested that some mechanism in addition to oxidation of lipid fuels was operating to reduce the proportion of active complex. Studies by Kerbey et al [20, 24, 32] showed that the effect of starvation and diabetes to decrease the proportion of active complex persists into mitochondria prepared from the heart. The difference is most obvious when mitochondria are incubated with inhibitors of PDH kinase (pyruvate, dichloroacetate) or when the phosphatase is activated with Ca^{2+}. Similar observations have been made with skeletal muscle mitochondria (S.J. Fuller & P.J. Randle, unpublished work) as shown in Fig. 2. This effect of starvation or diabetes persisting into isolated mitochondria is not explicable in terms of the concentration of known effectors of PDH kinase or phosphatase or of the oxidation of lipid fuels. The observation indicated the existence of a new mechanism of increasing PDH kinase activity or of decreasing PDH phosphatase activity or of both. Subsequent studies showed that both mechanisms operate i.e. increased activity of PDH kinase and decreased reactivation of phosphorylated complex by PDH phosphatase [32 - 34]. Current knowledge of the mechanisms involved is summarised below.

Kinase/activator protein. When rat heart mitochondria are incubated in the absence of respiratory substrate all of the PDH complex is converted into the active (dephosphorylated) form. The activity of PDH kinase can then be followed in extracts by measuring inactivation with ATP or ^{32}P incorporation. The concentrations of known metabolite effectors of the kinase are too low to interfere and the phosphatase is inoperative if EGTA is present. With this technique it was shown that the rate of the kinase reaction is increased in extracts of heart mitochondria from starved or diabetic rats relative to controls [33]. Similar observations have been made with skeletal muscle mito-

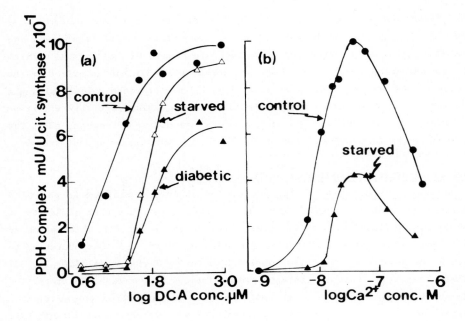

Fig. 2. Concentration of active PDH complex in mitochondria prepared with trypsin from rat leg muscles. Incubations/assays were as described in [24]. (a) effect of dichloroacetate (4µM – 1mM); $K_{0.5}$ were 15.5µM (control), 62µM (starved) and 72µM (diabetic). (b) effect of Ca^{2+} (6nM – 713nM achieved with Ca/4.5mM – EGTA buffers); $K_{0.5}$ for Ca^{2+} were 10nM (control) and 17.5nM (diabetic) and maximum values for proportion of active complex were 100% (control) and 50% (starved).

chondria (S.J. Fuller & P.J. Randle, unpublished work) as shown in Fig. 3.

More recent studies have shown that the factor which may be responsible for increased PDH kinase activity in extracts of heart mitochondria from starved or diabetic rats, can be separated from the PDH complex by precipitation of the latter at 150000g. Such high speed supernatant fractions (HSS) activate PDH kinase in mitochondrial extracts or in purified PDH complex denuded of kinase as described in [35,36]. Activation by HSS from mitochondria of starved or diabetic rats is substantially greater (threefold) than with HSS from fed normal rats [35]. The factor, termed kinase/activator, appears to be a protein (M_r approximately 100000 by gel filtration) or protein-

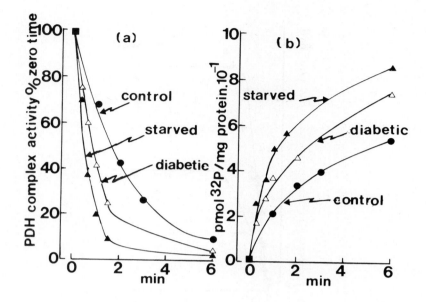

Fig 3. Inactivation of PDH complex with [γ-^{32}P]ATP in extracts of rat leg muscle mitochondria from fed, starved and diabetic rats; (a) inactivation (b) ^{32}P incorporation. The experimental protocol was essentially as in [33]. Apparent first order rate constants for inactivation were (min $^{-1}$) 0.51 (fed), 1.82 (starved) and 1.00 (diabetic).

associated factor. Studies with cycloheximide and puromycin in rats during starvation suggest that the factor is induced by cytoplasmic protein synthesis [36]. Disappearance of the factor on refeeding starved rats correlates with an increase in the proportion of active PDH complex [36]. More recent studies (A.L. Kerbey & P.J. Randle, unpublished work) of the HSS concentration / response relationship have indicated that the effect of starvation is not mediated solely by an increase in the concentration of a factor normally present in fed rats (see Fig. 4). This finding has delayed purification and characterisation because of the lack of a bulk source from slaughterhouse material. The factor may be an activator of the kinase or alternatively excess free kinase; i.e. PDH kinase that is not tightly bound to PDH complex. Attempts to distinguish between these two possibilities (eg. by use of affinity labels for the ATP binding site) have been unsuccessful, thus far.

It is important to determine which hormonal and nutritional factors

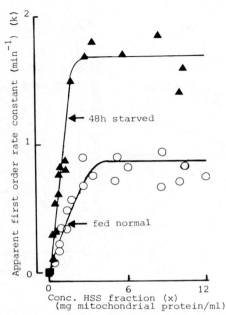

Fig. 4. Effect of concentration of a high speed supernatant fraction (HSS) prepared from rat heart mitochondria, on the rate of inactivation of pig heart PDH complex by ATP. Details of the preparation and assay of HSS are in [36]. V_{max} (expressed as apparent first order rate constant, min^{-1}) was 1.05 ± 0.05 (fed) and 2.34 ± 0.34 (starved) ($P<0.01$ for difference). The values for $K_{0.5(x)}$ did not differ significantly.

modulate induction of kinase/activator protein during starvation or the development of diabetes. In preliminary experiments attempts have been made to reverse the effects of starvation by perfusion of hearts for 6h with medium containing glucose, insulin and an amino acid mixture. Some preliminary evidence for partial reversal has been obtained (T.C. Vary, A.L. Kerbey & P.J. Randle, unpublished work). It is likely that tissue culture of myocytes which will allow longer in vitro incubation may be more successful. Full reversal in vivo, on refeeding starved rats requires 20 - 24h [36] and perfusions for this length of time are impractical.

<u>Reactivation by PDH phosphatase; regulation by multisite phosphorylation.</u> In a substantial number of studies in this laboratory it has been found that phosphorylation of sites 2 and 3 inhibits reactivation of the complex and dephosphorylation of site 1 by PDH phosphatase. The initial rate of

reactivation was reduced about fivefold by phosphorylation of sites 2 and 3 in purified pig heart complex [37-39]. It was important, therefore, to estimate the occupancy of phosphorylation sites in vivo and its relation to the proportion of complex in the inactive form. This has been achieved in mitochondria with $^{32}P_i$ and in rat heart in vivo either by back titration of unoccupied sites with [γ-^{32}P]ATP or by perfusion with $^{32}P_i$ [10, 40, 41].

The patterns of occupancy in mitochondria and in the heart in vivo were closely similar. Occupancy of site 1 was correlated linearly with the proportion of inactive complex. Relative to site 1 occupancies of sites 2 and 3 were constant at approximately 0.6 and 0.4 respectively when the proportion of complex in the inactive form was 70% or less (as in the fed normal rat). When the proportion of complex in the inactive form approached 100% (as in starved or diabetic rats) the occupancy of sites 2 and 3 approximated to that of site 1. The initial rate of reactivation of phosphorylated complex from hearts of fed normal rats was approximately three times that of phosphorylated complex from hearts of 48h starved or diabetic rats [41]. Representative data are shown in Table 1. It is suggested that this mechanism of regulation by multisite phosphorylation, which is kinase directed, is a hysteresis mechanism which restrains reactivation of phosphorylated complex in tissues of starved and diabetic animals. Knowledge of this mechanism is not necessarily complete as no effect of site 2 and 3 phosphorylations was found by Teague et al [42] with PDH phosphatase purified to apparent homogeneity. It is possible that some other factor is involved that is lost from highly purified phosphatase.

The activity of PDH phosphatase towards added phosphorylated pig heart PDH

TABLE 1

OCCUPANCY OF PHOSPHORYLATION SITES IN PDH COMPLEX IN RAT HEART IN VIVO
For further details see [41].

Rats	% of complex in inactive form	Occupancy of sites (%) in phosphorylated inactive complexes			Rate of reactivation by PDH phosphatase (min^{-1})[a]
		Site 1	Site 2	Site 3	
Fed, normal	70.5	99±0.9	56±1.2	37±1.6	0.51±0.04
Fed, normal[b]	37.5*	101±2.1	54±2.2	41±2.6	0.56±0.06
48h starved	99.4*	97±1.5	95±1.7*	86±1.9*	0.20±0.02*
Alloxan-diabetic	99.0	96±3.0	95±1.2*	90±1.0*	-

a. Apparent first order rate constant; b. exercised; all others resting.
Results are mean ± sem. * P<0.001 against fed normal control.

complex was not altered by starvation or diabetes [24, 32].

Complex activity in vivo. The proportion of active complex in the heart may be manipulated in the short term by factors which alter the concentration of metabolite effectors of PDH kinase and PDH phosphatase. It was of interest to examine the interaction of these short term manipulations with the effects of starvation and of diabetes. Based on the hypothesis outlined above, the expectation was that short term manipulations would not influence differences in the proportion of active complex between control and starved or diabetic. This expectation was fulfilled by the results shown in Table 2. High work increased the proportion of active complex presumably through activation of PDH phosphatase by increased mitochondrial $[Ca^{2+}]$ and inhibition of PDH kinase by decreased mitochondrial $[NADH]/[NAD^+]$ and [acetyl CoA]/[CoA] (column 3). The proportion of active complex was lowered by DL-3-hydroxybutyrate presumably as a result of increased mitochondrial [acetyl CoA]/[CoA] (column 4). Pyruvate increased the proportion of active complex by inhibiting PDH kinase (column 5). Prolonged perfusion in the absence of insulin increased the proportion of active complex presumably because PDH kinase was inhibited by lowering of mitochondrial $[NADH]/[NAD^+]$ (column 6). Ischemia decreased the proportion of active complex possibly because PDH kinase was activated by increasing mitochondrial $[NADH]/[NAD^+]$ (column 7). In each case the effect of starvation or diabetes to lower the proportion of active complex was retained i.e. the order of ranking (fed normal > starved > diabetic) was maintained.

TABLE 2 — PDH COMPLEX ACTIVITY IN PERFUSED RAT HEART

Rats	Proportion (%) of complex in active form (mean ± sem) for:-					
	Low work	High work	High work	Pyruvate	Low work	Ischemia
Normal fed	29 ± 4.2	78 ± 4.4	33 ± 3.4	69 ± 1.4	92 ± 4.5	32 ± 2.3
48h starved	12 ± 0.3	38 ± 5.9	5 ± 0.8	-	24 ± 2.9	14 ± 1.8
Diabetic	0.3 ± 0.2	6 ± 1.3	-	13 ± 3.0	8 ± 1.8	6 ± 0.03
Insulin(U/l)	20	20	20	20	-	-
Glucose(mM)	5.5	11	11	5.5	11	11
3-hydroxy-butyrate(mM)	-	-	5	-	-	-
Perfusion:- pressure(kPa)	8	16	16	8	8	-
time(min)	10 or 25	25	25	10	40	40
flow(ml/min)	15	35	35	15	15	1

CONCLUSIONS

A major problem in unravelling mechanisms of physiology and pathology in metabolic regulation is to know whether all of the mechanisms have been uncovered. Clearly in this particular case the identity and mechanism of action of kinase/activator, and the nature of the factors that may determine its formation in insulin deficient animals are outstanding problems amenable to investigation. Whether there are other factors involved which have yet to displace our ignorance remains to be seen.

ACKNOWLEDGEMENTS

These studies have been supported by grants from the Medical Research Council (U.K.), the British Diabetic Association and the British Heart Foundation. S.J.F. holds an MRC Research Studentship and T.C.V. is a British-American Research Fellow sponsored by the American Heart Association and the British Heart Foundation.

REFERENCES

1. Randle,P.J., Sugden, P.H., Kerbey, A.L., Radcliffe, P.M. and Hutson, N.J. (1978) Biochem. Soc. Symp. 43, 47
2. Linn, T.C., Pettit, F.H. and Reed, L.J. (1969) Proc. Natl. Acad. Sci. U.S.A. 64, 227
3. Wieland, O.H., Siess, E.A., Weiss, L., Loffler, G., Patzelt, C., Portenhauser, R., Hartmann, U. and Schirmann, A. (1973) Symp. Soc. Exp. Biol. 27, 371
4. Reed, L.J. (1981) Current Topics in Cell Regulation, 18, 95
5. Pettit, F.H., Humphreys, J. and Reed, L.J. (1982) Proc.Natl.Acad.Sci.U.S.A. 79, 3945
6. Teague, M., Pettit, F.H., Wu, T-L., Silberman, S.R. and Reed, L.J. (1982) Biochemistry, 21, 5585
7. Sugden, P.H., Kerbey, A.L., Randle, P.J., Waller,C.A. and Reid, K.B.M. (1979) Biochem. J., 181, 419
8. Sale, G.J. and Randle, P.J. (1981) Eur. J. Biochem. 120, 535
9. Reed, L.J. and Pettit, F.H. Cold Spring Harbor Conf. Cell Proliferation, 8, 701
10. Tonks, N.K., Kearns, A. and Randle, P.J. (1982) Eur. J. Biochem. 122, 549
11. Sale, G.J. and Randle, P.J. (1982) Biochem. J. 203, 99
12. Cooper, R.H., Randle, P.J. and Denton, R.M. (1974) Biochem. J. 143, 625
13. Whitehouse, S., Cooper, R.H. and Randle, P.J. (1974) Biochem. J. 141, 761
14. Pettit, F.H., Pelley, J.W. and Reed, L.J. (1975) Biochim. Biophys. Res. Commun. 65, 575
15. Cooper, R.H., Randle, P.J. and Denton, R.M. (1975) Nature, Lond. 257, 808

16. Kerbey, A.L., Radcliffe, P.M., Randle, P.J. and Sugden, P.H. (1979) Biochem. J., 181, 427
17. Denton, R.M., Randle, P.J. and Martin, B.R. (1972) Biochem. J., 128, 161
18. Randle, P.J., Denton, R.M., Pask, H.T. and Severson, D.L. (1974) Biochem. Soc. Symp., 39, 75
19. Hansford, R.G. (1976) J. Biol. Chem., 251, 5483
20. Kerbey, A.L., Radcliffe, P.M. and Randle, P.J. (1977) Biochem. J., 164, 509
21. Denton, R.M., McCormack, J.G. and Edgell, N.J. (1980) Biochem. J. 190, 107
22. Hennig, G., Loffler, G. and Wieland, O.H. (1975) FEBS Lett., 59, 142
23. Illingworth, J.A. and Mullings, R. (1976) Biochem. Soc. Trans., 4, 291
24. Kerbey, A.L., Randle, P.J., Cooper, R.H., Whitehouse, S., Pask, H.T. and Denton, R.M. (1976) Biochem. J., 154, 327
25. Hagg, S.A., Taylor, S.I. and Ruderman, N. (1976) Biochem. J., 158, 203
26. Caterson, I.D., Fuller, S.J. and Randle, P.J. (1982) Biochem. J., 208, 53
27. Vary, T.C. and Randle, P.J. (1984) in: Dhall, N.S. (Ed), Advances in Myocardiology, Volume 6. Plenum Press, New York. In Press
28. Denton, R.M. and Randle, P.J. (1967) Biochem. J., 104, 416
29. Denton, R.M. and Randle, P.J. (1967) Biochem. J., 104, 423
30. Randle, P.J., Garland, P.B., Hales, C.N., Newsholme, E.A., Denton, R.M. and Pogson, C.I. (1966) Recent Prog. Horm. Res., 22, 1
31. Pearce, F.J., Foster, J., DeLeeuw, G., Williamson, J.R. and Tutwiler, G.F. (1979) J. Mol. Cell. Cardiol., 11, 893
32. Hutson, N.J., Kerbey, A.L., Randle, P.J. and Sugden, P.H. (1978) Biochem.J., 173, 669
33. Hutson, N.J. and Randle, P.J. (1978) FEBS Lett., 92, 73
34. Baxter, M.A. and Coore, H.G. (1978) Biochem. J., 174, 553
35. Kerbey, A.L. and Randle, P.J. (1981) FEBS Lett., 127, 188
36. Kerbey, A.L. and Randle, P.J. (1982) Biochem. J., 206, 103
37. Sugden, P.H., Hutson, N.J., Kerbey, A.L. and Randle, P.J. (1978) Biochem. J. 169, 443
38. Kerbey, A.L., and Randle, P.J. (1979) FEBS Lett., 108, 485
39. Kerbey, A.L., Randle, P.J. and Kearns, A. (1981) Biochem. J., 195, 51
40. Sale, G.J. and Randle, P.J. (1980) Biochem. J., 188, 409
41. Sale, G.J. and Randle, P.J. (1982) Biochem. J., 206, 221
42. Teague, W.M., Pettit, F.H., Yeaman, S.J. and Reed, L.J. (1979) Biochem. Biophys. Res. Commun., 87, 244

Résumé

Dans les tissus de mammifères, l'oxydation du glucose est régulée par l'activité du complexe pyruvate déshydrogénase (PDH) et est réduite par le manque d'insuline en cas de jeûne et de diabète. Le complexe PDH est inactivé lorsqu'il est phosphorylé par l'ATP et est réactivé par déphosphorylation (PDH phosphatase).

Chez les animaux soumis au jeûne ou diabétiques le complexe PDH est inactivé par phosphorylation endéans les 48 heures. Les effets du jeûne sont inhibés par l'administration de glucose et ceux du diabète par un traitement à l'insuline in vivo. Ces effets ne sont pas inhibés par le glucose ÷ insuline in vitro. Les effets du jeûne et du diabète dépendent de l'oxydation des acides gras et sont inhibés par l'acide 2-tétradécylglycidique qui est un inhibiteur de la carnitine acyltransférase. L'effet de l'oxydation des acides gras est médié par l'activation de la PDH kinase provoquée par une augmentation des rapports [acétyl-CoA]/[CoA] et [NADH]/[NAD$^+$] dans la mitochondrie. La PDH kinase est aussi activée par une protéine ou par un facteur associé à une protéine appelé activateur de kinase dont on peut empêcher l'apparition en cas de jeûne au moyen de la cycloheximide ou de la puromycine. L'activité accrue de la PDH kinase provoque une augmentation de l'occupation de 2 sites de phosphorylation en plus du site d'inactivation. Ceci a pour conséquence d'inhiber la réactivation par la PDH phosphatase. L'insuline ne régule pas de manière aiguë la phosphorylation réversible du complexe PDH dans le muscle bien que ceci ait lieu dans les adipocytes.

REVERSIBLE PHOSPHORYLATION OF HORMONE-SENSITIVE LIPASE IN THE HORMONAL CONTROL OF ADIPOSE TISSUE LIPOLYSIS

PETER STRÅLFORS, GUDRUN FREDRIKSON, HÅKAN OLSSON AND PER BELFRAGE
Dept. of Physiological Chemistry 4, University of Lund, P.O.B. 750, S-220 07 Lund, Sweden

Excess energy intake is converted to fatty acids and stored as triacylglycerol in the adipose tissue. The mobilization of the fatty acids from adipose tissue is regulated by hormones and the sympathetic nervous system through control of the activity of the hormone-sensitive lipase (HSL). This enzyme catalyzes the rate-limiting step in adipose tissue lipolysis, the hydrolysis of the first ester bond of the triacylglycerol and, also, the hydrolysis of the produced 1,2-diacylglycerols to 2-monoacylglycerol. A separate monoacylglycerol lipase (1) catalyzes the hydrolysis of the monoacylgycerol.

HSL from rat adipose tissue has been extensively purified and partially characterized (2), its properties and regulation have recently been reviewed (3-5). The enzyme has a minimum M_r of 84000 (sodium dodecylsulphate polyacrylamide gel electrophoresis (SDS-PAGE)) and an apparent molecular size by gel chromatography of about 150000, which may indicate a dimeric structure. The enzyme has a marked preference for the 1(3)-ester bonds of the acylglycerol substrates, and hydrolyzes tri-:1,2-di-:2-monoacylglycerol:cholesterol ester at the maximal relative rates of 1:10:1:1.5. Like other lipases, e.g. pancreatic lipase and lipoprotein lipase, it has a high specific activity, 400 μmoles of fatty acids released per min per mg enzyme with 1,2-dioleoylglycerol as substrate, accounting for a high level of activity in adipose tissue in spite of a low tissue concentration (1-2 μg per g of tissue). The enzyme is inhibited by micromolar diisopropylfluorophosphate and cysteine-directed reagents (Hg^{2+}, N-ethylmaleimide) indicating a reactive serine group in the catalytic site, and one or several functional sulfhydryl groups.

It was recently demonstrated that the neutral, cytosolic cholesterol ester hydrolase in bovine adrenal cortex (6) and corpus luteum (7) was identical, or closely similar, to the hormone-sensitive lipase. The enzyme may thus have other important functions besides the control of adipose tissue lipolysis. In fact, it seems likely that hormone-sensitive lipase will turn out to be a hormone-activatable, multi-functional tissue lipase (for discussion, see ref. 4).

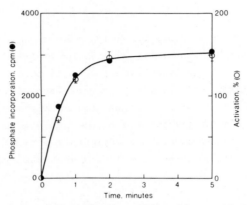

Fig. 1. Time-course of hormone-sensitive lipase phosphorylation and activation. The enzyme (approx. 30 nM) was incubated for 5 min with [γ-^{32}P] ATP-Mg^{2+} and the catalytic subunit of cAMP-PrK (0.3 μM). At the indicated time-points aliquots were withdrawn, and [^{32}P] phosphate incorporation into the enzyme protein and activation against an emulsified trioleoylglycerol substrate determined. Enzyme activation has been calculated as percent increase over control. ●, phosphate incorporation, mean of duplicates; ○, activation, mean of four determinations; vertical bars indicate S.E.

Regulation of hormone-sensitive lipase activity by reversible phosphorylation in vitro

HSL is rapidly phosphorylated and activated by cyclic AMP-dependent protein kinase (cAMP-PrK) (Fig. 1). Maximally one mol of phosphate per mol of enzyme protein is incorporated, into a single serine residue (8,9). A small, acidic phosphopeptide (approx. 10 amino acids) containing this site has been isolated by peptide mapping, after proteolytic degradation of ^{32}P-HSL with Staph. aur. V8 protease and trypsin (9). The rate of phosphorylation of the enzyme is comparable to that found in intact adipocytes (see below) (9). The specific inhibitor protein of cAMP-PrK arrests the phosphorylation immediately, demonstrating that cAMP-PrK acts directly on HSL without any intervening lipase kinase (2).

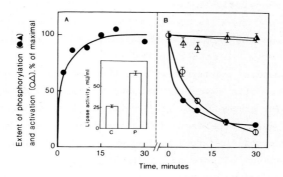

Fig. 2. Dephosphorylation and deactivation of hormone-sensitive lipase phosphorylated by cAMP-PrK. A. Phosphorylation of HSL, essentially as in Fig. 1., but at a lower concentration of the catalytic subunit of cAMP-PrK. AT 45 min the $[\gamma-^{32}P]$ ATP was removed by desalting and by hexokinase catalyzed phosphorylation of glucose. Inset: Activation of HSL, C=control, P=maximally phosphorylated enzyme; one mU represents the release of one nanomol of fatty acid per min. B. Time-course of dephosphorylation and deactivation of HSL. Partially purified protein phosphatase was added at arrow. At indicated time-points aliquots were withdrawn for determination of extent of HSL phosphorylation and activity as in Fig. 1. Controls (triangles) without protein phosphatase were run in parallel. Each point is the mean of three to five determinations, vertical bars are S.E.

Partially purified protein phosphatase from rat adipose tissue dephosphorylates and deactivates HSL, phosphorylated by the cAMP-PrK (Fig. 2) (H. Olsson, P. Strålfors and P. Belfrage, unpublished). The protein phosphatase preparation contained mainly protein phosphatase classified as type 2A (80%) and type 1 (see ref. 10). Protein phosphatases 1, 2A and 2C, purified from rabbit skeletal muscle, have also been found to catalyze the dephosphorylation of HSL. However, the relative importance of the different protein phosphatases in the dephosphorylation of the enzyme in the adipocyte is not yet known.

Phosphorylation of hormone-sensitive lipase in intact adipocytes

HSL is phosphorylated in intact rat adipocytes, as shown by the isolation of ^{32}P-labelled enzyme protein from adipocytes preincubated with [^{32}P]orthophosphate to label intracellular ATP (11). This phosphorylation occurs on two phosphorylation sites (8; P. Strålfors and P. Belfrage, unpublished). One of these is identical to the site phoshorylated by cAMP-PrK in vitro, as shown by peptide mapping of the phosphopeptides obtained from ^{32}P-HSL isolated from the fat cells and proteolytically degraded with V8 protease and trypsin. Since this site is only phosphorylated when the fat cells have been exposed to lipolytic hormones (e.g. catecholamines, ACTH, glucagon) and is directly involved in the control of HSL activity, it will be referred to as the 'regulatory' site in the following. The other site, referred to as the 'basal' site, is phosphorylated in the absence of hormonal stimulation of the fat cells. It is found in an about 20-amino acid phosphopeptide well separated by peptide mapping from the phosphopeptide containing the 'regulatory' site. In both sites, serine residues are the phosphate acceptors.

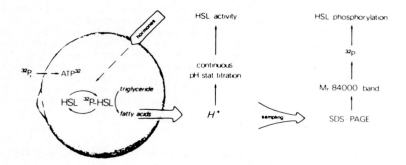

Fig. 3. Outline of methods for determination of the activity and the extent of phosphorylation of hormone-sensitive lipase in intact adipocytes.

Effect of hormones on hormone-sensitive lipase phosphorylation and activity in intact adipocytes.

Using a pH-stat titration technique HSL activity can be continuously monitored in a fat cell suspension as the release of fatty acids (Fig. 3) (12). Under the conditions used the reesterification of fatty acids is negligible. The extent of HSL phosphorylation in the intact fat cells can be determined by measuring the ^{32}P-radioactivity of the M_r 84000 ^{32}P-phosphopeptide band, obtained by SDS-PAGE of the total proteins of adipocytes preincubated with [^{32}P]orthophosphate (Fig. 3) (4,13,14).

Incubation of fat cells with [^{32}P]orthophosphate leads to incorporation of ^{32}P into HSL, to a steady state level after about 40 min (Fig. 4), with no measurable effect on the HSL activity. Recent analysis of the phosphorylatable sites has demonstrated that this phosphorylation is due to incorporation of phosphate in the 'basal' site only (8; P. Strålfors and P. Belfrage, unpublished). The phosphorylation of this site is not affected by exposure of the cells to insulin or to adenosine. Exposure of the cells to a fast-acting lipolytic hormone, e.g. noradrenaline (Fig. 4), rapidly increases HSL phosphorylation and, after a short time lag, the HSL activity (14). Half-maximal effect of noradrenaline is found at 20-30 nM (14). The rapid increase of the extent of HSL phosphorylation and activity could be readily reversed by the β-adrenergic antagonist propranolol. Analysis of the phosphorylated sites has shown that this reversible phosphorylation occurred at the 'regulatory' site, i.e. the site phosphorylated by cAMP-PrK in the isolated HSL. Approximately equal amounts of ^{32}P were incorporated into the 'basal' and 'regulatory' sites (see also Fig. 4), indicating that a single serine residue was phosphorylated in both sites.

Exposure of maximally noradrenaline-stimulated fat cells to 700 pM (100 μU/ml) insulin rapidly decreases the extent of HSL phosphorylation, accompanied by decreased HSL activity (Fig. 5) (13,14). When added shortly before, or together with noradrenaline, insulin (700 pM) completely prevents the anticipated increase of HSL phosphorylation and activity. Half-maximal inhibition is found at 20-30 pM insulin. Analysis of the phosphorylated sites showed that the insulin-induced decrease of HSL phosphorylation is due to net dephosphorylation of the 'regulatory' site (8; P. Strålfors and P. Belfrage, unpublished).

Modes of action of the fast-acting lipolytic hormones and of insulin

Exposure of the fat cells to millimolar concentration of dibutyryl cyclic AMP enhances the HSL phosphorylation and activity to the same extent as maximal noradrenaline stimulation (15). Noradrenaline enhances cellular cyclic AMP, HSL phosphorylation and activity in a dose-dependent

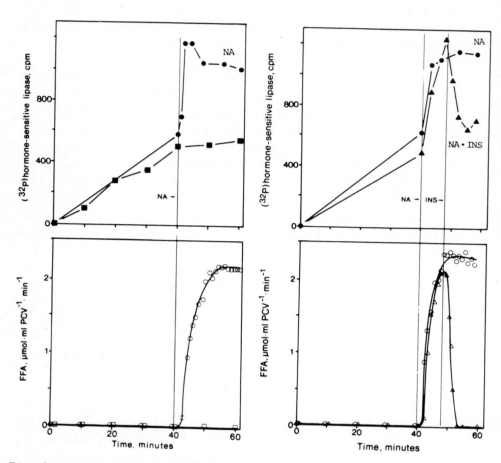

Fig. 4. Effect of noradrenaline on hormone-sensitive lipase phosphorylation and activity in intact adipocytes. Isolated rat adipocytes were incubated with [^{32}P]orthophosphate and the extent of HSL phosphorylation and activity determined as outlined in Fig. 3. Vertical line indicates addition of noradrenaline (NA); ■ , □ , control without noradrenaline; ●, ○, noradrenaline, approx. 0.3 μM. Reproduced from ref. 14, by permission.

Fig. 5. Effect of insulin on noradrenaline-stimulated hormone-sensitive lipase phosphorylation and activity in intact adipocytes. Conditions as in Fig. 4. Vertical bars indicate the addition of noradrenaline (NA), approx. 0.3 μM and insulin (INS), 700 pM, resp. ▲, △, noradrenaline followed by insulin. Reproduced from ref. 14, by permission.

manner (N.Ö. Nilsson, P. Strålfors, P. Björgell, J.N. Fain and P. Belfrage, unpublished). The cAMP-PrK activity ratio (-cAMP/+cAMP) in the fat cells increases from about 0.4 to 0.8 after maximal noradrenaline stimulation, with a time-course parallelling that of the phosphorylation of the HSL 'regulatory' site (P. Björgell and P. Belfrage, unpublished). Taken together these data establish that noradrenaline enhances HSL activity through cyclic AMP and cAMP-PrK mediated phosphorylation of the 'regulatory' site of HSL (Fig. 6), essentially as proposed in the 'lipolytic activation cascade' hypothesis a long time ago (16).

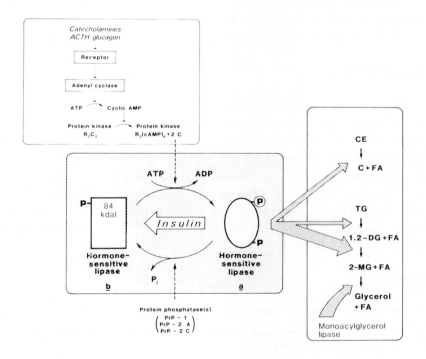

Fig. 6. Mechanisms for the hormonal control of adipose tissue lipolysis through reversible phosphorylation of hormone-sensitive lipase. Arrow, with insulin, indicates the net dephosphorylation of the 'regulatory' site, which accounts for insulin's anti-lipolytic effect. Encircled-P, phosphorylation of 'regulatory' site. CE, cholesterol ester; TG, DG, MG, tri-, di- and monoacylglycerol; FA, fatty acid.

In contrast, the mechanism(s) through which insulin causes net dephosphorylation of the 'regulatory' site of HSL (Fig. 6) cannot be established by the presented results. This insulin effect could have been mediated through a decrease of cAMP-PrK activity or an increase of the activity of one or several of the protein phosphatases involved in the dephosphorylation of HSL (see above). It has been found that prior exposure of fat cells to insulin substantially reduces the cellular cyclic AMP induced by subsequent maximal catecholamine stimulation (N.Ö. Nilsson, P. Strålfors, P. Björgell, J.N. Fain and P. Belfrage, unpublished). Cyclic AMP reduction, decreased cAMP-PrK activity, and thus a decreased rate of phosphorylation of HSL are certainly involved in the anti-lipolytic effect of insulin, but to an unknown extent.

ACKNOWLEDGEMENT

The contributions of Drs. Nils Östen Nilsson, Per Björgell and Staffan Nilsson to the work underlying the present review is gratefully acknowledged, as is the skilful technical assistance by Ingrid Nordh, Birgitta Danielsson, Aniela Szulezynski-Klein and Stina Fors. Financial support has been obtained from the following sources: A. Påhlsson, Malmö; T. and E. Segerfalk, Helsingborg; A.O. Swärd, Stockholm; Syskonen Svensson, Malmö; the Swedish Diabetes, Stockholm; Nordic Insulin, Copenhagen; P. Håkansson, Eslöv; The Medical Faculty, University of Lund and the Swedish Medical Research Council (project No. 3362).

REFERENCES

1. Tornqvist, H. and Belfrage, P. (1976) J. Biol. Chem. 251, 813.

2. Fredrikson, G., Strålfors, P., Nilsson, N.Ö. and Belfrage, P. (1981) J. Biol. Chem. 256, 6311.

3. Belfrage, P., Fredrikson, G., Strålfors, P. and Tornqvist, H. (1983) in: Borgström, B. and Brockman, H. (Eds.) Lipases, Elsevier/North Holland, Amsterdam, In press.

4. Strålfors, P. and Belfrage, P. (1983) in: Cohen, P. (Ed.)Molecular Aspects of Cellular Regulation. vol. 3. Recently Discovered systems of Enzyme Regulation by Reversible Phosphorylation. Part 2. , Elsevier/ North Holland, Amsterdam. In press.

5. Belfrage, P. (1983) in: Cryer, A. and Van, R.L.R. (Eds.) New Perspectives in Adipose Tissue Structure, Function and Development, Butterworths, London. In press.

6. Cook, K.G. and Yeaman, S.J.; Strålfors, P., Fredrikson, G. and Belfrage, P. (1982) Eur. J. Biochem. 125, 245.

7. Cook, K.G., Colbran, R.J., Snee, J. and Yeaman, S.J. (1983) Biochim. Biophys. Acta 752, 46.

8. Strålfors, P. (1983) Regulation of hormone-sensitive lipase through reversible phosphorylation. Role of cyclic AMP-dependent protein kinase. Thesis. University of Lund, Sweden. ISBN 91-722-627-7.

9. Strålfors, P. and Belfrage, P. (1983) J. Biol. Chem. In press.

10. Ingebritsen, T.S. and Cohen, P. (1983) Eur. J. Biochem. 132, 255.

11. Belfrage, P., Fredrikson, G., Nilsson, N.Ö. and Strålfors, P. (1980) FEBS Lett. 111, 120.

12. Nilsson, N.Ö. and Belfrage, P. (1981) Meth. Enzymol. vol. 72, p. 319. Acad. Press, New York.

13. Nilsson, N.Ö. (1981): Studies on the short-term regulation of lipolysis in rat adipocytes with special regard to the anti-lipolytic effect of insulin. Thesis, University of Lund, Sweden. ISBN 91-7222-402-9.

14. Nilsson, N.Ö., Strålfors, P., Fredrikson, G. and Belfrage, P. (1980) FEBS Lett. 111, 125.

15. Belfrage, P., Fredrikson, G., Nilsson, N.Ö. and Strålfors, P. (1981) Int. J. Obesity 5, 635.

16. Steinberg, D. and Huttunen, J.K. (1972) Adv. Cyclic Nucl. Res. 1, 47.

Résumé

    Les hormones lipolytiques à action rapide et l'insuline régulent la lipolyse dans le tissu adipeux au moyen du contrôle de l'activité de la lipase sensible aux hormones. Cet enzyme catalyse l'étape limitante de la lipolyse dans le tissu adipeux : l'hydrolyse du triacylglycérol de réserve. L'enzyme isolé peut être rapidement phosphorylé et activé par une phosphorylation catalysée directement par une protéine kinase AMP cyclique dépendante. Un maximum de 1 môle de phosphate par môle de sous-unité lipase peut être incorporé dans un seul résidu sérine. L'enzyme a été déphosphorylé et inactivé par les protéines phosphatases-1,-2A et -2C.. In vivo, dans les adipocytes isolés, l'enzyme incorpore du phosphate en l'absence de stimulation hormonale au niveau d'un site de phosphorylation spécifique de l'état "basal". La phosphorylation de ce site "basal" n'est pas associée à une augmentation

de l'activité de l'enzyme et n'est pas influencée par les hormones.
Les hormones lipolytiques à action rapide induisent la phosphorylation
d'un second résidu sérine au niveau d'un site de phosphorylation "de
régulation" qui est identique à celui phosphorylé in vitro par la kinase
AMP cyclique dépendante. Suite à la phosphorylation du site "de régulation", l'activité de l'enzyme et par conséquent la vitesse de la
lipolyse est augmentée d'environ 50 fois. L'insuline provoque de
manière spécifique une déphosphorylation rapide du site "de régulation"
inhibant ainsi l'activité de l'enzyme et la lipolyse. Des effets valant
la moitié de l'effet maximum sont obtenus sur les deux paramètres avec
une concentration d'insuline d'environ 25 pM.

KINASE F_A MEDIATED MODULATION OF PROTEIN PHOSPHATASE ACTIVITY

JACKIE R. VANDENHEEDE, SHIAW-DER YANG AND WILFRIED MERLEVEDE
Afdeling Biochemie, Faculteit Geneeskunde, Katholieke Universiteit Leuven, B-3000 Leuven (BELGIUM)

INTRODUCTION

The covalent modification of enzymes and regulatory proteins through reversible phosphorylation-dephosphorylation constitutes an important control mechanism for the regulation of many biological processes. The modulation of the enzymes involved in the hormonal control of glycogen metabolism is a prime example of a fine structured process which is largely governed by cyclic phosphorylation-dephosphorylation reactions. Figure 1 shows the major modifier enzymes in glycogen metabolism, and the interconversion between their active (a) and inactive (b) forms, catalyzed by protein kinases and phosphatase(s).

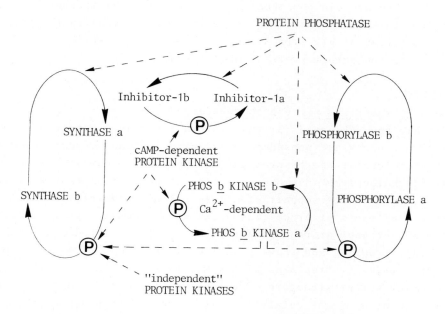

Fig. 1. Modifier enzymes involved in the regulation of glycogen metabolism.

The mode of action and regulation of the protein kinases has been intensively studied for several decades and this has resulted in the identification of several unrelated enzymes which phosphorylate their substrate proteins in a cyclic AMP- or Ca^{2+}-dependent or -independent way. For detailed information on the structure and function of the protein kinases, the reader is referred to a recent review by P. Roach (1). The enzymology of the protein phosphatases is only beginning to unravel, and has been the subject of substantial controversy and dispute. One reason for this is becoming clear as it is being realized that the protein phosphatase enzyme which is responsible for the major dephosphorylation reactions in glycogen metabolism, itself interconverts between an inactive- and an active conformation. Several proteins are involved in this phosphatase interconversion, and an intriguing aspect is that the activation of the protein phosphatase requires the action of an "independent" protein kinase (2).

The present report describes the recent advances made in the elucidation of the structure, function and regulation of this multisubstrate protein phosphatase. It will also be attempted to integrate this kinase mediated phosphatase modulation in the hormonal regulation of glycogen metabolism. We will restrict ourselves mostly to the rabbit skeletal muscle system.

RESULTS
The multisubstrate protein phosphatase

It has been observed (3) that in crude tissue preparations the major phosphorylase phosphatase activity is found associated with a high molecular weight protein (M_r=~260,000). However, this putative "holoenzyme" has proven to be extremely resistant to isolation, so that its subunit structure has not been elucidated. Harsh treatments of crude tissue extracts such as: precipitation with 80% ethanol at room temperature (4,5,6), limited proteolysis (3,7,8), freezing and thawing in the presence of 0.2 M 2-mercaptoethanol (8,9) or treatment with 4 M urea (10-12) destroy the high molecular weight structure of the phosphatase, and result in the creation of a M_r=35K catalytic unit, termed "phosphatase C" (3), which has a very broad substrate specificity (6,8). These M_r=35K protein phosphatases have been purified to homogeneity from several tissues and their properties have been described and compared in a comprehensive review by H.-C. Li (13).

Together with the isolation of "phosphatase C" came the discovery of heat stable proteins which were inhibitory for the phosphatase activity (14). Soon, two different heat stable protein inhibitors were distinguished (15,16) and termed inhibitor-1 and inhibitor-2 respectively. Inhibitor-1 had the exciting property of becoming a potent inhibitor of the phosphatase activity only after phosphorylation by the cyclic AMP-dependent protein kinase (15). Both inhibitor-1 (17-18) and inhibitor-2 (19-21) have been purified to homogeneity from several tissues. Inhibitor-1 could impose a hormonal regulation on the phosphatase activity (22-25) and inhibitor-2 has been identified (26) as a very important modulator protein for the interconversion of the multisubstrate protein phosphatase (see later). It has been proposed that the putative holoenzyme ($M_r = \sim 260,000$) represents a complex of protein phosphatase C and heat stable inhibitory units, which is dissociated by harsh purification procedures giving rise to various lower molecular weight phosphatase forms (3,27).

The inactive protein phosphatase. An inactive (ATP,Mg-dependent) form of protein phosphatase has been identified in various animal tissues (28-31), and the rabbit skeletal muscle enzyme has been extensively purified and characterized (26,32-36).

The inactive, rabbit skeletal muscle protein phosphatase is composed of two proteins: an inactive catalytic subunit (F_C) and a regulatory subunit which has been termed "modulator" or (M) and which is identical to the previously characterized inhibitor-2. The inactive phosphatase, represented as $[F_C M]$, is activated in a process involving ATP-Mg and an activating factor (F_A) which has been identified as an "independent" synthase kinase (37). This kinase F_A has subsequently been shown to be identical to the enzyme termed GSK-3 (38). The activated enzyme exhibits a broad substrate specificity, dephosphorylating among others phosphorylase, synthase and the β-subunit of phosphorylase kinase (33,39).

The purified enzyme shows one major protein staining band on non-denaturing polyacrylamide gels, which contains both the phosphatase activity (measurable after activation with kinase F_A and ATP,Mg) and the modulator activity (measured as heat stable inhibitor activity towards phosphatase C). This heat stable inhibitor protein (inhibitor-2) is strictly required for the reversible activation of the inactive phosphatase, and has therefore been termed "modulator protein" (26). The inactive enzyme reveals two major protein staining bands on SDS-PAGE corresponding to a $M_r=70K$ and $M_r=35K$ respectively. The smaller, $M_r=35K$ proteins contain the modulator activity. It has been suggested that the $M_r=70K$ proteins harbour the potential phosphatase activity (2), although no conclusive evidence for this is available.

The inactive enzyme migrates on gel filtration with an approximate $M_r=120K$, and on sucrose density gradient centrifugations as a $M_r=70K$ protein; the enzyme- and modulator activities perfectly coincide in both cases (40). It has recently been observed that the total modulator (inhibitor-2) activity, and the total protein phosphatase activity, present in a rabbit skeletal muscle extract, copurify in a constant ratio (40). Moreover, all through the purification procedure, the modulator is never available for inhibition of added phosphatase C, if the preparation is not first subjected to a boiling step, which dissociates the $[F_C-M]$-complex. This suggests that unboiled tissue extracts do not contain free modulator. Activation of the phosphatase by kinase F_A does not make the modulator readily available for inhibition of added phosphatase C (without a boiling step)(36). These results suggest that the modulator protein constitutes a regulatory subunit of the multisubstrate protein phosphatase, and that the kinase F_A-mediated activation of the $[F_CM]$-enzyme is not the result of the removal of an inhibitory moiety.

If these conclusions are valid, then one would expect the active multisubstrate protein phosphatase enzyme in the cell also to be an [enzyme-modulator]-complex (see section 2).

The inhibitor-2 protein is very susceptible to proteolysis which abolishes its activity (45). Similarly, limited proteolysis destroys the inactive phosphatase enzyme as well as the modulator activity it contains (37). The activated enzyme form, however, resists extensive proteolysis, although it is degraded from a $M_r=70K$ active enzyme to a $M_r=40K$ active fragment as judged by sucrose density gradient centrifugations, and has lost its modulator activity. Recombination of free modulator with this $M_r=40K$ active unit produces again an inactive "ATP,Mg-dependent" enzyme, and shifts the molecular weight to M_r=60K (27). This suggests that the modulator is only necessary for the kinase F_A-mediated activation of the inactive phosphatase, but is not required for the final expression of the activity of the (activated) enzyme. The inactive conformation of the $[F_CM]$-enzyme seems to be more susceptible to proteolysis, since addition of free modulator to proteolyzed inactive phosphatase preparations does not restore their capacity to be activated by kinase F_A. These observations are schematically represented in Fig. 2.

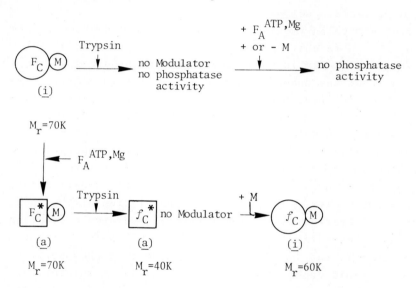

Fig. 2. Effect of trypsin on inactive and activated $[F_CM]$-enzyme. (i) and (a) represent (inactive) or (active) enzyme forms; f_C represents the proteolyzed catalytic subunit; the M_r are tentative values obtained in sucrose density gradient centrifugations (26).

The active protein phosphatase. A substantial part of the active multisubstrate protein phosphatase enzyme is found associated with the glycogen particle. The enzyme sediments together with the glycogen particle during high speed centrifugation, and can thus be separated from the cytosolic proteins. Further purification of this enzyme yields an active protein phosphatase (specific activity ± 1000 U/mg) which migrates in sucrose density gradient centrifugations as a M_r=70K complex of an active catalytic subunit and modulator protein (42). No low molecular weight phosphatase (M_r=40K) is observed. Limited proteolysis again destroys the modulator activity and converts the phosphatase to a M_r=~40K active fragment without measurable change in the phosphatase activity. Readdition of free modulator protein reconstitutes an inactive [phosphatase-modulator]-complex of M_r=60K, which can be activated by kinase F_A and ATP,Mg (Fig. 3). These results are analogous to those presented in section 1, and point to a structural and functional relationship between the kinase F_A-activated protein phosphatase and the "native" active enzyme which is associated with the glycogen particle.

The "native" active enzyme can also directly be converted to its inactive form by an incubation at 30° with excess free modulator protein. This conversion is time dependent, and produces an inactive enzyme without any change in molecular weight (M_r=70K). This suggests that the *in vitro* produced inactivation represents a substitution of a part of the active enzyme (possibly a modified modulator) by the free modulator. This process, which requires non-existing concentrations of free modulator, may not be operative *in vivo*, where the inactivation of the phosphatase could rather be the result of the reversal of the putative modulator modification. As we will discuss later, a transient phosphorylation of the modulator protein is observed during the kinase F_A-mediated activation of the inactive phosphatase, so that it is not unlikely that the modulator present in the "native" active enzyme is a phosphorylated species. A rapid dephosphorylation of the free modulator (after boiling) by phosphatase C may however prevent one from detecting the possible functional difference between the inhibitory properties of the phosphorylated and non phosphorylated modulator protein. This hypothesis however needs more extensive investigation.

Preparations of rabbit skeletal muscle protein phosphatase C (3) migrate in sucrose density gradient centrifugation with a M_r=40K, and form a M_r=60K inactive complex with added modulator which can be activated by kinase F_A (36). One distinguishing feature between the "reconstituted inactive enzymes" and the inactive protein phosphatase which is isolated as the [F_CM]-enzyme is the availability of the modulator activity in unboiled preparations. Where the un-

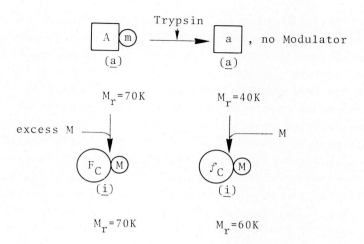

Fig. 3. Interconversion of protein phosphatase forms. Symbols are basically as in Fig. 2. In addition: A (square symbol) represents the active catalytic phosphatase subunit; a (square symbol) represents the proteolyzed active catalytic subunit; and m represents the putatively modified modulator present in the active phosphatase, whose activity can be measured after the boiling step.

boiled [F_CM]-phosphatase described in section 1 does not have modulator available for inhibition of exogeneously added protein phosphatase C, the unboiled reconstituted enzyme can substantially inhibit this low molecular weight phosphatase. This would point to a different, less efficient binding of modulator in these reconstituted enzyme complexes. Such a reconstituted, ATP,Mg-dependent phosphatase has been characterized by Cohen and coworkers (43,44) who do observe a dissociation of the modulator protein from the catalytic entity upon activation by kinase F_A (GSK-3). Their results may reflect a low affinity binding of the modulator in the reconstituted enzyme complex.

The mechanism of the kinase F_A mediated activation. The molecular mechanism of activation of the inactive multisubstrate protein phosphatase by kinase F_A has been thoroughly investigated in the last couple of years.

Since the inactive enzyme can be activated to some extent by metal ions such as Mn^{2+} and Co^{2+} (45) and the ATP or PP_i inhibition of active phosphatase forms is reversed by the same divalent metal ions, it has been proposed that the kinase F_A might possibly use ATP,Mg as a metal ion donor to activate an inactive demetallized form of phosphatase (33). The metal ion dependency or stabilisation of protein phosphatases is a generally observed phenomenon, which has been excellently reviewed by H.C. Li (13). So far, no evidence has been obtained for the incorporation of metal ions into the inactive phosphatase enzyme.

Initial attempts to incorporate [^{32}P]-phosphate into the inactive enzyme with kinase F_A and [γ-^{32}P]ATP,Mg failed as a fully activated [F_CM]-enzyme could be obtained without [^{32}P]-phosphate attached to it. The kinase F_A can use ATP[γ-S] to phosphorylate and inactivate glycogen synthase, but is unable to activate the phosphatase when ATP[γ-S] is used as nucleotide triphosphate (45). Since thiophosphorylated proteins are rather resistant to dephosphorylation by protein phosphatases (46), it was even suggested that a dephosphorylation step might be a requirement for the transition of the inactive phosphatase enzyme into an active conformation (2). These seemingly controversial results can however be explained if the kinase F_A mediated activation of the [F_CM]-enzyme involves a transient phosphorylation of the modulator subunit. This has been observed during the activation of the isolated [F_CM] enzyme (35,36,41) as well as with reconstituted inactive phosphatase forms (43,44). A mechanism of activation can be put forward which incorporates all the experimental results obtained so far. It is schematically presented in Fig. 4.

The kinase F_A initiates the activation of the inactive [F_CM]-enzyme by the incorporation of phosphate into the M-subunit. This phosphorylated modulator induces a conformational change in the inactive catalytic subunit F_C, which results in the activation of the phosphatase. During the initial period of activation, the degree of phosphorylation of the modulator protein may correlate well with the generation of phosphatase activity, if the second step, the conformational change is not rate limiting. Early in the activation of the phosphatase, the level of phosphate incorporated levels off and starts to decline, although the activation of the phosphatase continues, and ATP is being hydrolyzed at a constant rate all through the activation process. This suggests that the phosphorylated [F_CM]-complex is undergoing a dephosphory-

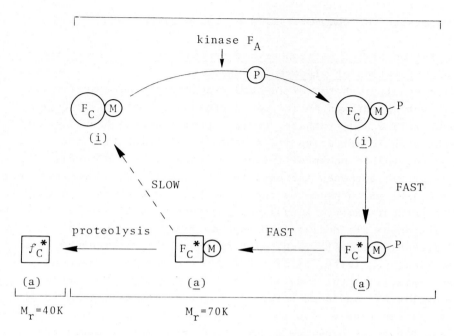

Fig. 4. The kinase F_A mediated activation of the $[F_CM]$-enzyme. All symbols are as in Fig. 2.

lation -possibly an autocatalytic reaction- and that the dephosphorylated enzyme slowly reverts back to the thermodynamically more stable inactive conformation. In other words, the phosphorylation of M initiates the transition of the inactive catalytic subunit, F_C, into an active conformation F_C^*, but the phosphoform of the modulator is not required at all times for the expression of the phosphatase activity. This explains the lack of correlation between the amount of phosphorylated modulator and the phosphatase activity generated during the later stage of the activation process. Proteolysis of the modulator subunit blocks this cyclic activation-inactivation process by converting the enzyme to a M_r=40K active fragment which maintains the active conformation and no longer reverts back to the inactive form, since the modulator subunit is missing (Fig. 4). These results do not agree with the hypothesis presented by Cohen and coworkers (43,44) who invoke a dissociation of the modulator (inhibitor-2) protein as a prerequisite for the activation of the phosphatase.

Integration of the multisubstrate protein phosphatase in the hormonal regulation of glycogen metabolism

Glycogenolytic hormones (α- and β-adrenergic hormones) exert their action through an increased phosphorylation of glycogen phosphorylase and synthase, thereby promoting the breakdown of glycogen while arresting its synthesis. A simplified scheme of these reversible phosphorylation reactions is given in Fig. 1. β-Adrenergic hormones increase the intracellular concentration of cyclic AMP, which promotes the sequential activation of the cyclic AMP-dependent protein kinase and phosphorylase kinase. α-Adrenergic responses are mediated through an increase in cytosolic Ca^{2+}, and probably reflect stimulation of the non-activated phosphorylase kinase activity, or of other Ca^{2+}-calmodulin-dependent protein kinase(s). For a detailed review, see (1). In the case of the cyclic AMP-mediated hormonal response, the cyclic AMP-dependent protein kinase also phosphorylates the heat stable inhibitor-1, and therefore provides a potential candidate to block the dephosphorylation process of phosphorylase a, synthase b and phosphorylase kinase (β-subunit)a by inhibiting the multisubstrate protein phosphatase activity. Although inhibitor-1 is a very potent phosphatase inhibitor *in vitro*, no *in vivo* correlation has ever been put forward between the phosphorylated state of inhibitor-1 and the activity of the multisubstrate protein phosphatase. In conclusion, both hormonal effects are mainly the result of an increased phosphorylation process since two intracellular second messengers, cyclic-AMP and Ca^{2+} have been identified which directly promote a kinase activation or stimulation. At the same time, the dephosphorylation processes are possibly being impaired through the cyclic AMP dependent phosphorylation of inhibitor-1.

The reversal of these hormonal effects is achieved by the dephosphorylation of glycogen phosphorylase and synthase, resulting in the arrest of glycogen breakdown and the promotion of glycogen synthesis. It has been demonstrated that a single multisubstrate protein phosphatase can dephosphorylate both of these rate limiting enzymes, and that this phosphatase interconverts between and active- and an inactive state (33). The modulation of this multisubstrate protein phosphatase activity may therefore constitute a major factor in determining the net effect that hormones have on the regulation of glycogen metabolism. The intricate regulation of this protein phosphatase activity is presently being elucidated.

The activity of the multisubstrate protein phosphatase described here can be modulated in two seemingly unrelated ways. One way, as already mentioned, is the inhibition of the phosphatase by the cyclic AMP-mediated phosphorylation of the heat stable inhibitor-1. The reversal of this inhibition is likely to involve the dephosphorylation of the phospho-inhibitor-1; a process which is not fully understood, and seems to be catalyzed by several unrelated enzymes (47-50). If the inhibition by inhibitor-1 occurs *in vivo*, it would be logical to assume that an inhibitor-1 phosphatase activity is responsible for the reversal of the cyclic AMP-mediated protein phosphorylations which promote glycogenolysis.

The second mechanism whereby the multisubstrate protein phosphatase is regulated is the kinase F_A-mediated cyclic interconversion between the active and inactive form of the enzyme, described in section 3. This process is dependent upon the presence of the modulator, as a non-dissociable regulatory subunit of the phosphatase, and on the activity of the kinase F_A, which is also a cyclic AMP- and Ca^{2+}-independent synthase kinase. This additional property of the kinase F_A places the enzyme in a very strategic situation, since it can regulate glycogen metabolism in two opposite directions. As a synthase kinase, it will inactivate synthase and stop glycogen synthesis, whereas in its role of activator for the multisubstrate protein phosphatase, it will promote exactly the opposite reaction. Clearly some regulation should be imposed on this kinase F_A, so that it readily distinguishes between its two functions. Several "independent" protein (synthase) kinase have been described (1) but no physiologically relevant regulatory factors have been proposed for them so that the potential regulation of these enzymes has not been implicated in the hormonal control of glycogen metabolism.

More and more evidence accumulates for an insulin mediated activation of a protein phosphatase which should reverse the cyclic AMP-dependent phosphorylations of the modifier enzymes in glycogen metabolism. Although concrete evidence to substantiate this hypothesis is still lacking, the stimulation of glycogen synthesis by insulin could easily be explained as the result of an increased activity of the multisubstrate protein phosphatase. As this protein phosphatase activity is possibly regulated by two separate mechanisms, so could insulin also increase the phosphatase activity in two different ways. Firstly it could stimulate the multisubstrate protein phosphatase indirectly by abolishing the inhibitor-1 mediated inhibition of the enzyme. This could be achieved by increasing the inhibitor-1 phosphatase activity. Suggestive evidence for this has been obtained in as much as the amount of phospho-inhibitor-1 was found to decrease after insulin administration (24,25). A small heat stable protein, termed "deinhibitor" has been reported to affect the inhibitor-1 phosphatase activity (50,51). Several laboratories have shown an adrenaline mediated increase in the phospho-form of the same protein (22-25), so that some of the antagonistic properties of insulin and adrenaline could be explained through the change in the phosphorylated state of the inhibitor-1 protein. Secondly, insulin could influence the activity of the multisubstrate protein phosphatase directly by promoting the kinase F_A mediated activation of the enzyme. The two opposite kinase F_A activities would be a likely target for hormonal regulation, since a coordinated control of glycogen metabolism could then be achieved by the selective involvement of this enzyme in either the activation of the multisubstrate protein phosphatase, causing glycogen synthesis, or in the phosphorylation of synthase, ultimately resulting in glycogen breakdown. Preliminary results have indicated that the phosphorylation sites on the synthase enzyme, which are the substrate sites for the kinase F_A (GSK-3), are dephosphorylated under the influence of insulin (52). An increased phospho-content of the same region in the synthase enzyme is observed after adrenaline administration (53). The increase of phosphate in any synthase site, caused by adrenaline might be the result of an overall decrease of the multisubstrate protein phosphatase activity, expected after cyclic AMP-mediated phosphorylation of inhibitor-1. The specific dephosphorylation of the kinase F_A-sites on synthase, seen after insulin administration could reflect an insulin mediated regulation of the kinase F_A-activity, so that it preferentially or exclusively acts as activator for the multisubstrate protein phosphatase, while its synthase kinase activity is blocked or impaired.

CONCLUDING REMARKS

In this report we have tried to illustrate the complex enzymology of the multisubstrate protein phosphatase which reverses most of the cyclic AMP-mediated protein phosphorylation reactions that regulate glycogen metabolism. The active enzyme dephosphorylates glycogen phosphorylase and synthase which are the rate limiting enzymes for glycogen breakdown or -synthesis respectively; it also inactivates phosphorylase kinase by dephosphorylating the enzyme's β-subunit. The activity of the protein phosphatase is controlled in a dual way: it interconverts between an active- and an inactive form, while the expression of its activity can furthermore be prevented by the phospho-inhibitor-1. The reversible interconversion of the multisubstrate protein phosphatase is made possible by the presence of the modulator protein, which constitutes the enzyme's regulatory subunit, and by the action of the kinase F_A which is responsible for the transition of the phosphatase catalytic subunit into its active conformation. It becomes more and more evident that the regulatory role of protein phosphatase(s) in the hormonal control of glycogen metabolism has been greatly underestimated because of our limited knowledge of the structure and function of the phosphatase enzymes themselves.

ACKNOWLEDGEMENTS

The original work carried out in our laboratory was supported by grants from the "Fonds voor Geneeskundig Wetenschappelijk Onderzoek" and the "Onderzoeksfonds K.U.Leuven". JRV is a Senior Research Associate of the "Nationaal Fonds voor Wetenschappelijk Onderzoek".

REFERENCES

1. Roach, P.J. (1981) Curr. Top. Cell. Regul. 20, 45-105.
2. Merlevede, W., Vandenheede, J.R., Goris, J. and Yang, S.-D. (1983) Curr. Top. Cell. Regul. 23 (in press).
3. Lee, E.Y.C., Silberman, S.R., Ganapathi, M.K., Petrovic, S. and Paris, H. (1980) Adv. Cycl. Nucl. Res. 13, 95-131.

4. Brandt, H., Killilea, S.D. and Lee, E.Y.C. (1974) Biochem. Biophys. Res. Commun. 61, 548-604.
5. Brandt, H., Capulong, Z.L. and Lee, E.Y.C. (1975) J. Biol. Chem. 250, 8038-8044.
6. Killilea, S.D., Brandt, H., Lee, E.Y.C. and Whelan, W.J. (1976) J. Biol. Chem. 251, 2363-2368.
7. Mellgren, R.L., Aylward, J.H., Killilea, S.D. and Lee, E.Y.C. (1979) J. Biol. Chem. 254, 648-652.
8. Khandelwal, R.L., Vandenheede, J.R. and Krebs, E.G. (1976) J. Biol. Chem. 251, 4850-4858.
9. Kato, K. and Sato, S. (1974) Biochim. Biophys. Acta 358, 299-307.
10. Goris, J., Defreyn, G. and Merlevede, W. (1977) Biochimie 59, 171-178.
11. Gratecos, D., Detwiler, T.C., Hurd, S. and Fischer, E.H. (1977) Biochemistry 16, 4812-4817.
12. Detwiler, T.C., Gratecos, D. and Fischer, E.H. (1977) Biochemistry 16, 4818-4823.
13. Li, H.C. (1982) Curr. Top. Cell. Regul. 21, 129-174.
14. Brandt, H., Lee, E.Y.C. and Killilea, S.D. (1975) Biochem. Biophys. Res. Commun. 63, 950-956.
15. Huang, F.L. and Glinsmann, W.H. (1975) Proc. Natl. Acad. Sci. U.S.A. 72, 3004-3008.
16. Huang, F.L. and Glinsmann, W.H. (1976) Eur. J. Biochem. 70, 414-426.
17. Nimmo, G.A. and Cohen, P. (1978) Eur. J. Biochem. 87, 341-351.
18. Nimmo, G.A. and Cohen, P. (1978) Eur. J. Biochem. 87, 353-365.
19. Foulkes, J.G. and Cohen, P. (1980) Eur. J. Biochem. 105, 195-203.
20. Yang, S.-D., Vandenheede, J.R. and Merlevede, W. (1981) FEBS Lett. 132, 293-295.
21. Khandelwal, R.L. and Zinman, S.M. (1978) J. Biol. Chem. 253, 560-565.
22. Khatra, B.S., Chiasson, J.-L., Shikima, H., Exton, J.H. and Soderling, T.R. (1980) FEBS Lett. 114, 253-256.
23. Tao, S.-H., Huang, F.L., Lynch, A. and Glinsmann, W.H. (1978) Biochem. J. 176, 347-350.
24. Foulkes, J.G., Jefferson, L.S. and Cohen, P. (1980) FEBS Lett. 112, 21-24.
25. Foulkes, J.G., Cohen, P., Strada, S., Everson, W.V. and Jefferson, L.-S. (1982) J. Biol. Chem. 257, 12493-12496.
26. Yang, S.-D., Vandenheede, J.R. and Merlevede, W. (1981) J. Biol. Chem. 256, 10231-10234.
27. Lee, E.Y.C., Brandt, H., Capulong, Z.L. and Killilea, S.D. (1976) Adv. Enz. Regul. 14, 467-490.

28. Merlevede, W. and Riley, G.A. (1966) J. Biol. Chem. 241, 3517-3524.
29. Goris, J., Defreyn, G. and Merlevede, W. (1979) FEBS Lett. 99, 279-282.
30. Yang, S.-D., Vandenheede, J.R., Goris, J. and Merlevede, W. (1980) FEBS Lett. 111, 201-204.
31. Goris, J., Doperé, F., Vandenheede, J.R. and Merlevede, W. (1980) FEBS Lett. 117, 117-121.
32. Yang, S.-D., Vandenheede, J.R., Goris, J. and Merlevede, W. (1980) J. Biol. Chem. 255, 11759-11767.
33. Vandenheede, J.R., Yang, S.-D. and Merlevede, W. (1981) J. Biol. Chem. 256, 5894-5900.
34. Vandenheede, J.R., Goris, J., Yang, S.-D., Camps, T. and Merlevede, W. (1981) FEBS Lett. 127, 1-3.
35. Jurgensen, S., Shacter-Noiman, E., Huang, C.Y., Chock, P.B., Vandenheede, J.R. and Merlevede, W. (1983) Fed. Proc. 42, 2026.
36. Jurgensen, S., Shacter-Noiman, E., Huang, C.Y., Chock, P.C., Yang, S.-D., Vandenheede, J.R. and Merlevede, W. (1983) manuscript in preparation.
37. Vandenheede, J.R., Yang, S.-D., Goris, J. and Merlevede, W. (1980) J. Biol. Chem. 255, 11768-11774.
38. Hemmings, G.A., Yellowlees, D., Kernohan, J.C. and Cohen, P. (1981) Eur. J. Biochem. 119, 443-451.
39. Stewart, A.A., Hemmings, B.A., Cohen, P., Goris, J. and Merlevede, W. (1981) Eur. J. Biochem. 115, 197-205.
40. Yang, S.-D., Vandenheede, J.R. and Merlevede, W. (1983) Biochem. Biophys. Res. Commun. 113, 439-445.
41. Ballou, L.M., Brautigan, D.L. and Fischer, E.H. (1983) Biochemistry 22, 3393-3399.
42. Vandenheede, J.R., Yang, S.-D. and Merlevede, W. (1983) Biochem. Biophys. Res. Commun. (in press).
43. Hemmings, B., Resink, T.J. and Cohen, P. (1982) FEBS Lett. 150, 319-324.
44. Resink, T.J., Hemmings, B.A., Tung, H.Y. and Cohen, P. (1983) Eur. J. Biochem. 133, 455-461.
45. Yang, S.-D., Vandenheede, J.R. and Merlevede, W. (1981) FEBS Lett. 126, 57-60.
46. Tabarini, D. and Li, H.-C. (1980) Biochem. Biophys. Acta 15, 1192-1199.
47. Cohen, P. (1978) Curr. Top. Cell. Reg. 14, 117-196.
48. Stewart, A., Ingebritsen, T.S., Manalan, A., Klee, C.B. and Cohen, P. (1982) FEBS Lett. 137, 80-84.
49. Yang, S.-D., Tallant, E.A. and Cheung, W.Y. (1982) Biochem. Biophys. Res. Commun. 106, 1419-1425.

50. Goris, J., Camps, T., Defreyn, G. and Merlevede, W. (1981) FEBS Lett. 134, 189-193.
51. Goris, J., Waelkens, E., Camps, T. and Merlevede, W. (1984) Adv. Enzyme Regul. 24 (in press).
52. Parker, P.J., Embi, N.,Cauldwell, F.B. and Cohen, P. (1982) Eur. J. Biochem. 124, 47-55.
53. Parker, P.J.,Cauldwell, F.B. and Cohen, P. (1983) Eur. J. Biochem. 130, 227-234.

Résumé

Cet exposé illustre l'enzymologie complexe de la protéine phosphatase à substrats multiples qui inverse la plupart des réactions de phosphorylation de protéines médiées par l'AMP cyclique qui contrôlent le métabolisme du glycogène. L'activité de la protéine phosphatase est régulée de deux manières : elle peut s'interconvertir en une forme active et une forme inactive; de plus, l'expression de son activité peut être inhibée par une protéine stable à la chaleur (inhibiteur-1). L'interconversion de la protéine phosphatase à substrats multiples est rendue possible par la présence d'une protéine modulatrice,qui constitue la sous-unité régulatrice de l'enzyme et par l'action d'une protéine activante, la kinase F_A qui est responsable de la transition de la sous-unité catalytique de l'enzyme en sa conformation active. L'expression physiologique de l'activité de cette protéine phosphatase à substrats multiples pourrait être soumise à une régulation hormonale.

Hormones and Cell Regulation, Volume 8
INSERM European Symposium
J. E. Dumont and J. Nunez editors
© 1984 Elsevier Science Publishers B.V.

DEIODINATION AND CONJUGATION OF THYROID HORMONE IN RAT LIVER

THEO J. VISSER, DURK FEKKES*, MARTEN H. OTTEN*, JAN A. MOL, ROEL DOCTER AND GEORG HENNEMANN

Department of Internal Medicine III and Clinical Endocrinology, Erasmus University Medical School, P.O.Box 1738, 3000 DR Rotterdam, The Netherlands.

INTRODUCTION

Thyroxine is the main secretory product of the follicular cells of the thyroid gland. Its intrinsic thyromimetic activity is, if anything, low compared with that of 3,3',5-triiodothyronine (T_3). Although some T_3 is also secreted by the thyroid, about 80% of its total body production is derived from the outer ring deiodination (ORD) or 5'-deiodination of T_4 in peripheral tissues (for reviews, see refs. 1-3). Extra-thyroidal production is an even more important source for 3,3',5'-triiodothyronine (reverse T_3, rT_3). This compound is formed by inner ring deiodination (IRD) or 5-deiodination of T_4, and it is probably devoid of biological activity. Further stepwise deiodination of T_3 and rT_3 lead to a series of diiodothyronines (T_2), monoiodothyronines (T_1) and finally thyronine (T_0). A central metabolite in this cascade is 3,3'-T_2, which is generated by both IRD of T_3 and ORD of rT_3 (Fig. 1).

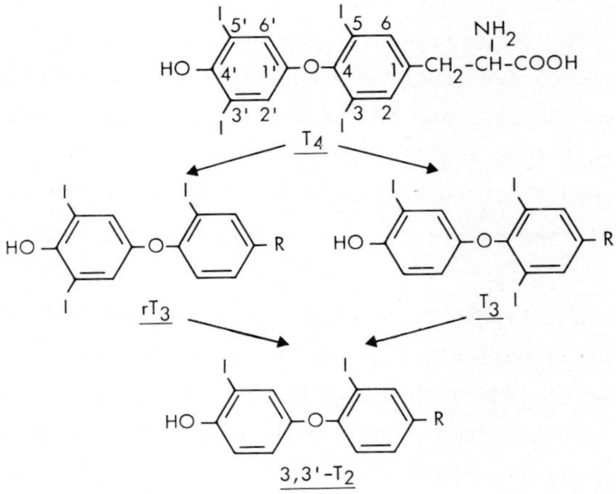

Fig. 1. Production of T_3, rT_3 and 3,3'-T_2 by stepwise deiodination of T_4.

* Present addresses: Durk Fekkes, Department of Pharmacology, Erasmus University Medical School, Rotterdam, The Netherlands; Marten H. Otten, Department of Gastroenterology, University of Nijmegen Medical School, Nijmegen, The Netherlands.

ORD may be regarded as an activating pathway since by this reaction T_4 is converted to T_3. Conversely, IRD is seen as an inactivating pathway as it diverts the metabolism of T_4 to rT_3 and mediates the degradation of T_3 to $3,3'-T_2$. The ultimate biological effect of T_4 is thus determined by the relative rates of these competing reactions. Consequently, the nature of the underlying enzymic mechanisms has been the subject of many investigations.

In healthy humans, mean normal plasma levels in nmol/l amount to about 100 for T_4, 2 for T_3 and 0.3 for rT_3. The metabolic clearance rates in l/day are on average 1 for T_4, 23 for T_3 and 100 for rT_3. The apparent production rates, therefore, approximate 100 nmol T_4, 45 nmol T_3 and 30 nmol rT_3 per day (1). Taking into consideration the thyroidal secretion of T_3, and the underestimation of the true total body production of T_3 and rT_3 by such calculations, it may be concluded that a) deiodination accounts for at least 75% of the metabolism of T_4, and b) roughly equal proportions of T_4 are converted to T_3 and rT_3.

Typical changes in the metabolism of thyroid hormone are observed in several clinical conditions, commonly referred to as the "low T_3 syndrome" (4). These include fasting, systemic illness, surgical stress, and the administration of certain drugs, i.e. propylthiouracil (PTU), propranolol, dexamethasone and X-ray contrast agents. In these situations plasma T_3 levels decrease and those of rT_3 increase while T_4 levels often remain constant. Kinetic studies have demonstrated that the production of T_3 is decreased but that its clearance is unaffected. Conversely, the production of rT_3 is apparently not changed but its metabolic clearance rate is decreased. These observations have led to the speculation that two enzymes are involved with the stepwise deiodination of T_4: an outer ring deiodinase (ORDase) and an inner ring deiodinase (IRDase) (2,5). The "low T_3 syndrome" would then result from a specific decrease in ORDase activity, leading to a diminished T_4-T_3 conversion and rT_3 degradation. The unchanged production rate of rT_3 and clearance rate of T_3 would point to a normal IRDase activity.

In the light of recent findings the above two enzymes-hypothesis is no longer tenable. Among others it neglects the importance of non-deiodinative pathways of thyroid hormone metabolism such as conjugation with sulfate or glucuronic acid. Furthermore, it has become increasingly clear that the mechanism of enzymatic deiodination differs among tissues. It is the purpose of this paper to review recent findings suggesting that a single enzyme catalyses both IRD and ORD of iodothyronines in the liver. The characteristics of this enzyme are compared with the specific IRDase's and ORDase's of other tissues. Evidence is also reviewed showing that the hepatic IRD of T_3 and the subsequent ORD of

$3,3'-T_2$ is enhanced by sulfate conjugation of the phenolic hydroxyl group.

ENZYMATIC DEIODINATION OF IODOTHYRONINES IN RAT LIVER

Essentially two methods have been used in the study of the deiodination of iodothyronines in tissue homogenates. Firstly, experiments have been done with radioiodine-labeled substrates, and the radioactive products have been characterized with a variety of chromatographic techniques. Secondly, tissue has been incubated with non-radioactive iodothyronines, followed by the measurement of the deiodination products with specific radioimmunoassays. Investigations of the deiodination of iodothyronines in rat liver preparations have utilized mainly the latter method with the following results.

<u>Subcellular location</u>. Both ORDase and IRDase activities are associated with the particulate fractions of liver homogenates, especially with the microsomes (6-11). There is controversy about the exact subcellular location. Most investigators reported on the co-sedimentation of deiodinase activity with marker enzymes of the endoplasmic reticulum (7-10). There is, however, one report that suggests a plasma membrane location (11) similar to the deiodinase activity of the kidney (12). These enzymes are integral membrane proteins that resist extraction with high-salt buffers but may be solubilized with selected detergents with retention of deiodinase activity (13,14).

<u>Cofactor requirement</u>. Deiodination by liver microsomes is stimulated by the addition of cytosol (6), suggesting the requirement of a soluble cofactor. Neither NADH nor NADPH support deiodination in the absence of cytosol (6) but NADPH may enhance T_4-T_3 conversion in homogenates (15-16). Thiols obviate the requirement for a cytosolic cofactor and stimulate microsomal IRD and ORD directly (6,17). Most potent in this respect are the dithiols dihydrolipoamide, dithiothreitol (DTT) and dithioerythritol, while 2-mercaptoethanol and 2,3-dimercaptoethanol are less active (6,18-20). Reduced glutathione is one of the least effective thiols tested (18,20). This raises the question concerning the nature of the endogenous cofactor. The high activity of dihydrolipoamide is counterbalanced by extremely low cytosolic concentrations. Conversely, reduced glutathione is the most abundant thiol in cells (21) but is of low potency. The identity of the physiological cofactor, therefore, remains obscure.

<u>Mechanism of enzymatic deiodination</u>. The way by which thiols stimulate enzymatic deiodination has been revealed in studies of the ORD of T_4 or rT_3 by rat liver or kidney microsomes (22-24). The results indicate that ORD follows ping-pong type reaction kinetics with DTT as the cofactor. This implies that there is no direct interaction between the iodothyronine substrate and the

thiol cofactor but that they react with different forms of the enzyme. Microsomal ORDase activity is inhibited by sulfhydryl-blocking reagents, in particular by micromolar concentrations iodoacetate (24,25). This suggests the participation of an enzyme SH group in the catalytic process. Especially revealing have been the investigations of the effects of PTU (22-24). These studies have demonstrated that PTU inhibits deiodination uncompetitively with substrate and competitively with cofactor. A thiourea derivative, PTU exhibits a high reactivity towards sulfenyl compounds producing stable mixed disulfides (26).

As a corollary of these findings, it has been proposed (22-24) that the mechanism of enzymatic deiodination in the liver and the kidneys is as set out in Fig. 2. In a first half reaction iodine is transferred (as I^+) from the substrate to the acceptor SH on the enzyme. Substitution with a proton yields the monodeiodinated product. In a second half reaction the enzyme sulfenyl iodide (E-SI) intermediate is reduced by cofactor with regeneration of native enzyme. The validity of this reaction model is supported by the observations that a) the persistent inactivation of enzyme activity by PTU is potentiated by ORDase substrates (25,27), b) incorporation of radioactive PTU in microsomal protein is stimulated by these same iodothyronines (25,28), and c) blockade of the essential SH group by formation of a mixed disulfide with PTU prevents the carboxymethylation of the enzyme (25).

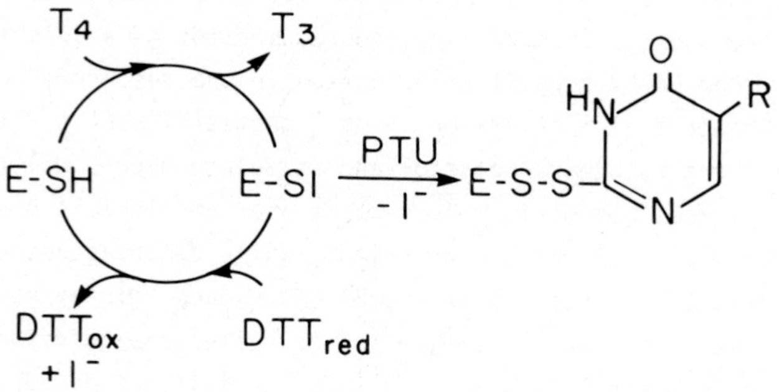

Fig. 2. Mechanism of the enzymatic deiodination of iodothyronines in rat liver, and inhibition by PTU. ORD of T_4 is shown as an example.

<u>Substrate and inhibitor specificities of IRDase and ORDase activities; evidence for a single enzyme.</u> We have investigated the following ORD reactions using rat liver microsomes and DTT: $T_4 \to T_3$ (6,17), $rT_3 \to 3,3'-T_2$ (17,29) and $3',5'-T_2 \to 3'-T_1$ (29,30). Similarly, studies have been made of the IRD reactions $T_4 \to rT_3$ (+ $3,3'-T_2$) (17) and $T_3 \to 3,3'-T_2$ (17,30). In addition, we also

measured the IRD of T_3 sulfate and the ORD of $3,3'-T_2$ and its sulfate (see below). The apparent K_m and V_{max} values of the various reactions are summarized in Table 1. As judged from the V_{max}/K_m ratio's, the susceptibility of the non-conjugated iodothyronines to undergo deiodination decreases in the order $rT_3 > 3',5'-T_2 > 3,3'-T_2 > T_4 > T_3$. Due to the high rate of rT_3 ORD, further degradation to $3,3'-T_2$ should be taken into account when the IRD of T_4 is measured (17).

Table 1
Mean kinetic parameters for the deiodination of several iodothyronines and sulfate conjugates by rat liver microsomes at pH 7.2, 37C and 3-5 mM DTT.

Substrate	Reaction	K_m [a]	V_{max} [b]	V_{max}/K_m	Ref.
T_4	ORD	2.3	30	13	17
T_4	IRD	1.9	18	9	17
rT_3	ORD	0.064	559	8734	17
T_3	IRD	6.2	36	6	17
T_3 sulfate	IRD	4.6	1050	228	42
$3',5'-T_2$	ORD	0.77	270	351	29
$3,3'-T_2$	ORD	8.9	188	21	39
$3,3'-T_2$ sulfate	ORD	0.34	353	1038	39

[a] µM; [b] pmol/min/mg protein

It has been widely recognized that ORD of the different iodothyronines is catalysed by the same liver enzyme. This is not only indicated by the mutual competitive inhibition by the substrates (1,17,29,31,32), but also by the equal sensitivity of the different reactions to a wide variety of inhibitors (29,33, 34). Fig. 3 shows an example of such a study comparing the ORD of rT_3 and that of $3',5'-T_2$. However, the following observations demonstrate also striking similarities between liver ORDase and IRDase activities, providing circumstantial evidence that both activities are intrinsic to a single enzyme.
1. The subcellular distribution of IRDase and ORDase activities is identical (see above). Moreover, concomitant purification of IRDase and ORDase activities was observed when microsomal protein was solubilized with detergent and subjected to a series of chromatographic steps (unpublished results). This proce-

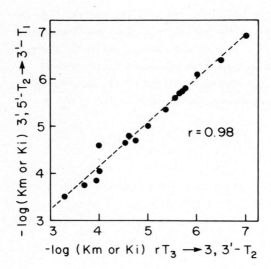

Fig. 3. Correlation between apparent K_m and K_i values for substrates and inhibitors of the ORD of rT_3 and that of $3',5'-T_2$ (data taken from ref. 29).

dure resulted in a 440-fold enrichment in specific enzyme activity (35).
2. Both IRD and ORD are reductive processes, and are driven by thiols like DTT (see above).
3. Not only enzymatic ORD (see above) but also IRD is inhibited by thiouracil derivatives such as PTU (36-38). In both cases this inhibition is uncompetitive with substrate (22-24, 36). This suggests that IRD and ORD follow the same reaction mechanism.
4. Reaction of microsomes with $3,5-T_2$, which can only undergo IRD, yields the sulfenyl iodide complex of the enzyme catalysing the ORD of other iodothyronines. This is evidenced by the persistent inactivation of ORDase activity in the presence of thiouracil (27).
5. The IRD of T_3 and the ORD of $3',5'-T_2$ or rT_3 are equally affected by several competitive substances covering a wide range of inhibitory potencies (30). Furthermore, T_3 is a competitive inhibitor of the 5'-deiodination of $3',5'-T_2$ and rT_3 with K_i values that are equal to the K_m value for T_3 in its IRD to $3,3'-T_2$. Conversely, the K_i value for $3',5'-T_2$ as a competitive inhibitor of T_3 IRD is equal to its K_m as an ORD substrate (30). A summary of these findings is illustrated in Fig. 4. Chopra and Chua Teco (38) made similar observations by comparison of the IRD of $3,5-T_2$ and the ORD of other iodothyronines.

These findings seriously challenge the hypothesis of two enzymes, i.e. an IRDase and an ORDase, being involved in the stepwise deiodination of T_4. The evidence is compelling that in rat liver a single enzyme is capable of carrying out both IRD and ORD.

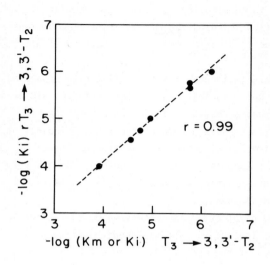

Fig. 4. Correlation of the apparent K_m value for T_3 and K_i values for inhibitors of T_3 IRD with the K_i values of these compounds as inhibitors of rT_3 ORD (see also ref. 30).

METABOLISM OF 3,3'-T_2 AND T_3 BY RAT HEPATOCYTES

<u>Metabolism of 3,3'-T_2</u>. The low yield of 3,3'-T_2 produced by ORD of rT_3 by hepatocytes prompted us to investigate the metabolism of 3,3'-T_2 by these cells. In these experiments, 3,[3'-^{125}I]T_2 or unlabeled 3,3'-T_2 were incubated in medium containing 10% fetal calf serum with hepatocytes in primary culture. The disappearance of unlabeled 3,3'-T_2 was analysed with a specific radioimmunoassay. The results were consonant with those obtained by analysis of parallel incubations with labeled 3,3'-T_2 using Sephadex LH-20 chromatography. The latter also allowed the quantitation of the labeled products, iodide and 3,3'-T_2 sulfate. Here we review our findings which have been reported in detail elsewhere (39,40).

The metabolism of 3,3'-T_2 by hepatocytes proved to be very fast. For instance, at a starting concentration of 10 nM, 80% of added 3,3'-T_2 is degraded in 60 min. Addition of thiouracil has little effect on this clearance, showing that deiodination is not a rate-limiting step. However, Sephadex chromatography revealed that iodide - derived from ORD - is produced almost exclusively in the absence of thiouracil. In the presence of the inhibitor, iodide formation is greatly reduced but this is accompanied by a reciprocal increase in the appearance of 3,3'-T_2 sulfate, explaining the unimpeded 3,3'-T_2 clearance. In an attempt to account for these findings we tested the possibility that in the hepatic metabolism of 3,3'-T_2 sulfation of the phenolic hydroxyl group precedes ORD. Evidence that indeed sulfation facilitates deiodination of 3,3'-T_2 may be summarized as follows (see also refs. 39,40).

1. The rate of 3,3'-T_2 deiodination in hepatocytes is similar to that of rT_3, which is in contradiction with the slow deiodination of 3,3'-T_2 in incubations with rat liver microsomes. This suggests an intermediate step that facilitates the ORD of 3,3'-T_2 in hepatocytes.

2. The accumulation of 3,3'-T_2 sulfate in incubations of 3,3'-T_2 with hepatocytes is only observed if deiodination is inhibited with, for instance, PTU or if deiodination is saturated at increasing medium 3,3'-T_2 concentrations. In both cases the proportion of 3,3'-T_2 metabolized remains constant. Deiodination is completely saturated at 1 µM medium 3,3'-T_2, whereas the apparent K_m of 3,3'-T_2 for the microsomal deiodinase amounts to 9 µM (see below). This is very suggestive of the formation of an intermediate, presumably 3,3'-T_2 sulfate, with an increased affinity for the deiodinase.

3. The rate of 3,3'-T_2 metabolism by hepatocytes is determined by the sulfate-transferring potential of the cells. The activity of the phenol sulfotransferases may be manipulated by variation of the medium SO_4^{2-} concentration, or by addition of competitive inhibitors, such as salicylamide, dichloronitrophenol and pentachlorophenol (41). Both approaches have been employed, and Fig. 5 illustrates the effects of varying medium SO_4^{2-} concentrations. In all instances the changes observed in the generation of iodide in the absence of PTU paralleled the changes in 3,3'-T_2 sulfate accumulation in the presence of this inhibitor. The release of iodide is, therefore, directly related to the rate of 3,3'-T_2 sulfation. This can only be explained if sulfation is the rate-limiting step preceding deiodination.

4. The sulfate ester of 3,3'-T_2 was shown to be a better substrate than non-conjugated 3,3'-T_2 for microsomal ORD in the presence of 5 mM DTT (Table 1). The difference in substrate behavior is mainly reflected in the apparent K_m values, i.e. 0.3 µM for 3,3'-T_2 sulfate and 9 µM for 3,3'-T_2, while V_{max} values differ less markedly (350 vs 190 pmol/ min/mg protein). ORD of rT_3 and that of 3,3'-T_2 sulfate are equally sensitive to a variety of inhibitory substances, suggesting that both reactions are catalysed by a single enzyme (39).

Metabolism of T_3. Similar evidence has been obtained showing that a major proportion of T_3 deiodinated in the liver in fact represents deiodination of the sulfate conjugate (39,42). In hepatocytes, T_3 is metabolized by glucuronidation, sulfation, and successive IRD and ORD. Iodide and T_3 glucuronide are the main radioactive products observed after incubation of $[3'-^{125}I] T_3$ with hepatocytes. However, in the presence of PTU, iodide formation is inhibited with a concomitant increase in the appearance of medium T_3 sulfate. Accumulation of T_3 sulfate in the presence of PTU and production of iodide in the absence of PTU are similar functions of the medium SO_4^{2-} concentration (Fig. 5). They are also inhibited to similar extents by the addition of salicylamide, dichloronitrophenol or pentachlorophenol (39).

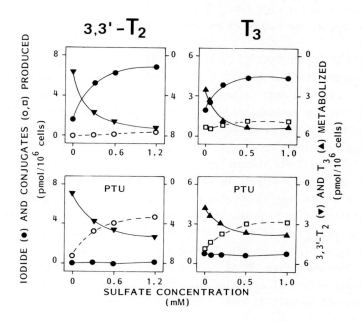

Fig. 5. Production of radioactive iodide and conjugates from outer ring labeled $3,3'$-T_2 and T_3 by rat hepatocytes as a function of the medium SO_4^{2-} concentration. Approximately 2×10^6 cells were incubated for 30 min at 37C with 10 nM $3,3'$-T_2 in Dulbecco's medium with 0.5% bovine serum albumin and 0-1.2 mM SO_4^{2-} (left panels), or for 180 min with 10 nM T_3 in Dulbecco's medium with 0-1 mM SO_4^{2-} but without albumin (right panels). Parallel incubations were done in the absence (upper panels) or presence (lower panels) of 100 μM PTU. Products were analysed by Sephadex LH-20 chromatography and by treatment with sulfatase or glucuronidase. $3,3'$-T_2 was exclusively sulfated but T_3 was also glucuronidated. T_3 glucuronide was the major conjugate in the absence of PTU; the sulfate-dependent increment of T_3 conjugates in the presence of PTU represents accumulation of T_3 sulfate. (Reprinted with permission from Otten et al, Science (1983) 221, 81).

There is a dramatic increase in the microsomal deiodination of T_3 sulfate compared with non-sulfated T_3. The initial IRD rates of T_3 sulfate may be compared with those of T_3 if precautions are taken to prevent subsequent ORD of the $3,3'$-T_2 sulfate generated. One then arrives at V_{max} values which are markedly higher for T_3 sulfate than for T_3 (1050 vs 36 pmol/min/mg protein), while apparent K_m values do not differ appreciably (Table 1). Interestingly, the IRD of T_3S and the ORD of rT_3 are equally affected by different inhibitory substances, including PTU (42). This corroborates the findings in Fig. 4, and is compatible with the view that IRD and ORD of iodothyronines as well as their sulfate conjugates are catalysed by one and the same PTU-sensitive enzyme.

DEIODINATION OF IODOTHYRONINES IN OTHER TISSUES

The characteristics of the ORD of T_4 and rT_3 in the rat kidney resemble those in the liver in many respects (33). However, distinctly different deiodinase activities are encountered in other tissues, i.e. brain (43), pituitary (44), placenta (45) and brown adipose tissue (46). With respect to IRDase activity, monkey hepatocarcinoma cells have also properties different from rat hepatocytes (47,48). These tissues contain membrane-bound but PTU-insensitive enzymes that selectively deiodinate either the inner or the outer ring of the substrate. Although they also require thiols, the reaction kinetics with DTT as the cofactor differ from the PTU-sensitive enzyme (46,49-51). Remarkable are also the low K_m values for the PTU-insensitive ORD of T_4 ($\sim$ 1 nM) and the IRD of T_4 and T_3 ($\sim$ 10 nM). Sulfate conjugation does not appear to enhance the PTU-insensitive IRD of T_3 in monkey hepatocarcinoma cells (47,48). The facilatory effect of sulfation may, therefore, be another distinction between PTU-sensitive and PTU-insensitive deiodination.

CONCLUSION

At least three different enzymes are involved with the peripheral deiodination of thyroid hormone in rats. Type I is the non-specific, PTU-sensitive deiodinase which is most abundant in the liver and the kidneys. Type II is the PTU-insensitive ORDase and type III the PTU-insensitive IRDase which have been observed in the other, abovementioned tissues. Not only their properties differ but these enzymes also appear to fulfill different purposes (3). Most of circulating T_3 in euthyroid subjects is derived from type I deiodination of T_4. However, a major proportion of circulating T_3 in hypothyroid subjects may be produced by type II deiodination of T_4 (52). An important role for enzymes type II and III in the brain and the pituitary seems to be the maintenance of local tissue T_3 concentrations within narrow boundaries even if thyroid function is abnormal. The findings of a close interaction between type I deiodination and sulfate conjugation is a further illustration of the complexity of the peripheral metabolism of thyroid hormone.

ACKNOWLEDGEMENTS

We appreciate the competent technical assistance of Ms. Ellen Kaptein, Ms. Marla van Loon and Ms. Jannet Blom, and the expert secretarial assistance of Ms. Yvonne van Dodewaard. This work was supported by grants from the Netherlands Organisation for the Advancement of Pure Research ZWO, from the Foundation for Medical Research FUNGO (nr. 13-34-108), and from the Division for Health Research TNO (nr. 13-34-110).

REFERENCES

1. Chopra, I.J., D.H. Solomon, U. Chopra, S.Y. Wu, D.A. Fisher and Y. Nakamura (1978) Rec. Progr. Horm. Res. 34, 521.
2. Visser, T.J. (1978) Mol. Cell. Endocrinol. 10, 241.
3. Larsen, P.R., J.E. Silva and M.M. Kaplan (1981) Endocrine Rev. 2, 87.
4. Hesch, R-D. (1981) The Low T_3 Syndrome, Academic Press, New York.
5. Schimmel, M. and R.D. Utiger (1977) Ann. Intern. Med. 87, 760.
6. Visser, T.J., I. van der Does-Tobé, R. Docter and G. Hennemann (1976) Biochem. J. 157, 479.
7. Fekkes, D., E. van Overmeeren-Kaptein, R. Docter, G. Hennemann and T.J. Visser (1979) Biochim. Biophys. Acta 587, 12.
8. Auf dem Brinke, D., R-D. Hesch and J. Köhrle (1979) Biochem. J. 180, 273.
9. Auf dem Brinke, D., J. Köhrle, R. Ködding and R-D. Hesch (1980) J. Endocrinol. Invest 3, 73.
10. Saito, K., K. Yamamoto, T. Takai and S. Yoshida (1980) J. Biochem. 88, 1595.
11. Maciel, R.M.B., Y. Ozawa and I.J. Chopra (1979) Endocrinology 104, 365.
12. Leonard, J.L. and I.N. Rosenberg (1978) Endocrinology 103, 274.
13. Fekkes, D., E. van Overmeeren, G. Hennemann and T.J. Visser (1980) Biochim. Biophys. Acta 613, 41.
14. Leonard, J.L. and I.N. Rosenberg (1981) Biochim. Biophys. Acta 658, 202.
15. Balsam, A., S.H. Ingbar and F. Sexton (1979) J. Clin. Invest. 63, 1145.
16. Sato, T., S. Maruyama and K. Nomura (1981) Endocrinol. Japon. 28, 451.
17. Visser, T.J., D. Fekkes, R. Docter and G. Hennemann (1979) Biochem. J. 179, 489.
18. Chopra, I.J. (1978) Science 199, 239.
19. Gavin, L.A., F. Bui, F. McMahon and R. Cavalieri (1980) J. Biol. Chem. 254, 49.
20. Goswami, A. and I.N. Rosenberg (1983) Endocrinology 112, 1180.
21. Meister, A. and S.S. Tate (1976) Ann. Rev. Biochem. 45, 559.
22. Leonard, J.L. and I.N. Rosenberg (1978) Endocrinology 103, 2137.
23. Visser, T.J. (1979) Biochim. Biophys. Acta 569, 302.
24. Leonard, J.L. and I.N. Rosenberg (1980) Endocrinology 107, 1376.

25. Leonard, J.L. and I.N. Rosenberg (1980) Endocrinology 106, 444.
26. Cunningham, L.W. (1964) Biochemistry 3, 1629.
27. Visser, T.J. and E. van Overmeeren-Kaptein (1981) Biochim. Biophys. Acta 658, 202.
28. Visser, T.J. and E. van Overmeeren-Kaptein (1979) Biochem. J. 183, 167.
29. Fekkes, D., G. Hennemann and T.J. Visser (1982) Biochem. Pharmacol. 31, 1705.
30. Fekkes, D., G. Hennemann and T.J. Visser (1982) Biochem. J. 201, 673.
31. Chopra, I.J. (1977) Endocrinology 101, 453.
32. Kaplan, M.M. and R.D. Utiger (1978) J. Clin. Invest. 61, 459.
33. Kaplan, M.M., J.B. Tatro, R. Breitbart and P.R. Larsen (1979) Metabolism 28, 1139.
34. Chopra, I.J. (1981) Endocrinology 108, 464.
35. Mol, J.A., T.P. v.d. Berg and T.J. Visser (1983) Ann. Endocrinol. 44, 28A.
36. Chopra, I.J., S.Y. Wu, Y. Nakamura and D.H. Solomon (1978) Endocrinology 102, 1099.
37. Visser, T.J., D. Fekkes, R. Docter and G. Hennemann (1978) Biochem. J. 174, 221.
38. Chopra, I.J. and G.N. Chua Teco (1982) Endocrinology 110, 89.
39. Otten, M.H., J.A. Mol and T.J. Visser (1983) Science 221, 81.
40. Otten, M.H., G. Hennemann, R. Docter and T.J. Visser (to be published).
41. Mulder, G.J. (1981) Sulfation of Drugs and Related Compounds, CRC Press, Boca Raton, Florida.
42. Visser, T.J., J.A. Mol and M.H. Otten (1983) Endocrinology 112, 1547.
43. Kaplan, M.M. and K.A. Yaskoski (1980) J. Clin. Invest 66, 551.
44. Kaplan, M.M. (1980) Endocrinology 106, 567.
45. Roti, E., S-L. Fang, L.E. Braverman and C.H. Emerson (1982) Endocrinology 110, 34.
46. Leonard, J.L., S.A. Mellen and P.R. Larsen (1983) Endocrinology 112, 1153.
47. Sorimachi, K. and J. Robbins (1979) Biochim. Biophys. Acta 583, 443.
48. Sato, K. and J. Robbins (1980) J. Clin. Invest. 68, 475.
49. Visser, T.J., J.L. Leonard, M.M. Kaplan and P.R. Larsen (1981) Proc. Nat. Acad. Sci. USA 79, 5080.
50. Visser, T.J., M.M. Kaplan, J.L. Leonard and P.R. Larsen (1983) J. Clin. Invest. 71, 992.
51. Kaplan, M.M., T.J. Visser, K.A. Yaskoski and J.L. Leonard (1983) Endocrinology 112, 35.
52. Silva, J.E., M.B. Gordon, F.R. Crantz, J.L. Leonard and P.R. Larsen (1983) Ann. Endocrinol. 44, 27A.

Résumé

Le foie est un site important de production extra-thyroïdienne de 3, 3'5-triiodothyronine (T_3) à partir de thyroxine (T_4), par un processus appelé désiodation du cycle externe (ORD). La désiodation du cycle interne (IRD) de la T_4 aboutit à la 3,3',5'-triiodothyronine (rT_3). La T_3 est la forme la plus active, si pas la seule active, parmi les hormones thyroïdiennes. Jusqu'à présent, il était établi que les désiodations IRD et ORD nécessitaient des enzymes différents. Une autre importante voie d'élimination des iodothyronines est la conjugaison du groupement hydroxyle phénolique avec un sulfate ou l'acide glucuronique. La désiodation et la conjugaison ont été considérées commes des réactions indépendantes avec différentes fonctions. Dans cet article nous présentons une revue des arguments montrant que dans le foie de rat les réactions IRD et ORD sont toutes les deux catalysées par une seule et même enzyme. De plus, dans des expériences réalisées avec des hépatocytes isolés, nous avons observé que l'IRD de la T_3 et l'ORD subséquente de la 3,3'-diiodothyronine (3,3'-T_2) est précédée et en fait accélérée, par une conjugaison avec du sulfate, la T_3-sulfate, et la 3,3'-T_2 sulfate s'avèrent être de meilleurs substrats pour la désiodase microsomale que les composés non sulfatés. Ces résultats montrent que l'IRD et l'ORD des iodothyronines, aussi bien que de leur conjugué sulfaté, sont catalysées au niveau d'un seul et même enzyme hépatique.

NEUROBIOLOGICAL CONTROLS
CONTRÔLES NEUROBIOLOGIQUES

FUNCTIONAL ASPECTS OF THE COEXISTENCE OF CLASSICAL NEUROTRANSMITTERS AND PEPTIDE NEUROTRANSMITTERS

ANITA WESTLIND, JANIS ABENS and TAMAS BARTFAI

Department of Biochemistry, Arrhenius Laboratory
University of Stockholm, 10691 Stockholm, Sweden

INTRODUCTION

The presence of various putative peptide neurotransmitters in neurons containing classical neurotransmitters such as acetylcholine and monoamines has repeatedly been demonstrated in the central and peripheral nervous systems of several species during the last five years (cf for references 1,2). Following the immunohistochemical identification of more than one neurotransmitter in certain neurons, release of both neurotransmitters and interactions at the postsynaptic membrane have also been demonstrated (3,4). Today the number of instances where the coexistence of several neurotransmitters have been demonstrated far exceeds the number of cases where any acute or chronic drug effects on these neurotransmitters have been studied.

In this paper we shall seek (admittedly teleological) answers to the functional significance of such neurotransmitter coexistence on the basis of data available for a few selected systems. We shall try to suggest some possible functions for the coexisting neurotransmitters in pre- and postsynaptic function during neuronal development, during normal function of these neurons as elements of a neuronal circuitry and during acute or chronic drug treatment which affects one or both of the coexisting neurotransmitters.

Gross anatomy of classical neurotransmitter- peptide neurotransmitter coexistence.

Table 1 gives some selected examples of the immunohistochemical findings indicating coexistence of classical neurotransmitters with putative peptide neurotransmitters. Table 1 shows that a large variety of the classical neurotransmitters ranging from the monoamines to GABA and acetylcholine have been found in coexistence with a peptide. It is also apparent that a variety of peptides are present in these neurons. Noradrenaline, for example, is found to coexist with somatostatin in the sympathetic ganglia of the guinea pig (10), and is found together with neurotensin in the cat adrenal medulla (12). This indicates that such coexistence situations are a widespread phenomenon, not restricted to one species, one cell type or one classical neurotransmitter - peptide neurotransmitter pair.

It should also be noted that both excitatory, inhibitory or excitatory and inhibitory classical neurotransmitters can occur together with peptides which themselves were found to exert excitatory, inhibitory or excitatory and inhibitory effects.

Table 1 also lists an example of a classical neurotransmitter occuring together with more than one putative peptide neurotransmitter; serotonin coexisting with both substance P and thyrotropin- releasing hormone (TRH) in cells of the rat medulla oblongata (22,23).

Examples of coexistence of two putative peptide neurotransmitters in neurons are also given; substance P and cholecystokinin (CCK) were found in neurons of the rat dorsal root ganglion (25). This situation is likely to be widespread since many of the peptide neurotransmitters are synthesized in the form of large pre-prohormone or prohormone precursors which are cleaved and chemically modified in the posttranslational process to several biologically active peptides. Examples of this phenomenon are abundant in the biosynthesis of opioid peptides where the proenkephalin sequence contains several met- and leu-enkephalin sequences in addition to two other biologically active peptides (28). Another illustration of this is the finding that when isolating the pre-prohormone for vasoactive intestinal polypeptide (VIP) it was found to contain the sequence for another putative peptide neurotransmitter, PHM-27 (27).

Table 1

Selected examples of classical neurotransmitter- peptide and peptide-peptide coexistence in neurons.

Transmitter	Peptide	Area	Refs.
Serotonin	Substance P (SP)	Medulla oblongata (rat)	5,6,7, 8,9
Noradrenaline	Somatostatin (SS)	Sympathetic ganglia (guinea pig)	10
		SIF-cells (cat)	11
Noradrenaline	Neurotensin	Adrenal medulla (cat)	12
Dopamine	Cholecystokinin (CCK)	Ventral tegmental area (A9,A10) (rat, human)	13
Dopamine	Neurotensin	Arcuate and periventricular hypothalamic neurons (rat)	14
Acetylcholine	Vasoactive intestinal polypeptide(VIP)	Autonomic ganglia exocrine glands (cat)	15,16
γ-aminobutyric acid (GABA)	Somatostatin	Nucleus reticularis thalami (cat)	17
γ-aminobutyric acid	Motilin	Cerebellar Purkinje cells (monkey,mouse)	18
L-glutamate	Proctolin	Slow skeletal motoneuron (cockroach)	19
Catecholamines*	Bovine pancreatic polypeptide (BPP)	Ventral bundle (rat)	20
		PNS and CNS (rat)	21
Serotonin	SP and thyrotropin releasing hormone (TRH)	Medulla oblongata (rat)	22,23

Peptide	Peptide		
Substance P	Cholecystokinin	Mesencephalic periaqueductal central gray (rat)	24
		Dorsal root ganglia (rat)	25
Substance P	Leu-enkephalin	CNS (avian)	26
Vasoactive intestinal polypeptide	PHM-27	Neuroblastoma cells	27

* The synthesizing enzyme Tyrosine hydroxylase was detected.

It is obvious that all neurons possess the genetic information to synthesize all the classical and peptide neurotransmitters of the given species and that the development and environment (innervation etc.) will decide which of these substances will in fact be synthesized. In view of this fact it is not astounding that several neurotransmitters coexist, and the actual maximum number of coexisting transmitters in a single neuron may be a good deal greater than three.

Cellular sites of synthesis and storage of coexisting classical neurotransmitters and peptide neurotransmitters.

As Hokfelt et.al. have pointed out in their important review (1) on peptidergic neurons, there are basic differences in the sites of synthesis of classical neurotransmitters and peptides. Classical neurotransmitters such as ACh and monoamines are most efficiently synthesized in the nerve terminals where their synthesizing enzymes are present with the highest specific activity. In the case of the monoamines, high-affinity uptake systems provide an additional factor in maintaining a high concentration of the classical neurotransmitter in the nerve endings. Classical neurotransmitters are also synthesized in the soma and dendrites since the synthesizing enzymes themselves originate from the soma and occur in their active forms there as well as in the axon during axonal transport to the nerve terminals.

According to presently available data, peptide neurotransmitters are synthesized on ribosomes in the soma and their presence in dendrites and axonal nerve endings is wholly dependent on axonal transport. (It should be noted that most of the immunohistochemical studies referred to in Table 1 used inhibitors of axonal transport to improve the chances of detecting the peptide, e.g. colchicine.)

The immunohistochemical studies (cf Table 1) and subcellular fractionation studies (29) have indicated that classical neurotransmitters and peptides are stored in different vesicle populations.

The result of the separation of the biosynthetic activities and the existence of discrete populations of storage vesicles for the classical neurotransmitters and the peptide neurotransmitters is that the relative concentrations of the two transmitters are widely different in various locations within one and the same neuron (cf Fig. 1). This uneven distribution of the coexisting neurotransmitters may serve as the basis of interesting release patterns and of "unexpected" changes in neurotransmitter level upon chronic drug treatments as discussed below.

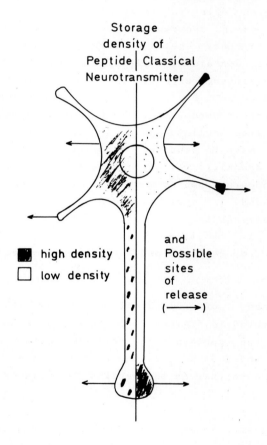

Fig.1

Schematic diagram of a neuron with coexisting classical (low molecular weight) and peptide neurotransmitters. Arrows indicate possible release sites, from bottom: axonal nerve terminals, dendrites and soma.

Presynaptic events at neurons with coexisting classical neurotransmitter and peptide neurotransmitter.

The question we shall examine here is : What are the differences in the release patterns of coexisting neurotransmitters? We shall present some alternative solutions to the problem of dual innervation of a postsynaptic cell. These alternatives, which exist in abundance in the peripheral and central nervous system are represented in Figs. 2A and 2B. In Fig. 2A the postsynaptic cells A,B,C,D are innervated by two neurons; I and II, containing a classical neurotransmitter and a peptide neurotransmitter, respectively. If coordinated activation of the postsynaptic cell D is needed it can be achieved either via a synchronizing interneuron between neurons I and II (neuron III), or via axon collaterals, or by the use of both arrangements. We have to consider that the recent findings (30) indicating that dendritic release or release of substances from the soma also occurs upon stimulation. Thus, when stimulated, both neurons I and II could via dendritic and somal release activate the postsynaptic cells A, B, and C. This activation would involve a simultaneous release of the classical neurotransmitter from neuron I and the peptide neurotransmitter from neuron II.

In the case of classical neurotransmitter- peptide coexistence indicated in Fig. 2B it is conceivable that the neuron releases onto the postsynaptic cell D both the classical neurotransmitter and the peptide while at the same time only the classical neurotransmitter or the peptide is released onto cells B and C.

There is evidence showing the different frequency dependencies for release of classical neurotransmitters and peptides. The release of the latter (VIP) from ACh/VIP neurons at the cat submandibular salivary gland, for example, requires a higher stimulation frequency than the release of ACh (3). Thus one may speculate on a "frequency code being translated into a chemical code" where either only one of the coexisting

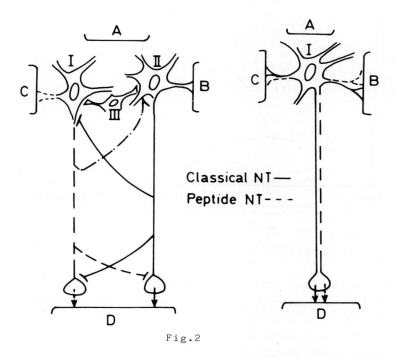

Fig.2

a) A,B,C,D are postsynaptic cells. Neurons I and II contain peptide and classical transmitter, respectively. Simultaneous action at cell D could be achieved via interneuron III or axon collaterals.

b) Neuron I can transmit to postsynaptic cells A,B,C,D by releasing either peptide or classical transmitter, or both.

neurotransmitters, or the other, or both are being released. As of now we have no indications or reasons to discuss why the coexistence situation or, two neurons with a synchronizing interneuron and/or axonal collaterals are the preferred arrangements in certain parts of the nervous system, but it is noteworthy that the same signaling problem (sending codes A, B, or AB where A is the classical neurotransmitter, B is the peptide) can be solved by one unit (neuron) instead of two or three.(This is by no means meant to suggest that "economics governs the development of the nervous system.")

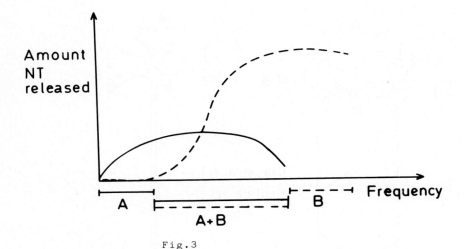

Fig.3

Schematic representation of the relative amounts of transmitter released as a function of stimulation frequency. The solid line represents the release of classical neurotransmitter; the dotted line represents the release of peptide. The symbols A,B and AB represent stimulation frequency ranges that would release classical transmitter (A) only, peptide transmitter (B) only, or both transmitters (AB) at once,

Developmental considerations and speculations.

It is known that sympathetic neurons can switch to synthesize ACh upon synapse formation with e.g. heart cells (31). It is thus conceivable that the neurons which contain coexisting neurotransmitters have not started to synthesize these neurotransmitters at the same time. The ontogenesis of the peptidergic component of the coexisting classical neurotransmitter - peptide pairs is practically unexplored . We may speculate that in coordination with the innervation of certain areas by the cholinergic or monoaminergic neuronal systems, the neuron which will also synthesize a peptide may begin to produce a specific peptide in response to its environment.

Regulation of release of classical neurotransmitter and coexisting peptide neurotransmitter.

As pointed out earlier the coexisting neurotransmitters seem to be synthesized at different sites, stored in different populations of vesicles and have different dependencies on stimulation frequency for their release. It is thus possible by certain treatments to influence the release of one but not the other, while other treatments will affect the release of both the classical neurotransmitter and the peptide. Reserpine, which induces depletion of monoamines caused depletion of noradrenaline but not of the coexisting bovine pancreatic peptide (BPP) in neurons of rat vas deferens (21). On the other hand, reserpine treatment depletes both serotonin and substance P from neurons in the rat ventral spinal cord, where serotonin and substance P coexist.

Regulation of release via autoreceptors.

It is well documented that many classical neurotransmitters such as noradrenaline(32) and acetylcholine (33) can inhibit their own release via autoreceptors located on the nerve terminals. If a nerve terminal with autoreceptors for the classical neurotransmitter contains a coexisting peptide, (Fig. 3) then stimulation conditions and drugs which influence the extent of autoinhibition may influence the release of not only the classical neurotransmitter but also of the coexisting peptide. Atropine, for example, acting on the postganglionic neuron at the cat submandibular salivary gland, in addition to enhancing release of ACh by blocking the inhibitory autoreceptors, enhances release of the costored peptide. This event is of importance in explaining why chronic atropine treatment depletes VIP in this tissue (34). Since the inhibitory regulation of VIP release by acetylcholine is disabled, VIP release is greater than normal. The replenishment of VIP stores is dependent on axonal transport of VIP-containing vesicles, which is a slower process than release itself. Thus the chronic presence of atropine depletes VIP at the salivary gland of atropine- treated rats. ACh levels are also lowered, but due to the on-site synthesis in the nerve terminal ACh is not as fully depleted as VIP, which virtually disappears from the tissue.

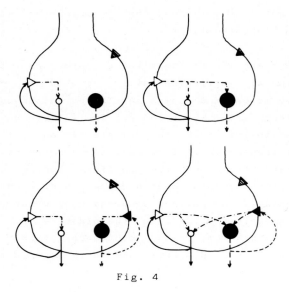

Fig. 4

Some possible presynaptic interactions between coexisting classical neurotransmitter (o) and peptide neurotransmitter (●)
△ = presynaptic autoreceptor for classical neurotransmitter
▲ = presynaptic autoreceptor for peptide neurotransmitter
▲ = presynaptic receptor

The above observations may be the basis for the findings that chronic atropine treatment caused supersensitivity of not only muscarinic cholinergic- but also of VIP-receptors in the rat submandibular salivary gland (34). The potential pharmacological relevance of such interactions (where blockage of a presynaptic receptor for one of the coexisting neurotransmitters affects the release of both costored transmitters) is readily understood if one considers that several of the clinically most often used psychopharmacological drugs modify monoaminergic transmission. Already in 1980 Hokfelt et.al.(1) pointed out that the dopamine-cholecystokinin coexistence may imply that antipsychotic agents which were previously thought to affect only dopaminergic transmission may also affect CCK transmission. Similarly the antihypertensive agents which change noradrenergic transmission may change sensitivity for the peptides which are costored with noradrenaline. Preliminary experiments also indicate that antidepressant drugs which affect serotonin turnover affect the tissue levels of substance P in the spinal cord where these neurotransmitters coexist (44).

Postsynaptic relevance of classical neurotransmitter-peptide coexistence.

Neurons in the CNS receive inputs from a large number of other cells and integration of signals by the postsynaptic cells is a common and continuously occuring phenomenon. Here we shall discuss the special case where two (or more) signal substances are released from the same presynaptic element. This fact may be of no consequence to the postsynaptic cell, i.e. that it may not distinguish between uni- and multicellular origin of the neurotransmitters (Fig. 2). Two neurotransmitters may influence each other's postsynaptic effect even when they are not released from the same neuron. Stimulation of adenylate cyclase, for example in membranes of the rat heart via occupancy of β-receptors is inhibited by acetylcholine occupying muscarinic receptors in the same membrane (37). Thus examples of postsynaptic interaction by no means imply that the two neurotransmitters must be released from the same presynaptic element. The only requirement is that they share a common site of action and are released approximately simultaneously. The common site can be the same ion channel, as suspected in the case of the bullfrog sympathetic ganglion, where muscarinic agonists and luteinizing hormone releasing factor activate the same K^+ channel (36). The postsynaptic interaction may be mediated by a common membrane protein such as the guanine nucleotide binding protein (38), which can bind to both receptors in membranes from the heart. (Guanine nucleotides affect both the β-receptor (39) and the muscarinic receptor (43) in the cardiac membrane.)

Specific cases of postsynaptic interactions of coexisting classical neurotransmitters and peptides.

Lundberg has shown in the cat submandibular salivary gland, where acetylcholine and VIP coexist in the postganglionic neuron, that although acetylcholine (released alone at low stimulation frequency) can cause salivary secretion, this secretion is potentiated by VIP (released at higher stimulation frequency)(3). Infused VIP has the same effect, i.e. in combination with low stimulation frequencies (which release only ACh) it potentiates salivary secretion. Biochemical studies indicate that VIP in nanomolar concentration affects muscarinic agonist binding in the cat submandibular salivary gland. The muscarinic receptor occurs in two agonist binding conformations; one with high affinity (0.5 uM) and one with low affinity (0.3 mM). When VIP binds to its receptor (Kd = 2nM) in the membrane, it induces a conversion of low affinity muscarinic agonist binding sites to high affinity agonist binding sites (40). Since the latter are coupled to the secretory response, the net effect is that VIP enhances the ACh- induced salivary secretion. The interactions between ACh receptors and VIP receptors are mutual as evidenced by the effect of acetylcholine (40) and carbachol on the VIP receptor- coupled adenylate cyclase. In the cat and rat salivary gland muscarinic agonists potentiate the stimulation of adenylate cyclase by VIP (40). These interactions between the ACh and VIP receptors are probably mediated via a guanine nucleotide binding protein.

Another example of postsynaptic interactions between coexisting classical neurotransmitter and peptide is the effect of substance P on binding to serotonin receptors in bulbo-spinal neurons (42). It is noteworthy that VIP does not affect ACh receptors in other tissues tested; nor does substance P affect serotonin binding in other tissues tested. This indicates that the prerequisite for interaction exists only in the systems innervated by neurons in which the classical neurotransmitter and peptide coexist (or in postsynaptic cells which are simultaneously innervated by both although without presynaptic coexistence).

Significance of coexistence of classical neurotransmitters and peptides.

The interest in the coexistence phenomenon derives from the fact that single cells can transmit different types of information to various synapses, and even to a single postsynaptic cell simultaneously. The ultimate functional significance of coexisting peptide- classical neurotransmitter pairs is shrouded in the same mystery as the reason for such a variety of transmitters in an essentially binary system. We know that the peptides can act independently, that they can act in concert with and can antagonize the effects of the classical transmitters. We can also speculate that peptides might have an entirely non- transmitter, trophic function in the nervous system, in addition to the neurotransmitter function discussed above.

Acknowlegments.

This study was supported by a grant from the Swedish Board for Planning of Research.

REFERENCES

1. Hokfelt, T., Johansson, O., Ljungdahl, A., Lundberg, J.M. and Schultzberg, M. (1980) Peptidergic neurons. Nature 284, 515-521.

2. Lundberg, J.M., Hedlund, B., Anggard, A., Fahrenkrug, J., Hokfelt, T., Tatemoto, K. and Bartfai, T. (1982) Costorage of peptides and classical transmitters in neurons. Medica Hoechst, 18,7.

3. Lundberg, J.M., (1981) Evidence for coexistence of vasoactive intestinal polypeptide (VIP) and acetylcholine in neurons of cat exocrine glands. Acta. Phys. Scand. Suppl. 496.

4. Adams, M.E. and O'Shea, M. (1983) Peptide cotransmitter at a neuromuscular junction, Science 221, 286-289.

5. Hokfelt, T., Ljungdahl, A., Steinbusch, H., Verhofstad, A., Nilsson, G., Brodin, E., Pernow, P. and Goldstein, M. (1978) Immunohistochemical evidence of substance P-like immunoreactivity in some 5-hydroxytryptamine-containing central neurons. Neuroscience 3, 517-538.

6. Chan-Palay, V., Jonsson, G. and Palay, S.L. (1978) Evidence for the coexistence of serotonin and substance P in single raphe cells and fiber plexuses; combined immunocytochemistry and autoradiography. Proc. natl. Acad. Sci. U.S.A 75, 1582.

7. Bjorklund, A., Emson, P.C., Gilbert, R.T.F. and Skageberg, G. (1979) Further evidence for the possible co-existence of 5-hy-

droxytryptamine and substance P in medullary raphe neurons of rat brain. Br. J. Pharmac. 66,112P

8. Chan-Palay, V. (1979) Immunocytochemical detection of substance P neurons, their processes and connections by in vivo microinjections of monoclonal antibodies: Light microscopy and electron microscopy. Anat. Embryol. 156(3),225-240.

9. Singer, E., Sperk, G., Placheta, P. and Leeman, S.E. (1979) Reduction of substance P levels in ventral cervical spinal cord of the rat after intracisternal 5,7-dihydroxytryptamine injection. Brain Res. 174, 326.

10. Hokfelt, T., Elven, L.G., Elde, R., Schultzberg, M., Goldstein, M. and Luft, R. (1977) Occurence of somatostatinlike immunoreactivity in some peripheral sympathetic noradrenergic neurons. Proc. natn. Acad. Sci. U.S.A. 74, 3587-3591.

11. Lundberg, J.M., Hokfelt, T., Anggard, A., Uvnas-Wallensten, K., Brimjoin, S., Brodin, E. and Fahrenkrug, J. (1980) Peripheral peptide neurons: Distribution, axonal transport and some aspects on possible function. In: Neural Peptides and Neuronal Communication (ed. E. Costa and M. Trabucchi), 25-36. Raven Press, New York.

12. Lundberg, J.m., Roekaeus, A., Hokfelt, T., Rosell, S., Brown, M.R. and Goldstein, M. (1979) Neurotensin-like immunoreactivity in the preganglionic sympathetic nerves and in the adrenal medulla of the cat. Proc. natn. Acad. Sci. U.S.A. 76, 4079.

13. Hokfelt, T., Rehfeld, J.F., Skirboll, L.R., Ivemark, B., Goldstein, M. and Markey, K.A. (1980) in Neural peptides and neuronal communication (eds. Costa, E. and Trabucchi, M.) (Raven, N.Y.).

14. Ibata, Y., Fukui, K., Okamura, H., Kawakami, T., Tanaka, M., Obata, H.L., Tsuto, T., Terubayashi, H., Yanaihara, C. and Yanaihara, N. (1983) Brain Res. 269, 177-179.

15. Lundberg, J.M., Anggard, A., Fahrenkrug, J., Johansson, O. and Hokfelt, T. (1979) Vasoactive intestinal polypeptide in cholinergic neurons of exocrine glands. Neuroscience 4, 1539.

16. Lundberg, J.M., Anggard, A., Fahrenkrug, J., Hokfelt, T. and Mutt, V. (1980) Vasoactive intestinal polypeptide in cholinergic neurons of exocrine glands: Functional significance of coexisting neurotransmitters for vasodilatation and secretion. Proc. Natl. Acad. Sci. U.S.A. 77, 1651-1655.

17. Oertel, W.H., Graybiel, A.M., Mugnaini, E., Elde, R.P., Schmechel, D.E. and Kopin, I.J. (1983) Coexistence of glutamic acid decarboxylase- and somatostatin-like immunoreactivity in neurons of the feline nucleus reticularis thalami. J. of Neuroscience 3(6), 1322-1332.

18. Chan-Palay, V., Nilaver, G., Palay, S., Beinfeld, M.C.,

Zimmermann, E.A., Wu, J.-Y. and O'Donohue, T.L. (1981) Chemical heterogeneity in cerebellar Purkinji cells:Existence and coexistence of glutamic acid decarboxylase-like and motilin-like immunoreactivities. Proc. Natl. Acad. Sci. U.S.A. 78, 7787-7791.

19. Adams, M.E. and O'Shea, M. (1983) Peptide cotransmitter at a neuromuscular junction. Science 221, 286-289.

20. Jacobowitz, D.M. and Olschowka, J.A. (1982) Coexistence of bovine pancreatic polypeptide-like immunoreactivity and catecholamine in neurons of the ventral aminergic pathway of the rat brain. Brain Res. Bull.

21. Jacobowitz, D.M. and Olschowka, J.A. (1982) Bovine pancreatic polypeptide-like immunoreactivity in brain and peripheral nervous system: Coexistence with catecholaminergic nerves. Peptides 3, 569-590.

22. Hokfelt, T., Rehfeld, J., Skirboll, L., Ivemark, B. and Goldstein, M.(1980) In: Neural peptides and neuronal communication (eds. E. Costa and Trabucchi, M.) (Raven press, New York).

23. Johansson, O., Hokfelt, T., Pernow, B., Jeffcoate, S.L., White, N., Steinbusch, H.W.M., Verhofstad, A.A.J., Emson, P.C., Spindel, E. (1981) Immunohistochemical support for three putative transmitters in one neuron: coexistence of 5-hydroxytryptamine, substance P and thyrotropin- releasing hormone- like immunoractivity in medullary neurons projecting to the spinal cord. Neuroscience 6(10) 1883-98

24. Skirboll, L., Hokfelt, T., Rehfeld, J., Cuello, A.C. and Dockray, G. (1982) Coexistence of substance P and cholecystokinin-like immunoreactivity in neurons of the mesencephalic periaqueductal central gray. Neuroscience Lett. 28, 159-163.

25. Dalsgaard, C.-J., Vincent, S.R., Hokfelt, T., Lundberg, J.M., Dahlstrom, A., Schultzberg, M., Dockray, G.J. and Cuello, A.C. (1982) Coexistence of cholecystokinin- and substance P-like peptides in neurons of the dorsal root ganglia of the rat. Neuroscience Lett. 33, 159-163.

26. Erichsen, J.T., Reiner, A. and Karten, H.J. (1982) Co-occurence of substance P-like and Leu-enkephalin-like immunoreactivities in neurons and fibers of avian nervous system. Nature 295, 407-410.

27. Itoh, N., Obata, K.-i., Yanaihara, N. and Okamoto, H. (1983) Human preprovasoactive intestinal polypeptide contains a novel PHI-27-like peptide, PHM-27. Nature 304, 547-549.

28. Rossier, J. (1982) Pro-enkephalin sequencing and the advent of cDNA technologies. Trends, NeuroSci. 6, 179-180.

29. Lundberg, J.M., Fried, G., Fahrenkrug, J., Holmstedt, B., Hokfelt, T., Lagercrantz, H., Lundgren, G. and Anggard, A. (1981)

Subcellular fractionation of cat submandibular gland: comparative studies on the distribution of acetylcholine and vasoactive intestinal polypeptide (VIP). Neuroscience 6, 1001-1010.

30. Hery, F., Soubrié, P., Bourgoin, S., Montastruc, J.L., Artaud, F. and Glowinski, J. (1980) Dopamine released from dendrites in the substantia nigra controls the nigral and striatal release of serotonin. Brain Res., 193, 143-151.

31. Patterson, P.H. (1978) Environmental determination of autonomic neurotransmitter functions. Annual Review of Neurosciences, Vol.1 , 1-17.

32. Starke, K. (1977) Regulation of noradrenaline release by presynaptic systems.Rev. Physiol. Biochem. Pharmacol. 77,1-124

33. Vizi, E.S. (1979) Presynaptic modulation of neurochemical transmission. Prog. in Neurobiol. 12, 181-290

34. Hedlund, B., Abens, J. and Bartfai, T. (1983) Vasoactive intestinal polypeptide and muscarinic receptor supersensitivity induced by long-term atropine treatment. Science 220, 519-521.

35. Brodin et.al. (1983) The effect of 5-hydroxydopamine uptake inhibitors on substance P levels in the rat CNS. In preparation.

36. Adams, P.R. and Brown, D.A. (1980) Luteinizing hormone-releasing factor and muscarinic agonists act on the same voltage sensitive K+-current in bullfrog sympathetic neurones. Br. J. Pharmacol. 68, 353-355.

37. Watanabe, A.M., McConnaughey, M.M., Strawbridge, R.A., Flemming, J.W., Jones, L.R. and Besch, H.R. (1978) Muscarinic cholinergic receptor modulation of β-adrenergic receptor affinity for catecholamines. J. Biol. Chem. 253, 4833.

38. Rodbell, M. (1980) The role of hormone receptors and GTP-regulatory proteins in membrane transduction. Nature 284, 17-22

39. Lefkowitz, R.J. and Hoffman,B.B.(1981) New directions in adrenergic receptor research Part 1. In Towards Understanding Receptors, Elsevier-North Holland.

40. Enyedi, P., Fredholm, B.B., and Lundberg, J.M. (1982) Carbacholine potentiates the cyclic AMP stimulating effect of VIP in cat submandibular salivary gland. Eur. J. Pharmacol. 79, 139-143.

41. Lundberg, J.M., Hedlund, B. and Bartfai, T. (1982) Vasoactive intestinal polypeptide enhances muscarinic ligand binding in cat submandibular salivary gland. Nature 295, 147-149.

42. Agnati, L.F., Fuxe, K., Benfenati, F., Zini, I. and Hokfelt, T. On the functional role of coexistence of 5-HT and substance P

in bulbospinal 5-HT neurons. Substance P reduces affinity and increases density of ^{3}H-5-HT binding sites. Acta. Physiol. Scand. 117, 299-301.

43. Berrie, C.P., Birdsall, N.J.M., Burgen, A.S.V. and Hulme, E.C. (1979) Guanine nucleotides modulate muscarinic receptor binding in heart. Biochem. Biophys. Res. Comm. 87(4), 1000-1005.

44. Gilbert, F.T., Bennett, G.W., Marsden, C.A. and Emson, P.C. (1981) The effects of 5-hydroxytryptamine-depleting drugs on peptides in the ventral spinal chord. Eur. J. Pharm. 76, 203-210.

Résumé

Les conséquences pré et postsynaptiques de la coexistence d'un neurotransmetteur "classique" de petit poids moléculaire avec un neurotransmetteur peptidique sont étudiées en choisissant l'exemple du couple acétylcholine - Peptide intestinal vasoactif (VIP). Ce travail montre qu'un traitement chronique avec des drogues qui, croit-on, n'affectent que le neurotransmetteur classique, peut amener des changements dans la composante peptidique de la transmission synaptique. Par exemple, un traitement chronique à l'atropine conduit à une hypersensibilité non seulement des récepteurs muscariniques mais également des récepteurs du VIP dans la glande salivaire sous maxillaire et dans le cortex cérébral de rat. Des interactions entre des processus médiés par ces neurotransmetteurs coexistants, apparaissent également au niveau postsynaptique : par exemple le VIP affecte la liaison d'un ligand sur le récepteur muscarinique.

HISTAMINE IN BRAIN : ACTIONS AND FUNCTIONS

MONIQUE GARBARG, JEAN-MICHEL ARRANG AND JEAN-CHARLES SCHWARTZ
Unité 109 de Neurobiologie, Centre Paul Broca de l'INSERM, 2 ter rue d'Alésia - 75014 Paris (France)

During the past decades, histamine (HA) has been involved as a messenger molecule regulating various physiological processes in peripheral organs as in brain (1, 2, 3).

In brain, the amine appears to be held in non-neuronal cells (probably mast cells) possibly involved in vascular control or immune responses as well as in a class of neurones (4). Neuronal HA is likely to act as a neurotransmitter : it is synthetized at high rate by decarboxylation of L-histidine under the action of a specific L-histidine decarboxylase, it is held in specific neuronal tracts evidenced by lesion studies, it can be released upon depolarisation by a calcium-dependent process, it affects the activity of target neurons and is efficiently inactivated under the action of histamine-N-methyltransferase (1).

Such a role generally implies that the messenger molecule is recognised by specific receptors and large efforts have been devoted during the last years to the characterization of the latters using biochemical, electrophysiological or physiological approaches (5).

Hence, the two classes of HA receptors (H_1 and H_2) previously characterized in peripheral organs (6) were shown to be present in the mammalian brain and recently a novel class of HA receptor modulating the neuronal HA release has been evidenced (7).

H_1-histamine receptors labeled with ^{3}H-mepyramine. Binding techniques which have been so fruitful during the last years in investigating the properties of a variety of CNS receptors have begun to be applied to HA receptors in the CNS.

Hill et al. (8) demonstrated that ^{3}H-mepyramine can be used as a selective ligand of H_1-receptors in homogenates of smooth muscle of the guinea pig ileum.

It has been thereafter shown that ^{3}H-mepyramine can be used to label H_1-receptors in particulate fractions from the brain of various species (9, 10).

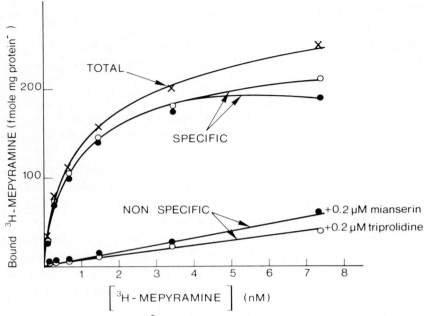

Fig. 1. Saturation curve of ^{3}H-mepyramine binding to guinea-pig cerebellum. The particulate fraction from guinea-pig cerebellum was incubated with increasing concentrations of ^{3}H-mepyramine alone or together with 0.2 µM of mianserin or tripolidine. Specific binding represents the difference between total and non specific binding. From Garbarg et al. (20).

In particulate fraction from guinea pig cerebellum, a region which contains a high density of H_1-receptor sites, the non-specific binding represents a small percentage of total binding and can be determined either with mianserin or triprolidine at 0.2-2 µM (Figure 1). Similar values are found in the two procedures and lead to the determination of a maximal capacity of 205 fmoles.mg protein^{-1} with a Kd value of 0.7 nM. This latter value is in good agreement with that obtained when considering the antagonism by mepyramine of the HA-induced contraction of the ileum , the reference biological system for an H_1-receptor mediated response (6). Moreover the pharmacological specificity of those binding sites, assessed by establishing the inhibitory potencies of a variety of histaminergic and non histaminergic antagonists, leaves little doubt that they represent the recognition moiety of H_1-receptors (11).

^{3}H-mepyramine can also be used to label H_1-receptors in the brain of the living mouse (12). Thus, a few minutes following i.v. administration of ^{3}H-mepyramine in low doses, a saturable binding occurs in vivo which presents characters of regional heterogeneity and pharmacological specificity paralleling those observed in binding studies in vitro. The density of ^{3}H-mepyramine binding sites in vivo is in good agreement with their density estimated in vitro.

A major interest in the in vivo test is that it shows that sytemic administration of most H_1-antihistamines in doses currently used in therapeutics to alleviate allergic symptoms, results in the occupation of a major fraction of H_1-receptors in brain. This observation strongly suggests that the well-known central effects of H_1-antihistamines (like sedation, "mental clouding", increased sleep duration) are, indeed, mediated by blockade of the actions of endogenous HA at H_1-receptors. Interestingly, mequitazine and terfenadine, two H_1-antihistamines devoid of sedative properties do not occupy H_1-receptors in brain, probably because they do not cross the blood brain barrier (13).

Biological responses to histamine mediated by H_1- and H_2-histamine receptors.
In brain slices, glycogenolytic actions can be easily investigated by measuring the changes in ^{3}H-glycogen levels in tissues preincubated in the presence of ^{3}H-glucose.

HA exerts a powerful glycogenolytic effect (EC_{50} = 3 µM) on slices from mouse cerebral cortex ; it appears to be selectively mediated by H_1-receptors as indicated by the relative potencies of various agonists (dimaprit being totally ineffective) and by the competitive antagonism by H_1-antihistamines but not by H_2-antihistamines. In addition, this preparation offers the opportunity to compare physiological responses to receptor occupancies : significantly lower HA concentrations are required to promote ^{3}H-glycogen hydrolysis than to prevent ^{3}H-mepyramine binding, suggesting the existence of "spare" H_1-receptors (14).

Neither the cellular localisation (neurons or glial cells ?) nor the functional role of glycogenolytic response is established. It might however be hypothetised that HA neurons control the energy supply of cerebral cells, a fonction shared with other classes of aminergic neurons (noradrenergic or serotoninergic) which also appear to project in a diffuse manner to telencephalic area.

HA receptors mediate also the formation of 3'-5' cyclic AMP formation in a concentration-dependent manner. Both classes of HA receptors are involved in a complex manner. In cell-free preparations, a HA-sensitive adenylate cyclase has been characterized, the enzyme being coupled to typical H_2-receptors strictly identified (15). In contrast, the cyclic AMP response to HA in intact brain cells is mediated by both H_1- and H_2-receptors (16).

TABLE 1

INFLUENCE OF CALCIUM ON CYCLIC AMP ACCUMULATION IN SLICES FROM GUINEA PIG HIPPOCAMPUS

The pooled slices were incubated for 30 min in complete Krebs-Ringer bicarbonate medium (NaCl 120 mM ; KCl 5 mM ; $CaCl_2$ 2.6 mM ; $MgSO_4$ 0.67 mM ; KH_2PO_4 1.2 mM ; $NaHCO_3$ 27.5 mM and D-glucose 5.9 mM) under a constant stream of O_2 : CO_2 (95 : 5). Then the slices were washed four times and resuspended either with complete medium (Ca^{2+} 2.6 mM) or with the same medium in which Ca^{2+} was omitted. From Schwartz et al. (21).

	cAMP (pmoles/mg protein)	
	2.6 mM Ca^{2+}	Ca^{2+}-free
Control	6.7 ± 0.8	6.0 ± 0.8
Dimaprit (0.1 mM)	24.6 ± 1.7	28.7 ± 1.5
Dimaprit (0.1 mM) + TEA (1 mM)	75.7 ± 6.4	54.2 ± 3.7[a]
Difference	51.1	25.5

[a] $p < 0.05$ when comparing with the cAMP accumulation induced by the combination of dimaprit and 2-thiazolylethylamine (TEA) in normal medium (Ca^{2+} 2.6 mM).

In this way, there is an additive response to dimaprit, a selective H_2-receptor agonist and 2-thiazolylethylamine (TEA) a predominantly H_1-receptor agonist (Table 1). Because this latter agent possesses significant agonist activity at the two classes of receptors, the response to TEA alone reflects activation of both H_1 and H_2 receptor sites but when H_2-receptors are already stimulated in the presence of a supramaximal concentration of dimaprit, the response to TEA is selectively mediated by H_1-receptors. H_1-receptors have never been reported to be directly coupled with an HA-sensitive adenylate cyclase in any biological system and several biological responses to H_1-receptors, such as the contraction of smooth muscles, appear to involve a translocation of Ca^{2+} ions. This might also be the case for the accumulation of cyclic AMP in hippocampal slices, since the response mediated by H_1-receptors is largely prevented in the absence of extracellular calcium. At the same time, the cyclic AMP accumulation elicited by H_2-receptor stimulation is not modified (Table 1). Interestingly, the HA-induced increase in cyclic GMP and the HA-induced glycogenolysis are also mediated by H_1-receptors and similarly involve calcium (14,17). Hence it is tempting to speculate that in a large number of biological systems, H_1-receptors might be associated with entry of calcium (5) (Figure 2)

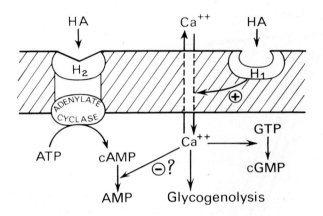

Fig. 2. Intracellular responses mediated by H_1- and H_2-histamine receptors in neurons tissues. From Schwartz (5).

Auto-inhibition of histamine release from rat brain slices : evidence for a new class of histamine receptors. It is well established that several neurotransmitters affect neuronal activity in the CNS through stimulation not only of post synaptic receptors, but also of receptors located presynaptically which often display distinct pharmacological specificity and by which they may control their own release (18, 19).

Neuronal HA stores of rat cerebral cortex were labelled by preincubation of cortical slices in the presence of ^{3}H-histidine. After extensive washings to eliminate excess ^{3}H-labelled precursor and to obtain a low constant basal efflux of ^{3}H-HA, the endogenously synthetised ^{3}H-amine was released upon depolarisation by 30 mM K^+, a process calcium-dependent.

The depolarisation-induced release of ^{3}H-HA is partially inhibited in the presence of 1 µM exogenous HA added in the incubation medium (7).
The ^{3}H-HA release induced by another depolarizing agent, veratridine is also significantly inhibited in the presence of 1 µM exogenous HA.

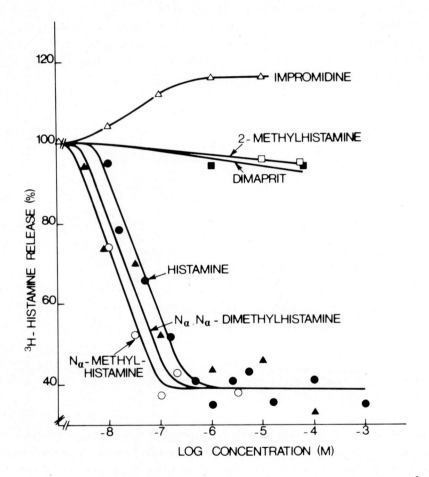

Fig. 3. Effect of various histamine agonists on the K^+-evoked release of ^{3}H-histamine from slices of rat cerebral cortex.
Slices were allowed to synthetize ^{3}H-HA from ^{3}H-histidine during a 30 min incubation. After several washing steps, HA or agonists were added and 5 min later, ^{3}H-HA was released by a 2 min application of 30 mM k^+. ^{3}H-HA was estimated in pellet and medium after purification by ion-exchange chromatography. Results are expressed as the percentage of ^{3}H-HA released by K^+ in the absence of exogenous HA or agonists. The spontaneous efflux of ^{3}H-HA into the medium in basal conditions represented 3 % of the total ^{3}H-amine present in tissue plus medium and the efflux of ^{3}H-HA in the presence of 30 mM K^+ was 13 %. From Arrang et al. (7).

The inhibitory effect of exogenous HA is a concentration-dependent process occuring with an EC_{50} value of 40 nM (Figure 3), quite lower than that required to stimulate H_1- or H_2-receptors in the slice preparation.

It is unlikely that these effects of exogenous HA result from an impaired synthesis of ^{3}H-HA via end-product inhibition because no significant ^{3}H-HA formation is occuring in the slices at the time when the depolarizing stimuli are applied, as shown by the unmodified tissue levels of ^{3}H-HA in the presence of α-hydrazinohistidine, an L-histidine decarboxylase inhibitor ; moreover the inhibitory effect of exogenous HA persists in the presence of this synthesis inhibitor.

The hypothesis that the inhibitory effect of exogenous HA in slices is a receptor-mediated process is substantiated by its characters of saturability, reversibility and high pharamcological specificity. Moreover, it is antagonisable in an apparently competitive manner.

The inhibitory action of HA is mimicked by its two $N\alpha, N\alpha$-methylderivatives but other compounds displaying H_1- or H_2-receptor agonist activity are devoid of significant inhibitory effect (Figure 3). Actually, impromidine shows a facilitatory action, suggesting that it may act as an antagonist towards endogenous HA released by the depolarising stimulus, thus revealing a normal negative feedback modulation of HA release. Indeed, impromidine is found to be a competitive antagonist of exogenous HA with a pA_2 value of 7.5, in agreement with concentration of impromidine required for the half maximal facilitatory effect.

While H_1-antihistamines are ineffective at concentrations at which they block H_1-receptors, several H_2-antihistamines antagonise in a concentration-dependent and surmountable manner the HA-induced inhibition of release. However, it is clear that these latter effects cannot be attributed to blockade of H_2-receptors since the potency of the various compounds markedly differ from that they display at either peripheral or cerebral H_2-receptors (Table 2).
For instance, tiotidine is a very weak antagonist whereas burimamide is quite potent. The pA_2 value of 7.5 determined for this latter compound by Schild plot analysis of its antagonist effect, is higher by two orders of magnitude than its value on H_2-receptors (pA_2 = 5.1). Hence, from the relative potencies of HA agonists and from the apparent dissociation constants of HA antagonists, as well as from the lack of effect of antagonists of other neurotransmitters, it can be concluded that the auto-inhibition of HA release in brain is mediated by a novel class of histamine receptors which might be called H_3.

TABLE 2

INHIBITION OF ^{3}H-HISTAMINE RELEASE FROM SLICES OF RAT CEREBRAL CORTEX BY VARIOUS H_1- AND H_2-HISTAMINE RECEPTOR ANTAGONISTS.

Potential antagonists were added in at least four different concentrations and five minutes later, the ^{3}H-HA neosynthetized by the slices of rat cerebral cortex was released by 30 mM K^+ in the presence of 1 μM HA. Apparent dissociation constants (K_B values) were calculated assuming a competitive antagonism according to the equation $K_B = \dfrac{IC_{50}}{1 + \dfrac{S}{EC_{50}}}$ where S represents the concentration of exogenous HA (1 μM) and EC_{50} the amine concentration eliciting a half maximal inhibitory effect on K^+-evoked release of ^{3}H-HA.

	Antagonist activity (K_B, M)		
	Inhibition of ^{3}H-histamine release	Ileum contraction (H_1-receptors)	Atrium rate (H_2-receptors)
Mepyramine	> 6 x 10^{-8}	4.4 x 10^{-10}	
D-Chlorpheniramine	> 6 x 10^{-8}	5 x 10^{-10}	
L-Chlorpheniramine	> 6 x 10^{-8}	1.5 x 10^{-8}	
Burimamide	7 x 10^{-8}		7.8 x 10^{-6}
Metiamide	2.5 x 10^{-6}		9.2 x 10^{-7}
Cimetidine	3.3 x 10^{-5}		7.9 x 10^{-8}
Ranitidine	> 10^{-6}		6.3 x 10^{-8}
Tiotidine	> 10^{-5}		1.5 x 10^{-8}

Preliminary experiments suggest a presynaptic localisation of these autoreceptors because the inhibitory action of exogenous HA on ^{3}H-HA release is not prevented by tetrodotoxin (0.2 μM) and is also observed on depolarised cortical synaptosomes.

The presence of HA autoreceptors with a pharmacological specificity clearly distinct from that of the classical H_1- and H_2-receptors may open new possibilities to unravel the role of histaminergic neurons in brain.

REFERENCES

1. Schwartz, J.C., Pollard, H. and Quach, T.T. (1980) Histamine as a neurotransmitter in mammalian brain, J. Neurochem. 35 (1) : 26-33.
2. Green, J.P. (1983) in : Iversen, L., Iversen, S.D. and Snyder, S.H. (Eds.), Handbook of psychopharmacology, Plenum Publishing Corporation, pp. 385-420.
3. Haas, H.L. and Wolf, P. (1977) Central actions of histamine : microelectrophoretic studies, Brain Res. 122 : 269-279.
4. Schwartz, J.C. (1977) Histaminergic mechanisms in brain, Ann. Rev. Pharmacol. Toxicol. 17 : 325-339.
5. Schwartz, J.C. (1979) Histamine receptors in brain, Life Sci. 25 : 895-912.
6. Ganellin, C.R. (1982) in : Ganellin, C.R. and Parsons, M.E. (Eds.), Pharmacology of histamine receptors, Wright PSG, pp. 10-102.
7. Arrang, J.M., Garbarg, M. and Schwartz, J.C. (1983) Auto-inhibition of brain histamine release mediated by a novel class (H_3) of histamine receptor, Nature 302 : 832-837.
8. Hill, S.J., Young, J.M. and Marrian, D.H. (1977) Specific binding of ^{3}H-mepyramine to histamine H_1-receptors in intestinal smooth muscle, Nature 270 : 361-363.
9. Hill, S.J., Emson, P.C. and Young, J.M. (1978) The binding of [^{3}H]mepyramine to histamine H_1 receptors in guinea-pig brain, J. Neurochem. 31 : 997-1004.
10. Tran, V.T., Chang, R.S.L. and Snyder, S.H. (1978) Histamine H_1 receptors identified in mammalian brain membranes with [^{3}H]mepyramine, Proc. Natl. Acad. Sci. (USA) 75 : 6290-6294.
11. Schwartz, J.C., Barbin, G., Duchemin, A.M., Garbarg, M., Llorens, C., Pollard, H., Quach, T.T. and Rose, C. (1982) In : Ganellin, C.R. and Parsons, M.E. (Eds.) Pharmacology of histamine receptors, Wright PSG, pp. 351-391.
12. Quach, T.T., Duchemin, A.M., Rose, C. and Schwartz, J.C. (1980) Labeling of histamine H_1-receptors in the brain of the living mouse, Neurosci. lett. 17 : 49-54.
13. Quach, T.T., Duchemin, A.M., Rose, C. and Schwartz, J.C. (1979) In vivo occupation of cerebral histamine H_1-receptors evaluated with ^{3}H-mepyramine may predict sedative properties of psychotropic drugs, Eur. J. Pharmacol. 60 : 391-392.
14. Quach, T.T., Duchemin, A.M., Rose, C. and Schwartz, J.C. (1980) ^{3}H-glycogen hydrolysis elicited by histamine in mouse brain slices : selective involvement of H_1-receptors, Mol. Pharmacol. 17 : 301-308.
15. Kanof, P.D., Hegstrand, L.R. and Greengard, P. (1977) Biochemical characterization of histamine-sensitive adenylate cyclase in mammalian brain, Arch. Biochem. Biophys. 182 : 321-334.
16. Palacios, J.M., Garbarg, M., Barbin, G. and Schwartz, J.C. (1978) Pharmacological characterization of histamine receptors mediating the stimulation of cyclic AMP accumulation in slices from guinea-pig hippocampus, Mol. Pharmacol. 14 : 971-982.
17. Richelson, E. (1978) Histamine H_1-receptor-mediated guanosine 3',5'-monophosphate formation by cultured mouse neuroblastoma cells, Science 201 : 69-71.
18. Langer, S.Z. (1977) Presynaptic receptors and their role in the regulation of transmitter release, Brit. J. Pharmacol. 60 (4) : 481-498.
19. Starke, K. (1978) Presynaptic receptors, Ann. Rev. Pharmacol. Toxicol. 21 : 7-30.

20. Garbarg, M., Pollard, H., Quach, T.T. and Schwartz, J.C. (in press) in : Najatsu, T. (Ed.), Methods in biogenic amines research.
21. Schwartz, J.C., Barbin, G., Duchemin, A.M., Garbarg, M., Palacios, J.M., Quach, T.T. and Rose, C. (1980) in Pepeu, G., Kuhar, M.J. and Enna, S.J. (Eds.) Receptors for neurotransmitters and peptide hormones, Raven Press, New-York, pp. 169-182.

RESUME

L'histamine est un neurotransmetteur putatif dans le cerveau des mammifères. Des études biochimiques et électrophysiologiques ont montré la présence de neurones histaminergiques se projettant de façon diffuse dans tout le télencéphale. Les deux classes de récepteurs à l'histamine, définies dans les systèmes périphériques ont été caractérisées et identifiées sur des cellules cibles par des études de liaison et des réponses fonctionnelles. Les récepteurs H_1 ont été marqués sélectivement par un antagoniste tritié, la mépyramine, dans des fractions particulaires de cerveau et chez la souris entière. Ces récepteurs sont impliqués dans l'action glycogénolytique de l'histamine et participent par une action indirecte à l'accumulation d'AMP cyclique induite par l'amine. La présence de calcium est nécessaire pour ces deux réponses. Les récepteurs H_2 sont couplés à une adénylate cyclase. Récemment, des récepteurs à l'histamine qui modulent sa propre libération à partir de coupes de cortex de rat dépolarisées, ont été mis en évidence.

La pharmacologie de ces autorécepteurs est clairement distincte de celle des récepteurs H_1 ou H_2. Par exemple, les agonistes sélectifs ne reproduisent pas l'effet inhibiteur de l'amine. Les antagonistes des récepteurs H_1 sont dépourvus d'action. Les antagonistes des récepteurs H_2 réversent de façon compétitive l'effet de l'histamine, mais avec des constantes apparentes de dissociation différentes de celles attendues pour des récepteurs H_2 typiques. La nomenclature H_3 est proposée pour ces autorécepteurs.

ROLE OF SPECIFIC NEURO-NEURONAL AND NEURO-GLIAL INTERACTIONS IN THE IN VITRO DEVELOPMENT OF DOPAMINERGIC NEURONS FROM THE MOUSE MESENCEPHALON

ALAIN PROCHIANTZ
Chaire de Neuropharmacologie, INSERM U 114, Collège de France, 11, pl. Marcelin Berthelot, 75231 Paris cedex 05

INTRODUCTION

Dopaminergic (DA) cell bodies can be visualized in the mouse brain mesencephalic flexure between the 12th and the 13th day of embryonic life (1). These neurons grow intrinsic and extrinsic neurites, some of the lattest being visible in one of their targets (namely the striatum) already in 15 day-old embryos. However complete maturation of the cells is only achieved 4 weeks after birth. During this period DA neurons go through several developmental processes including 1) the maturation of cell bodies with the increased expression of catecholaminergic traits 2) the outgrowth and differenciation of neurites, with in the case of axons their progression towards frontal brain regions 3) the recognition and invasion of target territories followed by the establishment of synaptic connections. It is likely that numerous trophic influences involving diffusible compounds and specific cellular interactions are involved in all stages of the maturation of DA pathways. Such interactions can be most advantageously studied *in vitro*, since, although disrupting the tissue architecture cell culture techniques enables one to proceed analytically. In this short review of our work I shall show how, using such an approach, the possible roles of neuro-neuronal and neuro-glial interactions in DA maturation can be envisaged.

RESULTS

1) Biochemical maturation of DA neurons in culture and co-culture.

When mesencephalic cells dissociated from 13 day-old mouse embryos are cultured for several days in the presence of fetal calf serum, the DA population can be shown to survive and develop quite correctly . This can be proved both by their morphological appearance as evidenced by autoradiographic, immunocytochemical and histochemical techniques (2) as well as by biochemical criteria such as their ability to take up synthetize and release tritiated DA (^{3}H-DA) (3,4). Following these characteristics allows to conclude that maturation occurs in vitro even in the absence of cells prepared from target areas such as the striatum.

However these maturation indexes can also be followed when target cells are added to the mesencephalic populations. Indeed it is clear that this addition accelerates both the levels of ^{3}H-DA uptake and synthesis after 8 and 15 days in co-culture (4). It was checked that these ameliorations of biochemical capacities are not due to a better cell survival and therefore probably reflect an enhanced maturation of DA cells.

In order to elucidate the respective influences of glial and neuronal striatal cells on DA maturation, the co-culture experiments were performed in conditions allowing only a very restricted growth of non-neuronal elements. These conditions consist in replacing fetal calf serum by an appropriate mixture of hormones, proteins and salts (5). In this medium the amount of neuronal and non neuronal cells was estimated after 8 days in culture. This was achieved using antibodies against the 200 KD subunit of the neurofilament triplet and the vimentin protein as specific markers of the 2 cellular categories. It was shown that 95% at least of the cells were neuronal in nature. Mesencephalic and mesencephalic plus striatal cells cultures were therefore done in these conditions. Again it was shown that striatal cells stimulate DA maturation (5). Although this result strongly suggests that DA development is enhanced through neuro-neuronal interactions it does not eliminate a possible role of glial cells, it does not even completely rule out that the observed phenomenon could exclusively be triggered by the small population of remaining glia.

· If we assume that neuro-neuronal interactions have some influence on DA maturation, such an influence can result from the secretion by the target neurons of soluble trophic factors or/and from the occurrence of direct cellular contacts. The addition to mesencephalic cultures of media conditioned either on culture or on co-culture never induced any specific increase in DA uptake or synthesis. It can therefore be proposed that the presence of direct contacts is likely to be the basis of the phenomenon observed, although it is true that a role for highly unstable soluble trophic compounds cannot be precluded.

2) Influence of target and non target membranes
     on in vitro DA maturation

The hypothesis of a crucial function of direct cellular interactions prompted us to test the influence of the addition of membrane preparations from several brain areas dissected at various developmental periods on the in vitro ability of dopaminergic cells to take up ^{3}H-DA. These studies brought some evidence that during the 2nd and the 3rd week of postnatal development striatal cells produce antigenic determinants which stimulate the growth rate of DA cells. Membranes prepared from hippocampus, parietal cortex or cerebellum were not active. This effect seems therefore specific of the target and the competent age nicely coincides with the period of maximal innervation of the striatum by the afferent DA terminals (7).

3) Morphological analysis of DA neurons
     - Neuro-neuronal interactions

Another way of looking for the presence of specific interactions between the DA terminals and the striatal neurons is to examine the morphological aspect of DA nerve fibers in the presence or not of striatal target cells.

Mesencephalic cells were therefore plated at low cell density with an without striatal cells or cerebellar cells . Cerebellar cells do no constitute a physiological target of DA neurons and consequently serve as control. After a few days in culture, autoradiography was accomplished and the morphology of DA neurites studied. It is quite appearant at first sight that DA terminals stopped on nerve cells, mainly in the case of the co-culture with striatal targets and that consequently the DA fibers are shorter in these conditions. The results are that the mean lengths of DA neurites are similar after 4 days _in vitro_ when mesencephalic cells are cultured alone or co-cultured with cerebellar cells. On the contrary, DA neurites mean lenght is reduced by twofold in the presence of target cells indicating the occurrence of specific transient arrest of growth cones on target elements. The use of specific neuronal surface markers such as an anti BSP-2 antiserum (a kind gift of Dr C. Goridis) indicates that the targets are neuronal in nature (8).

- Neuro-glial interactions

It is well known that neuro-glial interactions occur at different developmental or adult stages (9). In order to study whether glial cells from the target region are different from those constituting the immediate environment of cell bodies plus dendrites, monolayers of glial cells from both regions (striatum and mesencephalon) were prepared. Embryonic mesencephalons and striata were dissociated and grown during 3 weeks with frequent medium changes until an homogeneous monolayer was obtained, the absence of neuronal or fibroblastic cells being checked with antibodies against neurofilament proteins or fibronectin. Embryonic mesencephalic cells were then added and the morphology of DA neurons studied 2 days later. On striatal glia DA neurons exhibit one main neurite, long, thin and hardly branched. On mesencephalic glia, 40% of the population presents this same poorly branched phenotype, however 60% of the remaining neurons have a very different aspect , showing several highly branched and varicose neurites. This second population does not seem to represent a subset of DA cells unable to develop on striatal glia since the total number of DA neurons is identical on both substrata. We therefore have proposed two alternative explanations. Firstly a maturation growth factor might be associated with the surface of a subpopulation of mesencephalic glia, second, the type of fiber growing on either substratum is different, for example dendrites would not be able to grow on striatal glial cells (10).

DISCUSSION

In this report, several lines of evidence have been brought showing that specific neuro-neuronal and neuro-glial interactions are implied in the <u>in vitro</u> formation of neuronal circuits. These interactions seem to have important consequences both on the morphological and the biochemical maturation of DA neurons. We do not know whether the same phenomena demonstrated <u>in vitro</u> also happen <u>in vivo</u>, however it is likely that at least some of them are common to both situations. The work presented is much in favour of an important function for direct intercellular interactions through most probably membrane bound macromolecules, the existence of which remains to be directly demonstrated. Indeed the non finding of any diffusible trophic and/or survival diffusible factors do not mean that they do not exist, they might be very unstable or work only at very high concentration in specific intercellular spaces, two properties that are not easily reproducible in an <u>in vitro</u> approach. Whatever the role of soluble compounds, our results on the existence of specific membrane interactions between nerve cells and their environment are in good accordance with other studies. For example Hemmendinger et al., have demonstrated that co-culture of DA neurons with mesencephalic cells provokes the development of dendrites, while axons growth is promoted in the presence of the targets (11). It has also be shown that tectal cell surface or tectal cell membranes can interact specifically with growth cones from ganglionic cells from the retina (12,13). On another hand several cell surface proteins have been shown to regulate neuronal interactions in vitro (14,15). There is therefore hope that models similar to the one developed in our laboratory will lead to the discovery of some molecular mechanisms relevant to the establishment of ordered neuronal ciruits during development.

REFERENCES

1. Golden, G.S. (1973) Postnatal development of the biogenic amine systems of the mouse brain. Devl. Biol. 33, 300-311.

2. Berger, B., Di Porzio, U., Daguet, M.C., Gay, M., Vigny, A., Glowinski, J. and Prochiantz, A. (1982) Long-term development of mesencephalic dopaminergic neurons of mouse embryos in dissociated primary cultures : morphological and histochemical characteristis. Neurosci. 7, 193-205.

3. Prochiantz, A., Di Porzio, U., Kato, A., Berger, B. and Glowinski, J. (1979) In vitro maturation of mesencephalic dopaminergic neurons from mouse embryos in enhanced in presence of their striatal target cells. Proc. Natl. Acad. Sci. USA. 76, 5387-5391.

4. Daguet, M.-C., Di Porzio, U., Prochiantz, A., Kato, A. and Glowinski, J. (1980) Release of dopamine from dissociated mesencephalic dopaminergic neurons in primary cultures in absence or presence of striatal target cells. Brain Res. 191, 514-518.

5. Di Porzio, U., Daguet, M.-C., Glowinski, J. and Prochiantz, A. (1980) Effect of striatal cells on in vitro maturation of mesencephalic dopaminergic neurones grown in serum-free conditions. Nature 288, 370-373.

6. Prochiantz, A., Delacourte, A., Daguet, M.-C. and Paulin, D. (1982) Intermediate filament proteins in mouse brain cells cultured in the presence or absence of fetal calf serum. Exp. Cell Res. 139, 404-410.

7. Prochiantz, A., Daguet, M.-C., Herbet, A. and Glowinski, J. (1981) Specific stimulation of in vitro maturation of mesencephalic dopaminergic neurones by striatal membranes. Nature 293, 570-572.

8. Denis-Donini, S., Glowinski, J. and Prochiantz, A. (1983) Specific influence of striatal target neurons on the in vitro outgrowth of mesencephalic dopaminergic neurites, a morhological quantitative study. J. Neurosci. In the press.

9. Varon, S.S. and Somjen, G.C. (1979) Neuron-Glia interactions. Neurosci. Res. Program Bull. vol 17, MIT press.

10. Denis-Donini, S., Glowinski, J. and Prochiantz, A. (Submitted) Glial heterogeneity might define the three-dimansional shape of mesencephalic dopaminergic neurons.

11. Hemmendinger, L.M., Garber, B.B., Hoffamn, P.C. and Heller, A. (1981) Target neuron-specific process formation by embryonic mesencephalic dopaminergic neurones in vitro. Proc. Natl. Acad. Sci. USA. 78, 1264-1268

12. Bonhoeffer, F., and Huf, J. (1980) Recognition of cell types by axonal growth cones in vitro. Nature 288, 162-164

13. Halfter, W., Claviez, M. and Schwarz, U. (1981) Preferential adhesion of tectal membranes to anterior chick retina neurites. Nature 292, 67-70

14. Hausman, R.E. and Moscona, A.A. (1976) Isolation of retina-specific cell-aggregating factor from membranes of embryonic neuronal retina tissue. Proc. Natl. Acad. Sci. USA. 73, 3594-3598

15. Brackenbury, R., Thiery, J.P., Rutishauser, U. and Edelman, G.M. (1977) Adhesion among cells of the chick embryo. I. An immunological assay for molecules involved in cell-cell binding. J. Biol. Chem. 252, 6835-6840.

Résumé

Dans cette revue sont rapportées des expériences dont le but est d'analyser in vitro le rôle des interactions cellulaires neuro-neuronales et neuro-gliales au cours du développement. Le système modèle choisi est celui de la voie dopaminergique nigrostriatale chez la souris. Il est démontré que la maturation biochimique ainsi que la morphologie des neurones dopaminergiques sont spécifiquement affectées par la présence des cellules neuronales cibles et par celle des cellules gliales soit de la région cible soit de la région où se développe l'arborisation dendritique dopaminergique. La signification de ces résultats est discutée au regard de ce qui est connu sur l'importance des interactions membranaires dans l'élaboration des circuits neuronaux.

BIOSYNTHESIS AND DEGRADATION OF CARNOSINE AND RELATED DIPEPTIDES BY BRAIN CELLS IN PRIMARY CULTURE AND PURIFIED ENZYME PREPARATIONS

KARL BAUER[1], MICHAEL SCHULZ[1], NORBERT KUNZE[1], HORST KLEINKAUF[1], KLAUS HALLERMAYER[2] AND BERND HAMPRECHT[2]

[1]Institut für Biochemie und Molekulare Biologie, TU Berlin, Franklinstraße 29, 1 Berlin 10 (Berlin West) and [2]Institut für Physiologische Chemie, Universität Würzburg, 8700 Würzburg (FRG)

INTRODUCTION

Already in 1900, carnosine was isolated from Liebig's meat extract (1). Subsequently, this substance was identified as β-alanyl-histidine (2,3) and meanwhile various ω-aminoacyl-amino acids such as β-alanyl-1-methylhistidine (anserine) (4,5,6), homocarnosine (γ-aminobutyryl-histidine) (7), homoanserine (γ-aminobutyryl-1-methylhistidine (8) etc. have been isolated from excitable tissues, brain and muscle. Apparently limited by the restricted availability of the precursors, the β-alanine-containing peptides are mainly found in muscles while the γ-aminobutyryl-containing peptides are exclusively present in the Central Nervous System. In both cases, the ω-aminoacyl-amino acids represent major products of histidine metabolism.

The physiological functions of these peptides, however, are still unknown. There have been many suggestions concerning the role of carnosine in glycolysis, oxidative phosphorylation, muscle contraction (for review see ref. 9) and wound healing (10). More recently, it has been suggested that carnosine may represent the principal neurotransmitter of the mammalian olfactory pathway (for review see ref. 11,12). This hypothesis is based on the demonstration that carnosine is mainly present in the olfactory bulb and that it is selectively lost after peripheral deafferentation. This interpretation may not be ultimately conclusive, since the degeneration of one cell type might conceivably also cause severe damage to other, functionally closely related cells. Moreover, due to conflicting electrophysiological studies (13,14,15), the physiological function of carnosine in the mammalian olfactory system remains subject to further investigations. As yet, there is also no information available concerning the physiological function of homocarnosine, the major ω-aminoacyl-amino acid in the brain. We hope that intensive biochemical studies on the biogenesis, metabolism and distribution of these peptides may offer a clue as to their function.

Abbreviations used: GABA, γ-aminobutyric acid; DBCA, $N^6,O^{2'}$-dibutyryl adenosine 3':5'-cyclic monophosphate.

BIOSYNTHESIS OF CARNOSINE AND RELATED PEPTIDES IN VITRO

Studies on intact animals and with tissue slices provided first evidence that carnosine is directly synthesized from its constituent amino acids. Subsequent studies with cell-free extracts clearly demonstrated that this dipeptide is formed enzymatically by an enzyme termed carnosine synthetase (16,17,18) and indicated that β-alanyl-adenylate might be formed as an intermediate. Therefore, it seemed that the overall reaction of carnosine synthesis may follow the scheme:

$$\beta\text{-alanine} + \text{histidine} + \text{ATP} \xrightarrow{Mg^{++}} \text{carnosine} + \text{AMP} + \text{PP}_i$$

However, neither β-alanine dependent incorporation of [^{32}P]pyrophosphate into ATP, nor β-alanine dependent AMP formation has convincingly been demonstrated so far (18,19). Owing to the istability of the enzyme, only crude enzyme preparations could be used for these studies and therefore, the exact mechanisms of peptide bond formation could not be elucidated.

Recently, we succeeded in purifying carnosine synthetase from chick pectoral muscle more extensively. With this enzyme preparation we observed β-alanine dependent hydrolysis of ATP yielding ADP and inorganic phosphate. These results strongly indicate that the mechanisms of carnosine synthesis resembles that of glutathione rather than the mechanisms of amino acid activation by aminoacyl-t-RNA synthetases. It is interesting to note that synthesis of carnosine is strongly inhibited by the products of ATP hydrolysis, ADP and P_i as well as by pyrophosphate while AMP inhibits the reaction only slightly at high concentrations. Further studies with homogeneous enzyme preparations are clearly needed to delineate the exact mechanisms of carnosine synthesis.

Most interestingly, the earlier studies indicated a broad specificty of the peptide synthesizing system. Kalyankar and Meister (18) observed that the enzyme catalyzes not only the stereospecific synthesis of carnosine, but also the synthesis of anserine, β-alanyl-lysine, β-alanyl-ornithine and β-alanyl-arginine, as well as γ-amino-butyryl peptides. With our highly purified enzyme preparation we could confirm these findings and therefore it seems unlikely that the different peptides are synthesized by different enzymes.

Although anserine can be formed directly by carnosine synthetase in vitro, the earlier in vivo studies indicated that anserine is formed by methylation of carnosine. McManus (20) subsequently succeeded in purifying an enzyme from chick pectoral muscle which catalyzes the transfer of the methyl group of S-adenosyl-methionine to the imidazole ring in carnosine. This enzyme appeared to be specific to carnosine as acceptor and thus has been termed carnosine-N-methyl transferase.

DEGRADATION OF CARNOSINE AND RELATED PEPTIDES

Since various ω-aminoacyl-amino acids can be synthesized by carnosine synthetase and are found in various tissues, we might expect that several dipeptide hydrolyzing enzymes may exist, which either fulfill a physiological function for the inacvation of the biologically active peptide(s) or may function as a proof-reading system to hydrolyze peptides produced as byproducts of carnosine synthesis. In the kidney, such enzymes may additionally serve an important function in the hydrolysis of dietary dipeptides and thus may be especially important for the resorption of dietary histidine.

In most studies on the degradation of carnosine and related peptides mainly kidney extracts have been used. Hanson and Smith (21) were the first who described a carnosine degrading enzyme from swine kidneys. According to these authors, the metal dependent enzyme exhibits broad substrate specificity. It is activated by Mn^{++} or Zn^{++} and inhibited by chelating agents. Activation of carnosinase from porcine kidney by Cd^{++} and Mn^{++} was also observed by Rosenberg (22). Lenney (23) described 2 metal dependent carnosinases from the same material, differing in their substrate specificity. According to his report, one enzyme hydrolyzes carnosine, anserine and Gly-His but not homocarnosine, L-Ala-His or N-acetyl-His while all these substances and other dipeptides are hydrolyzed by another carnosinase with very broad specificity. Since only partially purified enzyme preparations have been used in this study, it cannot be excluded that the observed broad substrate specificity is mainly due to contamination of their enzyme preparation by other enzymes.

Margolis et al. (24) recently succeeded in purifying a mouse kidney carnosinase to apparent homogeneity. Besides carnosine, this enzyme also hydrolyzes Gly-His, L-Ala-His and anserine but not homocarnosine or histidinyl-β-alanine. This carnosinase is inhibited by metal-chelating agents, slightly activated by Ca^{++} or Mg^{++} but inhibited by heavy metal ions. Mn^{++} inhibits the enzyme at low concentrations of carnosine but stimulates above 1-2 mM carnosine concentrations thereby increasing V_{Max} about 50% and the K_M value from 65 µM in the absence of Mn^{++} to 2.5 mM at 500 µM $MnCl_2$. In part, these results may explain the differences between the previous studies with regard to the effect of divalent cations. According to these authors, the Mn^{++} independent carnosinase of kidney cytosol is also present in nasal olfactory epithelial tissue but differs antigenically from that of CNS, which is abolutely dependent on Mn^{++} for activity (25).

From our own studies, however, it seems unlikely that the two different forms of carnosinases are strictly tissue specific. After fractionation of rat brain extracts by ion exchange chromatography on DEAE cellulose and hydrophobic interaction chromatography on Phenyl-Sepharose we could separate two different forms of carnosinases which exhibit distinctly different properties. One enzyme preferentially hydrolyzes carnosine and also degrades anserine but not γ-amino-butyryl-containing peptides. This enzyme is activated by Ca^{++} and Mg^{++} and inhibited by metal chelating substances as well as by heavy metal ions including Mn^{++}. It is not inactivated by agents reacting with sulfhydryl- or serine groups such as diisopropylfluorophosphate or iodoacetamide. This enzyme seems to be very similar to the cytosolic kidney enzyme described by Margolis et al. (24). The other carnosinase from rat brain exhibits distinctly different characteristics. This enzyme is inactivated by-SH reactive substances, stabilized by dithioerythritol and dependent on Mn^{++} ions for full activity. Interestingly, this enzyme hydrolyzes β-alanyl-arginine about fifty times more effectively than carnosine but does not degrade the corresponding γ-aminobutyryl-dipeptides. Therefore, this enzyme might be characterized as a β-Ala-Arg hydrolase rather than a carnosinase.

It remains to be elucidated whether other ω-aminoacyl-amino acids are also hydrolyzed by highly specific enzymes. At least we should expect that a highly specific homocarnosinase exists, since this dipeptide is not effectively hydrolyzed by the enzymes described as yet.

SYNTHESIS OF CARNOSINE BY RAT-GLIAL C-6 CELLS

So far, antibodies against the different peptides or the synthesizing enzyme are not yet available and therefore, little is known about the localization of synthetic processes. As an approach to gain further information, Ng et al. (26) studied the metabolism of ω-aminoacyl amino acids using neuronal and glial cell-enriched fractions. Their studies indicated that in the brain, carnosine synthetase is preferentially localized in neurons. We extended this approach by screening various homogeneous cell lines for their ability to synthesize carnosine. In these experiments [^{3}H]β-alanine was used as tracer. After incubation for 24 hours, extracts of the cells and the culture medium were prepared and subjected to ion exchange chromatography. Incorporation of the radiolabeled precursor into carnosine and related peptides could not be observed with various hypothalamic cell lines (which have been characterized as primitive neurosecretory cells (27) and also not by various, well characterized neuroblastoma cells. However, when the rat-glial C-6 cells were used, the tracer incorporation studies clearly demonstrated that carnosine and related peptides are actively synthesized by these cells. Moreover, carnosine, homocarnosine as well as β-alanyl-ornithine and γ-aminobutyryl-ornithine could also be isolated from these cell extracts as well as from rat brain(28).

These studies strongly indicate that carnosine may not exclusively be synthesized by neurons but may also be synthesized by glial cells. However, since transformed cells may display some neuronal properties (for review see Ref. 29) it was important to repeat these studies with primary cultures of neurons and glial cells.

BIOSYNTHESIS OF CARNOSINE BY GLIAL CELLS IN PRIMARY CULTURE

Astroblasts. Two-week-old primary cultures from newborn mouse brain consist mainly of a confluent layer of flat, epitheloid cells containing glial fibrillary acidic protein (30,31,32). Therefore, they are regarded as astroblasts. Tracer incorporation studies clearly demonstrated that carnosine is actively synthesized by glial cells in primary culture (33). In addition, these studies also revealed considerable metabolic differences between glial cells in primary culture and the C-6 glioma cells.

First of all, there is significantly more carnosine synthesized by the glial cells in primary culture than by the C-6 glioma cells. While the C-6 cells represent an undifferentiated glial cell (with the potential to differentiate into astrocytes or oligodendrocytes), the glial cells in primary culture express mainly the characteristics of astroglial cells. Therefore, it will be interesting to study whether synthesis of carnosine can be attributed to a specific cell type.

Furthermore, there are considerable differences in the metabolism of GABA. Synthesis of homocarnosine could be demonstrated with the C-6 glioma cells but not by glial cells in primary culture. While GABA is slowly metabolized by the C-6 cells, it is quickly and completely degraded by glial cells in primary culture and synthesis of homocarnosine could only be observed by inhibiting γ-aminobutyrate-α-ketoglutarate transaminase, the key enzyme in the major pathway of GABA degradation.

Astrocytes. Upon treatment with N^6, O^2-dibutyryl adenosine 3':5'-cyclic monophosphate (DBCA) or brain extracts (34,35,36) astroblasts are known to differentiate and assume the morphology of astrocytes. After treatment of glial cells in primary culture with 1 mM DBCA for 7 days we observed that carnosine synthesis was almost completely inhibited (Fig. 1).

As can be seen from Fig. 1. this effect cannot be explained by reduced uptake of the radiolabeled precursor or increased metabolization of β-alanine to acidic, as yet unidentified products. This effect can also not be due to

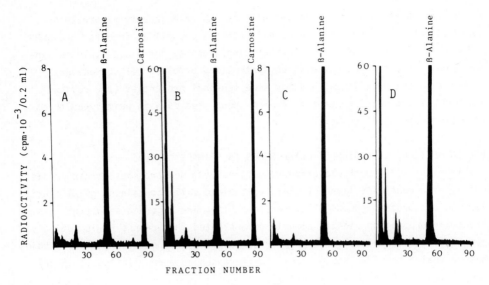

Fig. 1. Metabolism of radiolabeled β-alanine by astroblast and astrocyte-like cells in primary culture. Two-week-old cultures of newborn mouse brain (37), untreated (A,B) or treated for the last 7 days with 1 mM DBCA (C,D), were used. Radiolabeled β-alanine (10 µCi/ml) was added to fresh culture medium supplemented (C,D) or not (A,B) with 1 mM DBCA. After incubation for 24 hours, extracts were prepared from the cells (A,C) and the culture media (B,D). The extracts were resolved by ion exchange chromatography on Aminex Q-15S resin essentially as described previously (33). 25% of the column effluent was separated by stream splitting and continuously monitored for radioactivity by scintillation counting.

increased degradation of the enzymatically formed carnosine, because in both cultures radiolabeled carnosine is taken up at comparable rates but is not significantly degraded (Fig. 2).

When the cells were treated for short periods of time (7 hours) with 1 mM DBCA or noradrenalin (1 µM) and isoproterenol (1 µM) in the presence of 0.5 mM 3-isobutyl-1-methylxanthin (IBMX) as phosphodiesterase inhibitor, significant inhibition of carnosine synthesis could not be observed. Therefore, these results may suggest that the inhibition of carnosine synthesis might be related to the processes leading to the morphological differentiation of glial cells in primary culture. Further studies, however, are clearly needed to verify this working hypothesis.

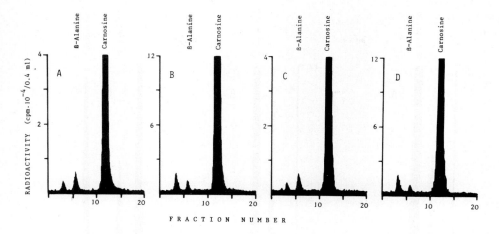

Fig. 2. Metabolism of radiolabeled carnosine by astroblast and astrocyte-like cells in primary culture. Two-week-old cultures of newborn mouse brain, untreated (A,B) or teated for the last 7 days with 1 mM DBCA (C,D), were used. Carnosine, radiolabeled in the β-alanine moiety, was added to fresh culture medium supplemented (C,D) or not (A,B) with 1 mM DBCA. After incubation for 24 hours, extracts of the cells (A,C) and the culture media (B,D) were prepared and resolved by ion exchange chromatography on Aminex Q-15S resin. Pyridine-acetate buffer (0.625 N; pH 4.6) was used for isocratic elution. The column effluent was monitored for radioactivity as in Fig. 1.

Neuronal cell cultures. Compared to the glial cultures, major differences could be observed when primary neuronal cell cultures were incubated with radiolabeled β-alanine (Fig. 3A,B) or carnosine (Fig. 3C). As with glial cells, β-alanine was found to be rapidly taken up by neuronal cells and to some extent (about 10%) the precursor was also found to be metabolized to acidic products. In contrast to the glial cells, however, synthesis of carnosine could not be observed with the neuronal cells. Furthermore, while carnosine was not effectively degraded by glial cells (see Fig. 2), considerable hydrolysis of the dipeptide could be observed after incubation of radiolabeled carnosine for 24 hours (Fig. 3C). The degradation of carnosine, however, does not explain the results of the tracer incorporation studies. If carnosine synthesis would occur, we should be able to detect at least trace amounts of the dipeptide. This is obviously not the case and therefore we conclude that under these conditions carnosine is not synthesized by neuronal cells in primary culture.

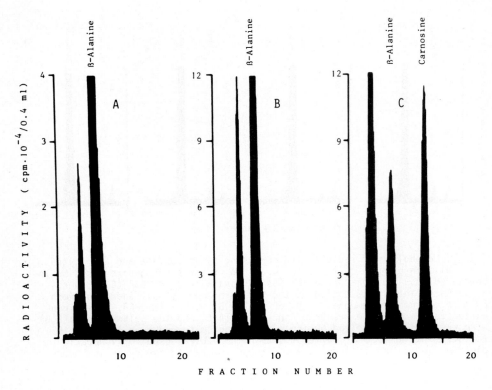

Fig. 3. Metabolism of radiolabeled β-alanine (A,B) and carnosine (C) by neuronal cells in primary culture. Neuronal cell cultures were prepared as described (38). Radiolabeled β-alanine (A,B) or carnosine (tritium labeled in the β-alanine moiety) (C) were added to fresh culture medium. After incubation for 24 hours, extracts were prepared from the cells (A), the culture medium (B) or the total incubation mixture (C). The extracts were resolved by ion exchange chromatography as described in Fig. 2.

CONCLUSIONS

The biochemical studies clearly demonstrate that carnosine and related peptides can be synthesized by undifferentiated glial cells in primary culture but not by differentiated glial cells as obtained by treatment with DBCA and also not by primary neuronal cell cultures. These results, however, should be interpreted with great caution. The present data do not mean that in vivo carnosine is exclusively synthesized by glial cells and not by neurons. Since GABA is degraded extremely rapidly by glial cells in primary culture, the present studies can not explain the fact that in the brain homocarnosine is found in high concentrations. Moreover, if the model of DBCA-induced differentiation of glial cells would reflect the natural processes leading to differentiation, the present results would even not explain the presence of any ω-aminoacyl-amino acid at all. However, since the structural and functional integrity of cell interaction in brain is not conserved in primary cultures of neurons and glial cells, a perfect correlation of the results from the in vivo and in vitro studies should not be expected. Nevertheless, as model systems, cell cultures provide exciting possibilities for studying some of the biochemical properties of the nervous system. Since carnosine most likely serves discrete biological functions (as other peptides but in contrast to some neuroactive substances which also serve other metabolic functions) we might have a sensitive chemical marker to investigate how cellular functions are expressed in biochemical events. This might be especially true for carnosine and related peptides which are not synthesized by general, constitutive enzymes but by a particular enzyme, carnosine synthetase. It is hoped that future investigations on carnosine metabolism by cultured cells may further our insight into the function of astroglia cells in the nervous system, the mechanisms of cellular communication and the processes of cellular differentiation. Studies along these lines together with extensive biochemical, histochemical as well as cytochemical studies will hopefully also help to attain an understanding of the yet unknown physiological functions of ω-aminoacyl-amino acids.

ACKNOWLEDGEMENTS

We thank Uta Heinrich for excellent technical assistance, Fridolin Löffler for the neuronal cell cultures and Bernhard Horsthemke for stimulating discussions. We also thank the Deutsche Forschungsgemeinschaft for financial support.

REFERENCES

1. Gulewitsch, W. and Amiradzibi, S. (1900) Ber. Dt. Chem. Ges. 33, 1902.
2. Barger, G. and Tutin, F. (1918) Biochem. J. 12, 402.
3. Baumann, L. and Ingwadsen, T. (1918) J. Biol. Chem. 35, 263.
4. Ackermann, D., Timpe, O. and Poller, K. (1929) Z. Physiol. Chem. 183, 1.
5. Linnewh, W., Keil, A.W. and Hoppe-Seyler, F.A. (1929) Z. Physiol. Chem. 183, 11.
6. Behrens, O.K. and Du Vigneaud, V. (1937) J. Biol. Chem. 120, 517.

7. Pisano, J.J., Wilson, J.D., Cohen, L., Abraham, D. and Udenfriend, S. (1961) J. Biol. Chem. 236, 499.
8. Nakajima, T., Wolfgram, F. and Clark, W.G. (1967) J. Neurochem. 14, 1107.
9. Crush, K.G. (1970) Comp. Biochem. Physiol. 34, 3.
10. Fitzpatrick, D.W. and Fisher, H. (1982) Surgery 91, 56.
11. Margolis, F.L. (1978) Trends Neurosci. 1,42.
12. Margolis, F.L. (1980) in: Barker, J.L. and Smith, T.(Eds.), Role of Peptides in Neuronal Function, Marcel Dekker, Inc., New York, pp. 545-572.
13. MacLeod, N.K. (1978) Trends Neurosci. 1, 69.
14. MacLeod, N.K. and Straughan, D.W. (1979) Brain Res. 34, 183.
15. Nicoll, R.A., Alger, B.E. and Jahr, C.E. (1980) Proc. R. Soc. Lond. Biol. Sci. 210, 133.
16. Winnick, R.E. and Winnick, T. (1957) Biochim. Biophys. Acta 23, 649.
17. Winnick, R.E. and Winnick, T. (1959) Biochim. Biophys. Acta 31, 47.
18. Kalyankar, G. and Meister, A. (1959) J. Biol. Chem. 234, 3210.
19. Stenesh, J.J. and Winnick, T. (1960) Biochem. J. 77, 575.
20. McManus, I.R. (1962) J. Biol. Chem. 237, 1207.
21. Hanson, H.T. and Smith, E.L. (1949) J. Biol. Chem. 179, 789.
22. Rosenberg, A. (1960) Biochim. Biophys. Acta, 45, 297.
23. Lenney, J.F. (1976) Biochim. Biophys. Acta 429, 214.
24. Margolis, F.L., Grillo, M., Grannot-Reisfeld, N. and Farbman, A.I. (1983) Biochim. Biophys. Acta 744, 237.
25. Margolis, F.L., Grillo, M., Brown, C.E., Williams, T.H., Pitcher, R.G. and Elgar, G.J. (1979) Biochim. Biophys. Acta 570, 311.
26. Ng, R.H., Marshall, F.D., Henn, F.A. and Sellström, A. (1977) J. Neurochem. 28, 449.
27. Tixier-Vidal, A. and de Vitry, F. (1976) Cell Tissues Res. 171, 39.
28. Bauer, K., Salnikow, J., de Vitry, F., Tixier-Vidal, A. and Kleinkauf, H. (1979) J. Biol. Chem. 254, 6402.
29. Pfeiffer, S.E., Betschart, B., Cook, J., Mancini, P. and Morris, R. (1977) in: Federoff, S. (Ed.) Cell, Tissue and Organ Cultures in Neurobiology, Academic Press, New York, pp. 287-346.
30. Shein, H.M. (1965) Exp. Cell Res. 40, 554.
31. Booher, J. and Sensenbrenner, M. (1972) Neurobiology 2, 97.
32. Wilson, S.H., Schrier, B.K., Farber, J.L., Thompson, E.J., Rosenberg, R.N., Blume, A.J. and Nirenberg, M.W. (1972) J. Biol. Chem. 247, 3159.
33. Bauer, K., Hallermayer, K., Salnikow, J., Kleinkauf, H. and Hamprecht, B. (1982) J. Biol. Chem. 257, 3593.
34. Lim, R., Mitsunobu, K. and Li, W.K.P. (1973) Exptl. Cell Res. 79, 243.
35. Shapiro, D.C. (1973) Nature 241, 203.
36. Moonen, G. and Sensenbrenner, M. (1975) Experentia 32, 40.
37. Van Claker, D., Müller, M. and Hamprecht, B. (1978) J. Neurochem. 30, 713.
38. Reiser, G., Löffler, F. and Hamprecht, B. (1983) Brain Res. 261, 335.

Résumé

Déjà en 1900, la carnosine (β-alanyl-histidine) était isolée à partir d'extraits de viande Liebig et depuis lors différents ω-amino-acyl acides aminés tels que l'ansérine (β-alanyl-l-méthylhistidine), l'homocarnosine (γ-amino--butyryl-histidine) etc...ont été isolés à partir de tissus excitables, cerveau et muscle. Le rôle physiologique de ces peptides, cependant, demeure une énigme.

Par l'étude de différentes lignées cellulaires, nous avons pu démontrer que la carnosine est synthétisée par la lignée cellulaire C_6 de gliome de rat mais pas par différentes lignées cellulaires hypothalamiques qui ont été définies comme cellules neurosécrétoires primitives, ni par différentes lignées cellulaires de neuroblastomes testées. Des études d'incorporation de traceur, réalisées sur des cultures primaires de cellules ont démontré que la carnosine est également synthétisée par des cellules gliales en culture primaire. Après synthèse, la carnosine est rapidement libérée dans le milieu. Aucune dégradation de carnosine par ces cellules n'a pu être observée. Par contre, la carnosine n'est pas synthétisée par des cellules neuronales en culture primaire mais elle est rapidement hydrolysée par ces cellules. Il est connu qu'en présence de dibutyryl AMP cyclique les astroblastes se différencient morphologiquement en cellules de type astrocytaire. Après traitement de cultures primaires de cellules gliales par du dibutyryl AMP cyclique (1 mM) pendant 7 jours, la synthèse de carnosine est presque complètement inhibée. Cet effet n'est pas observé après des traitements de courte durée avec du dibutyryl AMP cyclique, de la noradrénaline ou de l'isoprotérénol en présence d'IBMX en tant qu'inhibiteur de phosphodiestérase. Ces résultats suggèrent que l'inhibition de la synthèse de carnosine pourrait être liée au processus conduisant à la différenciation morphologique de cellules gliales en culture primaire.

CONTROLS AT THE GENE LEVEL
CONTRÔLES AU NIVEAU DU GÈNE

GENES INVOLVED IN DEVELOPMENT AND DIFFERENTIATION CONTROL IN PLANTS

JEFF SCHELL[1,2], MARC VAN MONTAGU[2,3], JOACHIM SCHRÖDER[1], GUDRUN SCHRÖDER[1], DIRK INZE[2], ROLF DEBLAERE[3], JEAN-PIERRE HERNALSTEENS[3] AND MARC DE BLOCK[2]
[1]Max-Planck-Institut für Züchtungsforschung, Köln (FRG)
[2]Laboratorium voor Genetika, Rijksuniversiteit Gent, 9000 Gent (Belgium)
[3]Laboratorium voor Genetische Virologie, Vrije Universiteit Brussel, St.-Genesius-Rode (Belgium)

INTRODUCTION

The study of cellular development and of organ differentiation, both normal and abnormal, will undoubtedly make important progress if one could identify and isolate the genes involved in the genetic molecular control of these processes. One would hope to be able to study the products of these genes and their mechanism of action. With the advent of suitable host gene vectors, with which one can reintroduce isolated genes into living cells, it has now become possible to modify isolated genes in vitro and to reintroduce them in living organisms to study their specific effect. The same approach could be used to study the mechanism of action of genes controlling both normal and abnormal cellular development and differentiation.

The difficulty however is to devise means to identify such genes. One system has turned out to be particularly suited for such an approach. Indeed the formation of so-called "Crown-gall" tumors on many dicotyledonous plants by the soil pathogen Agrobacterium tumefaciens is the direct result of the introduction into the nuclear genome of plant cells of a number of "oncogenes" that regulate cellular development and organ differentiation.

Agrobacterium tumefaciens is a soil microorganism that is capable of infecting a broad assortment of dicotyledonous plants after they have been wounded (1). As a result of this infection, the wound tissue begins to proliferate as a neoplastic growth referred to as a crown gall tumor. Once induced, the tumors no longer require the presence of bacteria to continue growing (2). Among the most important of the new properties of these transformed cells are, that they can grow axenically in vitro without the growth factors normally required by plant cell cultures, and that they can synthesize a variety of compounds unique to tumors. The latter compounds, which are termed opines, can be metabolized specifically by the bacteria responsible for inciting the tumor (3).

The study of these plant cancers has recently made very considerable progress because it was found that the genes responsible for hormone-independent neoplasmic growth as well as for the ability to induce opine biosynthesis are not normally present in plant cells e.g. in an inactive form, but are infact encoded by the so-called Tumor-inducing Ti-plasmid of A.tumefaciens (4-6) and are introduced into the nuclear genome of the transformed plant cells.

Crown-gall is therefore the first well documented instance of natural genetic engineering.

A specific portion of the Ti-plasmid, the T-DNA, is transferred from the bacteria to the nucleus of a susceptible plant host (7). There the DNA is integrated into plant chromosomes as a unit with discrete end points (8,9) which contains the genes responsible for opine biosynthesis and for tumor growth (5,6,10-12).

IDENTIFICATION OF TI-PLASMID SEQUENCES RESPONSIBLE FOR THE NEOPLASMIC MODE OF GROWTH

Agrobacteria from which the extrachromosomal Ti-plasmid has been eliminated, no longer induce tumors (13).

By means of transposon mutagenesis (5,6,10) and more recently, by analysis of the effects of substantial deletion mutations (14,15), it has been possible to demonstrate that the Ti-plasmid contains two distinct and separate regions that are essential to produce transformed cells (Fig. 1). The first of this segments has been called the T-DNA region for Transferred DNA region, because it is completely homologous with DNA sequences found in most established tumor lines. The second portion of the Ti-plasmid with sequences essential for tumor formation has been termed the "virulence-region". The DNA sequences of the vir-region have not been detected in established tumor lines and are therefore not essential for tumor maintenance.

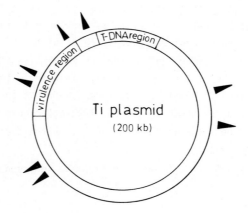

Fig. 1.

It is not clear how the vir-function contribute to the formation of Crown-gall tumors. Circumstancial evidence suggests that the vir-region codes for functions involved in (1) establishing cellular contacts between bacterial and plant cells (2) promoting transfer of DNA from the bacteria to the plant nucleus and (3) promoting integration of the T-DNA region in the plant nuclear DNA.

The T-DNA's in transformed plant cells represent an active form of inserted foreign DNA.

The number, sizes, and location of the transcribed T-DNA segments were studied in different tumors (16). Tumor-specific RNAs were detected and mapped by hybridization of ^{32}P-labeled-Ti plasmid fragments to RNA which had been seperated on agarose gels, and then transferred to DBM paper.

These transcripts differ in their relative abundance, and in their sizes. They all bind to oligo(dT)-cellulose, indicating that they are polyadenylated. The direction of transcription was determined, and the location of the approximate 5'- and 3'-ends were mapped on the T-DNA.

All RNAs mapped within the T-DNA sequence. This, and the observation that transcription is inhibited by low concentrations of α-amanitin suggests that each transcript is determined by a specific promoter site on the T-DNA recognized by plant RNA polymerase II. The results available so far are consistent with the assumption that each gene on the T-DNA has its own signals for transcription in the eukaryotic plant cells. Six transcripts were found to be derived from a "common" or "core" segment of the T-region. These transcripts were found to be identical in tumors induced by different types of Ti plasmids (16, 17).

These six transcripts are arranged on the T-DNA in the order 5, 2, 1, 4, 6a, 6b (Fig. 2)

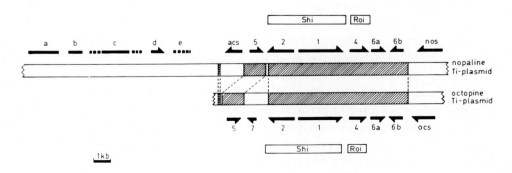

Fig. 2

Specific mutations were introduced in the T-DNA regions of Ti plasmids to produce transformed plant cells in which one or more T-DNA-derived transcripts would not be expressed. By observing the phenotypes of the plant cells harboring such partially inactivated T-DNAs, it was possible to assign functions to most of the different transcripts (12, 18). It was found that none of the T-DNA transcripts was essential for the transfer and stable maintenance of T-DNA segments in the plant genome. Essentially, two different functions were found to be determined by T-DNA transcripts.
(i) Transcripts coding for opine synthase: octopine tumors contain either one or two of such genes. One of them codes for octopine synthase (ocs) (Fig. 2) Nopaline tumors contain at least two transcripts coding for different opines. One is located at the right end of the T-DNA and codes for nopaline synthase (nos), whereas the other is located in the left part of the T-DNA, and codes for agrocinopine (acs) (see Fig. 2).
(ii) Transcripts (probably after translation into proteins) that are directly or indirectly responsible for tumorous growth: these transcripts are found to be derived from the "common" or "core" region of the T-DNA. Remarkably, all T-DNA functions affecting the tumor phenotype were located in this "common" region of the T-DNA. Several of these transcripts act by suppressing plant organ development. It was observed that shoot and root formation are suppressed independently and by different transcripts. Two transcripts (1 and 2) were identified that specifically prevent shoot formation. The effect of these T-DNA gene products is in many ways analogous to that of auxin-like plant growth hormones since the effect of these genes is similar to that observed for calli from normal plant cells with artificially increased auxin level. Another gene (transcript 4) was found to prevent specifically root formation, and the effect of this T-DNA gene can, therefore, be compared to the effects observed when normal plant cells are grown in the presence of high concentrations of cytokinins (12, 18).

It is important to note that the products of genes 1 and 2 not only suppress shoot formation but also stimulate root formation. Reciprocally, the product of gene 4 not only inhibits root formation, but also stimulates shoot formation. This conclusion is based on recent observations indicating that Ti plasmids from which genes 1 and/or 2 as well as gene 4 were eliminated by deletion-substitution, could transfer their modified T-DNA to plant cells, but did not promote either shoot or root development at the site of infection. Shoot formation therefore requires both the inactivity of genes 1 and/or 2, and the activity of gene 4, whereas root formation requires inactivity of gene 4 and activity of genes 1 and 2. This conclusion was further strengthened by experiments (unpublished data from the authors laboratory) involving Ti plasmids from which all of the T-region genes were removed by deletion (Zambryski et al. EMBO J. In Press). Individual T-DNA genes were subsequently reintroduced in the deleted T-region. Thus Ti plasmids which will only introduce genes 1 or 2 or 4 or a defined combination of these genes, are available and were used to demonstrate that the introduction in plant cells of the combination of genes 1 and 2 will result in the formation of small tumors from which roots proliferate. Shoot formation however is suppressed. The introduction of only gene 1 or only gene 2 does not yield a transformed phenotype.

On the other hand the introduction in plant cells of gene 4 will result in a tumorous growth from which both normal and abnormal (teratoma) shoots abundantly proliferate, whereas root formation is completely suppressed.

The neoplasmic or tumorous mode of growth of these transformed plant cells is therefore due primarily to a suppression (or negative control) of normal differentiation. It is however remarquable that the gene-products (1 and 2) that prevent shoot formation also stimulate root formation when the root inhibiting gene (4) is inactive or absent and that reciprocally the gene product (4) that prevents root formation will stimulate shoot differentiation when this is not prevented by the activity of genes 1 and 2. These observations therefore strongly support the notion that the major developmental systems in plants are organized and controlled in an interrelated and mutually exclusive way.

To differentiate into shoots, plant cells could rely on a single system that both suppresses root specific genes and stimulates shoot specific genes and reciprocally roots could rely on a single system that suppresses shoot specific genes while stimulating root specific genes.

Whereas these observations are consistent with the idea that the products of genes 1, 2, and 4 directly determine the auxin-cytokinin levels in the transformed cells, and that these hormone levels in turn would be responsible for the observed tumor phenotypes, they do not prove this point. It is, for instance, still conceivable that the products of genes 1, 2, and 4 could act directly at the level of gene regulation, and that the alterations in growth hormone levels would be the consequence, rather

than the cause, of the observed tumor morphology. It is essential to
isolate these gene products in order to determine in detail their
mechanism of action. As a preliminary step towards this goal both genes
1 and 2 were separately cloned, transferred and expressed in E.coli.
It was shown (Schröder et al. European J. Biochem., submitted)that the
product of gene 2 catalyzes the conversion of α-indol acetamide into
α-indol acetic acid which is a well known Auxin growth hormone.
Independently genetic evidence was obtained (D. Inzé et al., submitted to
MGG) showing that gene 1 might be involved in the biosynthesis of α-indol
acetamide. This evidence therefore supports the notion that genes 1 and 2
are directly involved in auxin biosynthesis. An analogous approach
is used to test whether or not gene 4 is similarly involved in cytokinin
biosynthesis.

In addition to this hormone-like activity, the T-DNA core codes for at
least three other transcripts, 5, 6, and 6a. Transcript 5, was found to
inhibit the organization of transformed cells into leaf bud structures.
Elimination of this transcript, along with the shoot inhibiting auxin-like
genes (genes 1 and 2) resulted in transformed cells organizing themselves
as teratomas (12, 18).

Whereas the hormone-like effect of genes 1, 2, and 4 results in a
suppression of regeneration by both non-T-DNA-containing as well as by
T-DNA-containing plant cells mixed in the same crown gall tissue, the
effect of gene 5 seems to be restricted to the plant cells in which this
gene is present and active.

We are uncertain about genes 6a and 6b because no important phenotypic
change has thus far been correlated with their inactivation (18).

In the previous literature there had been observations suggesting that
the T-DNA could not pass through meiosis. Seeds obtained by self-ferti-
lization of octopine positive plants, however, produced new plants with
active T-DNA-linked genes, demonstrating that genes introduced in plant
nuclei, via the Ti plasmid, can be sexually inherited (20, 21). A series
of sexual crosses were therefore designed to study the transmission
pattern of the T-DNA-specified genes. The results of these crosses
demonstrate very convincingly that the T-DNA-linked genes are transmitted
as a single Mendelian factor both through the pollen and through the eggs
of the originally transformed plant. These crosses also showed that the
original transformed plant was a hemizygote containing T-DNA only on one
of a pair of homologous chromosomes. By these crosses tobacco plants

homozygotes for the altered T-DNA were obtained (20). When regenerants from different transformations are crossed the two T-DNA loci segregate independently (21).

In view of the fact that in the previous experiments the opine positive plants did not contain any of the tumour controlling genes, it could still be argued that active tumour genes such as genes 1, 2 or 4 could not go through meiosis e.g. because they would prevent early seed embryo formation.

This possibility was studied in the following way: shoots sprouting from tumours containing genes 4, 6a, 6b and octopine synthase (see Fig. 2) were grafted onto normal tobacco, resulting in the formation of flowering shoots. Sexual crosses between different grafts and selfing produced two different types of seedlings. Some developed into normal, untransformed plants whereas others failed to develop any roots and remained very small. The abnormal seedlings were shown to consist of transformed cells and expressed the octopine synthase gene as well as gene 4. The phenotypes of root suppression coded for by gene 4 and of opine synthesis (coded for by the ocs gene, see Fig. 2) were 100% linked in the offspring (Wöstemeyer et al., in press).

These experiments therefore demonstrate that tumour genes (such as gene 4) can be transmitted sexually to offspring plants and that they do not prevent seed formation and embryo development but specifically interfere with root differentiation.

GENERAL CONCLUSIONS

Ti-plasmids of Agrobacteria have provided us with an unexpected and extraordinarily potent system to study the genetic control of plant organ development. Indeed, we found that these plasmids contain DNA sequences which are transferred and integrated into chromosomes of plant cells. These so-called T-DNA sequences contain genes that are expressed in plant cells. Some of these genes make products that specifically suppress either shoot or root formation. These observations indicate that plants have separate genetic programs for shoot development and for root development and that both of these programs must be internally coordinated since they can be controlled by the products of single genes.

Furthermore, the T-DNA remarkably has two independent but complementary mechanisms to suppress plant organ development. Each of these mechanisms alone may be sufficient to suppress the development of transformed cells. One of these mechanisms appears to function via a growth hormone-like mechanism and therefore suppresses both transformed and untransformed cells, thus explaining why uncloned primary tumors, consisting of a mixture of T-DNA-containing and normal untransformed cells, nevertheless grow as a uniformly undifferentiated callus. This mechanism might well play a role in the production of auxin- and cytokinin-like hormones. The other mechanism is even more intriguing since it appears to suppress transformed plant cells only.

REFERENCES

1. Braun, A.C. (1982) in: Kahl, G. and Schell, J. (Eds.), Molecular Biology of Plant Tumors, Academic Press, New York, pp. 155-210.

2. Braun, A.C. (1943) Amer. J. Bot., 30, 674.

3. Goldmann-Ménagé, A. (1970) Ann. Sci. Nat. Bot. (12° Sér.), 11, 233; Lejeune, B. (1973) Ph.D.Thesis CNRS, Paris; Tempé, J. and Petit, A. (1982) in: Kahl, G. and Schell, J. (Eds.), Molecular Biology of Plant Tumors, Academic Press, New York, pp. 451-459.

4. Zaenen, I., Van Larebeke, N., Teuchy, H., Van Montagu, M. and Schell, J. (1974) J. Mol. Biol., 86, 109;
Van Larebeke, N., Engler, G., Holsters, M., Van den Elsacker, S., Zaenen, I., Schilperoort, R.A. and Schell, J. (1974) Nature (London) 252, 169;
Van Larebeke, N., Genetello, C., Hernalsteens, J.P., Depicker, A., Zaenen, I., Messens, E., Van Montagu, M. and Schell, J. (1975) Mol. Gen. Genet. 152, 119;
Watson, B., Currier, T.C., Gordon, M.P., Chilton, M.-D. and Nester, E.W. (1975) J. Bacteriol. 123, 255.

5. Holsters, M., Silva, B., Van Vliet, F., Genetello, C., De Block, M., Dhaese, P., Depicker, A., Inzé, D., Engler, G., Villarroel, R., Van Montagu, M. and Schell, J. (1980) Plasmid 3, 212.

6. De Greve, H., Decraemer, H., Seurinck, J., Van Montagu, M. and Schell, J. (1981) Plasmid 6, 235.

7. Chilton, M.-D., Saiki, R.K., Yadav, N., Gordon, M.P., Quetier, F. (1980) Proc. Natl. Acad. Sci. USA 77, 4060;
Willmitzer, L., De Beuckeleer, M., Lemmers, M., Van Montagu, M., and Schell, J. (1980) Nature (London) 287, 359.

8. Lemmers, M., De Beuckeleer, M., Holsters, M., Zambryski, P., Depicker, A., Hernalsteens, J.P., Van Montagu, M. and Schell, J. (1980) J. Mol. Biol. 144, 353;
Zambryski, P., Holsters, M., Kruger, K., Depicker, A., Schell, J., Van Montagu, M. and Goodman, H.M. (1980) Science 209, 1385.

9. Thomashow, M.F., Nutter, R., Montoya, A.L., Gordon, M.P., Nester, E.W. (1980) Cell 19, 729;
De Beuckeleer, M., Lemmers, M., De Vos, G., Willmitzer, L., Van Montagu, M. and Schell, J. (1981) Mol. Gen. Genet. 183, 283.

10. Garfinkel, D.J. and Nester, E.W. (1980) J. Bacteriol. 144, 732.

11. Garfinkel, D.J., Simpson, R.B., Ream, L.W., White, F.F., Gordon, M.P. and Nester, E.W. (1981) Cell 27, 143.

12. Leemans, J., Deblaere, R., Willmitzer, L., De Greve, H., Hernalsteens, J.P., Van Montagu, M. and Schell, J. (1982) EMBO J. 1, 147.

13. Hamilton, R.H. and Fall, M.Z. (1971) Experientia 27, 229;
Engler, G., Holsters, M., Van Montagu, M., Schell, J., Hernalsteens, J.P. and Schilperoort, R.A. (1975) Mol. Gen. Genet. 138, 345.

14. Van Haute, E., unpublished results.

15. Hille, J., Klasen, I. and Schilperoort, R.A. (1982) Plasmid 7, 107.

16. Willmitzer, L., Simons, G., and Schell, J. (1982) EMBO J. 1, 139;
Bevan, M. and Chilton, M.-D. (1982) J. Mol. Appl. Genet. 1, 539;
Willmitzer, L., Dhaese, P., Schreier, P.H., Schmalenbach, W., Van Montagu, M. and Schell, J. (1983) Cell 32, 1045.

17. Chilton, M.-D., Drummond, M.H., Merlo, D.J. and Sciaky, D. (1978) Nature (London) 275, 147;
Depicker, A., Van Montagu, M. and Schell, J. (1978) Nature (London) 275, 150;
Engler, G., Depicker, A., Maenhaut, R., Villarroel-Mandiola, R., Van Montagu, M. and Schell, J. (1981) J. Mol. Biol. 152, 183.

18. Joos, H., Inzé, D., Caplan, A., Sormann, M., Van Montagu, M. and Schell, J. (1983) Cell 32, 1057.

19. Ooms, G., Hooykaas, P.J., Moleman, G. and Schilperoort, R.A. (1981) Gene 14, 33.

20. Otten, L., De Greve, H., Hernalsteens, J.P., Van Montagu, M., Schieder, O., Straub, J. and Schell, J. (1981) Mol. Gen. Genet. 183, 209.

21. De Greve, H., Leemans, J., Hernalsteens, J.P., Thia-Toong, L., De Beuckeleer, M., Willmitzer, L., Otten, L., Van Montagu, M. and Schell, J. (1982) Nature (London) 300, 752.

Résumé

Les plasmides Ti d'Agrobacterium nous ont fourni un système à la fois inattendu et extraordinairement puissant pour étudier le contrôle génétique du développement des plantes. En effet, nous avons trouvé que ces plasmides contiennent des séquences de DNA qui sont transférées et intégrées dans les chromosomes des cellules de plantes. Ces séquences appelées séquences T-DNA contiennent des gènes qui sont exprimés dans les cellules de plantes. Les produits de certains de ces gènes suppriment spécifiquement la formation soit des pousses, soit des racines. Ces observations montrent que les plantes ont des programmes génétiques différents pour le développement des pousses et des racines,et que ces deux programmes doivent être coordonnés étant donné qu'ils peuvent être contrôlés par les produits de gènes uniques.

De plus, le T-DNA, de façon remarquable, possède deux mécanismes indépendants mais complémentaires pour supprimer le développement d'organes de la plante. Chacun de ces mécanismes, isolément, peut être suffisant pour supprimer le développement des cellules transformées. Un de ces mécanismes semble fonctionner via une substance type hormone de croissance et réprime donc à la fois les cellules transformées et non transformées. Ceci explique pourquoi des tumeurs primaires non clonées, composées d'un mélange de cellules contenant du T-DNA et des cellules normales non transformées, poussent néanmoins comme un callus uniformément indifférencié. Ce mécanisme pourrait bien jouer un rôle dans la production d'hormones de type auxine et cytokinine. L'autre mécanisme est même plus étonnant étant donné qu'il semble ne réprimer que les cellules de plantes transformées.

AMPLIFICATION OF MOUSE MAMMARY TUMOR VIRUS (MMTV) DNA IN MAMMARY TUMORS OF
GR/A MICE, A MODEL FOR HORMONE SENSITIVE MAMMARY TUMORS

R. Michalides
The Netherlands Cancer Institute, Dept. of Virology, Plesmanlaan 121, 1066 CX
Amsterdam, The Netherlands

INTRODUCTION

The mouse mammary tumor viruses represent a set of variants (MMTV) belonging to the family of retroviruses and induce mammary tumors in susceptible strains of mice. The animal retroviruses can be divided in two classes: the acute transforming retroviruses with transforming (or onco-) genes, whose action initiate and maintain the neoplastic phenotype of the cell, and the slowly transforming retroviruses without such onc genes (1, 2). MMTV belongs to the slowly transforming category, inducing mammary tumors after a long latency period of approximately one year, and, although the majority of the mammary gland cells become infected, only one or a few mammary tumors appear in an infected animal. The virus is transmitted via the milk from mothers to offspring as an exogenous MMTV. The route of infection in the animal is not quite understood, but it is assumed that the virus finds its way through the gut wall and is transferred via white blood cells to the mammary gland cells. The onset and incidence of mammary tumors is, furthermore, influenced by the virulence of the MMTV variant, the histocompatibility genes of the host and the hormonal status of the animal (3).

Since retroviruses replicate through a DNA provirus intermediate integrated in the DNA of the host cell, only cells infected with exogenous MMTV contain proviral DNA of exogenous MMTV. All inbred mouse strains tested so far contain also some MMTV-DNA copies integrated in the germ line cells, and thereby also in all their cells. These socalled endogenous MMTVs are the result of infrequent insertions of infectious MMTV genome into the germ line during evolution of the species (4).

Two unique interactions have endeared the MMTV to molecular biologists and endocrinologists, viz. the interaction between the MMTV genome and a dexamethasone-receptor complex resulting in a hormone-induced expression of the MMTV provirus, and the interaction between MMTV and mammary gland cells leading to mammary tumors including hormone dependent tumors. The hormonal stimulation of MMTV expression became evident from studies measuring the production of MMTV in mammary tumor cell lines or MMTV infected non murine cells, such as rat hepatoma cells. This expression is 10-20 fold increased by the addition of the synthetic glucocorticoid dexamethasone (5). The effect of dexamethasone on MMTV expression is a direct one in that the dexamethasone-receptor complex binds to a defined region of the MMTV genome resulting in an increase of transcription. This direct effect of an interaction between a hormone receptor complex and a MMTV provirus region provides an excellent system to study hormone-stimulated transcription in cells.

Recent studies, which indicate that the oncogenicity of MMTV is best explained by the insertional mutagenesis model, have increased our understanding of the interaction between MMTV and mammary gland cells and the resulting mammary tumor development.

The present paper deals with these two processes.

STRUCTURE AND EXPRESSION OF THE MMTV GENOME

The MMTV-RNA genome contains 7800 nucleotides and is present in virus particles as a dimer. MMTV-RNA replicates via a DNA intermediate using an RNA-dependent DNA polymerase and becomes integrated in the cellular DNA in a provirus DNA form, which is bordered by long terminal repeats (LTR), see Fig. 1). These LTRs are generated at transcription of RNA into DNA (6), in such a way that specific fragments of the MMTV-RNA genome are duplicated within the LTR at the opposite site of the provirus. An unique region of MMTV-RNA at the 5' end, U5, becomes duplicated in the right LTR and an unique region at the 3' end, U3, duplicates in the left LTR, each LTR therefore contains an U3 and an U5 region. In between the LTRs are the genes for the structural internal, *gag*, and envelope, *env*, MMTV proteins and for *pol*, the RNA-dependent DNA polymerase.

Two major MMTV-RNAs are detected in MMTV producing cells: a full-length, genome sized, 7.8 Kb RNA and a 3.8 Kb RNA (7, 8). The 7.8 Kb MMTV-RNA instructs the synthesis of the *gag* proteins and of the RNA-dependent DNA polymerase, whereas the 3.8 Kb MMTV-RNA codes for the *env* proteins (9). The 3.8 Kb MMTV-RNA shares a 5' leader sequence of 287 nucleotides with the 7.8 Kb MMTV-RNA, from which it is derived by a splicing mechanism removing the *gag-pol* information. A third, minor, MMTV-RNA has recently been found in mammary tumors of GR and C3H mice and, in larger amounts, in BALB/c mammary glands (10, 11). The 5' end of this 1.7 Kb MMTV-RNA maps at a position indistinguishable from 7.8 and 3.8 Kb RNA, about 130 nucleotides upstream from the right end of the L-LTR. It has a 5' splice site identical to that of the 3.8 Kb MMTV-RNA, but is resumed just before the right-LTR where it continues to the transcription termination site. This small MMTV-RNA, therefore, contains information of the U3 part of the LTR.

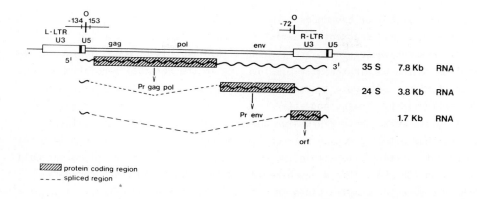

Fig. 1. Expression of the MMTV provirus. L-LTR, left long terminal repeat, R-LTR right long terminal repeat. The numbers at the L-LTR indicate the 5' transcription start site of MMTV-RNA (-134 nucleotides from the right of the L-LTR). Position -72 at the R-LTR indicates the 3' splice site (-72 nucleotides form the left of the R-LTR) of the 1.7 Kb RNA. Protein coding regions are marked on the RNAs.

This U3 part of the LTR contains an open reading frame (orf) of 960 nucleotides with a coding potential for a basic 36,700 dalton protein, which is found in different MMTVs. The conservation of orf in various MMTVs suggests some role for the orf protein. The orf protein was detected in in vitro translation assays, but has not yet been found in animal tissues (12-16). The orf protein could be involved in the integration of MMTV proviruses in the cellular genome, play a role in the interaction between mammatropic hormones and the MMTV genome, or be involved in the processing of MMTV precursor proteins.

INTERACTION WITH THE DEXAMETHASONE-RECEPTOR COMPLEX

MMTV is unique among retroviruses in that transcription of MMTV proviral DNA in cultured cells is stimulated by glucocorticoid hormones, and has therefore become a model for studying steroid hormone action (18). The modulation of MMTV expression is mediated by hormone-specific intracellular receptor proteins that associate with nuclear sites upon binding to the glucocorticoid dexamethasone. Following the binding of the dexamethasone-receptor complex the transcription of MMTV is rapidly increased. This hormonal regulation is also observed in cells transfected with molecularly cloned MMTV-DNA, either of endogenous (19,20) or exogenous origin (21, 22). Transfections with combinations of subgenomic MMTV fragments and marker genes (18, 23, 24) indicated that the hormone responsive sequences are located within the LTR of MMTV. To further define the region within the MMTV-LTR involved in the hormonal regulation of MMTV transcription, LTRs were partially deleted, combined with a marker gene such as thymidine kinase, and after transfections assayed for their ability to stimulate marker gene expression upon addition of dexamethasone. These studies showed that a region between 105 and 204 nucleotides upstream from the transcription start site within the MMTV-LTR is involved in the glucocorticoid regulation (25, 26). This area co-resides with a domain to which the dexamethasone-receptor complex binds in vitro. In vitro binding of partially purified hormone-receptor complexes to certain cloned MMTV-DNA fragments showed receptor binding domains in various regions of the MMTV provirus, but one binding domain in the LTR 400 base pairs upstream of the RNA transcription start is generally agreed upon (27, 29).

This glucocorticoid response element is distinct from sequences that are required for initiation of transcription, it functions also at larger distances from the promoter (25), and its orientation can be reversed (30). These properties are comparable with those of "enhancer elements" that reside within other viral genomes, such as SV40 (31) and Moloney murine sarcoma virus (32).

The hormonal regulation of MMTV expression may therefore operate via an enhancer element located just upstream of the RNA transcription start site in the LTR. Since MMTV proviruses become integrated more or less at random sites in the cellular genome, the introduction of such enhancer sequences in new sites may well affect the expression of other new genes.

INTEGRATION OF MMTV AND TUMORIGENESIS

Strains of mice which carry an exogenous milk-borne MMTV have a high incidence of mammary cancer. The exogenous virus infects the mammary gland cells of the offspring and is transcribed into viral DNA which integrates in the host cell DNA. Insertion of MMTV proviral DNA is therefore a requisite for mammary tumor formation, since mice devoid of milk-borne MMTV have a much lower mammary tumor incidence, with the exception of the GR strain discussed below. The genome of MMTV does not carry a transforming oncogene and is hereby related to other retroviruses, such as avian leukosis virus and murine leukemia virus, whose genomes only contain information for viral proteins. This group of slowly transforming retroviruses shares the following characteristics (see Fig. 2):
- tumors appear late after infection of the animals;
- the tumors produced by these viruses are clonal or semiclonal in origin.

SLOWLY TRANSFORMING RETROVIRUSES

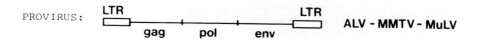

PROPERTIES:
- induce late appearing tumors in vivo
- tumors are clonal in respect with integration of extra proviruses
- no transformation in vitro
- at random integration of extra proviruses
- chance integration in crucial region influences expression cellular oncogene

Fig. 2. Structure and properties of slowly transforming retroviruses.

Many cells of the target organ become infected with these retroviruses, but only one or a few of these infected cells grow out into a tumor. Integration of proviral DNA into the host DNA seems to occur at random sites (33), but integrations at particular genomic sites are apparently required to promote tumorigenesis. Integration of proviral DNA at such particular site then activates a cellular oncogene. In case of ALV (Avian Leukosis Virus) induced lymphoid leukosis in chicken, a cellular oncogene, *c-myc*, is activated due to the insertion of ALV-DNA in the vicinity of the *c-myc* gene (34, 35). The ALV-DNA provides a promoter in the long terminal repeat (LTR) sequence for the transcription of the juxtaposed *c-myc* gene (see Fig. 3). *c-Myc* activation is also involved in chicken lymphomas induced by chicken syncytial virus (36), whereas *c-erb B* is activated by the insertion of ALV-LTR in ALV induced erytrhoblastosis (37). These findings suggest that a single retrovirus can activate different cellular oncogenes, and that the same oncogene *(c-myc)* can be activated by different retroviruses.

Activation of cellular oncogenes appears also to be implicated in the development of MMTV induced mammary tumors. Using an analogous strategy as used to reveal activation of cellular oncogenes by proviral insertion in chicken

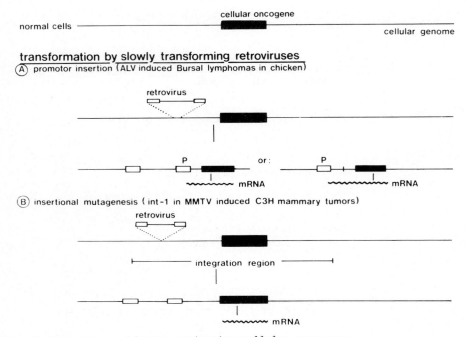

Fig. 3. Insertion models for activating cellular oncogenes.

lymphomas, domains of the mouse genome have been identified which are important for mammary tumorigenesis. The following approach was taken to identify these domains: a molecular clone containing an exogenous MMTV-DNA/cellular DNA junction fragment was isolated from a mammary tumor. The cloned cellular DNA fragment next to the MMTV provirus was then used to study whether other mammary tumors contained extra MMTV proviruses integrated in the same cellular DNA domain. In this way two viral DNA integration regions were identified: the *int-1* locus spanning a 35 Kb cellular DNA region, which was found to be occupied by an exogenous MMTV in the majority of MMTV induced C3H mammary tumors (38), and a similar region of 10 Kb, *int-2*, in MMTV induced BR6 mammary tumors (39). Both integration regions are not homologous to any of the known cellular oncogenes, and they are also unrelated to one another and located on different chromosomes.

A part of the *int-1* locus is transcribed into a 2.6 Kb mRNA in some mammary tumors, but is not in mammary glands. This part may well correspond with the cellular oncogene activated by insertion of MMTV in the *int-1* domain.

These studies show that exogenous MMTV is integrated in certain cellular domains in the majority of MMTV induced mammary tumors. Integration of MMTV proviruses in mammary gland DNA, however, takes place at multiple, possible random, sites. The MMTV induced mammary tumors therefore originate most likely from a clonal outgrowth of a mammary gland cell following the integration of MMTV in an *int*-domain. In case of integrations in *int-1*, the integration event also results in expression of an *int-1* 2.6 Kb RNA.

This mechanism of activating cellular oncogenes by insertion of proviruses is explained by the insertional mutagenesis model (40). This model proposes that chance integration of proviral DNA in the vicinity of a crucial cellular gene perturbs the expression of that gene. It is possible that the mouse genome contains multiple of such integration domains where integration of MMTV may trigger tumorigenesis. Another possibility is that different cellular oncogenes are involved in various pathological types of mammary cancer.

In ALV induced bursal lymphomas, activation of *c-myc* is most commonly achieved by promoter insertion (see Fig. 3), in which the LTR of the inserted ALV provirus provides a transcriptional promoter for the downstream located *c-myc* gene. This promoter insertion mechanism does not apply for expression of *int-1* since extra MMTV proviruses are positioned at both sides of the gene corresponding with the *int-1* RNA. Expression of *int-1* may well be mediated through "enhancer" like sequences in the LTR of the inserted MMTV provirus as described above.

THE GR MOUSE STRAIN, A MODEL SYSTEM FOR HORMONE SENSITIVE MAMMARY TUMORS

Practically all females of the strain GR develop mammary cancer at around one year of age. Breeding GR females develop pregnancy dependent mammary tumors, which regress upon parturition (42), these so called plaque type tumors are composed of ductal and alveolar-like elements. Hormone dependent as well as hormone independent mammary tumors appear after multiple pregnancies. Pregnancy dependent mammary tumors of GR are therefore considered a precursor state of the later appearing mammary tumors. Growth of the pregnancy dependent tumors requires prolactin, progesterone and estrone. Hormone dependent tumors can be induced experimentally in ovariectomized GR females by treatment with a combination of progesterone and estrone (41), or with the steroid compound 17 ethynyl-19-nortestosterone (ANT) (42). Mammary tumor incidence in virgins of GR is as high as in breeders; the tumors appear only a little later, are adenocarcinomas, and their growth is not influenced by steroids (44).

The appearance of these various types of mammary tumors in GR is controlled by the Mtv-2 locus, located on chromosome 18, which also controls expression of MMTV in this strain (45, 46). This Mtv-2 locus contains one endogenous MMTV provirus out of the five present in GR (47). Therefore, expression of an endogenous MMTV provirus, associated with the Mtv-2 locus, is linked with mammary tumorigenesis in GR. The expression of another endogenous MMTV provirus in mouse strain C3Hf is similarly linked with the development of late appearing mammary tumors in that strain (47,48).

The various stages of GR mammary tumor progression can be studied in lines of transplanted mammary tumors. Primary tumors from pregnant GR females are then transplanted into castrated GR males or females, with or without treatment with estrone and progesterone. When a graft yields an outgrowth in a hormone-treated mouse, but not (within 3 months) in hormone-untreated castrated mice, the tumor is designated hormone dependent (HD). If the outgrowth appears sooner in the hormone-treated castrated mice than in the untreated ones, it is termed hormone responsive (HR), whereas tumors which grow equally well are classified as hormone independent (HI). Transplant lines which initially behave as HD tumors usually convert into HI tumors after multiple passages (49). Primary tumors and early HD passages of such transplant lines contain extra MMTV provirus DNA which is lost at the conversion of hormone dependency (50).

This suggests that primary HD tumors of GR consist of different cell clones, each characterized by a difference in number and localization of extra MMTV-DNA fragments in the cellular genome. HD tumors of GR thus contain hormone dependent as well as hormone independent cells, the latter persisting upon multiple transplantation. The various histological patterns seen in HD tumors and the heterogeneous distribution of estrogen receptors in HD GR mammary tumors (51, 52) are in line with the heterogeneous composition of these tumors.

It is of interest to know whether hormone dependency of GR mammary tumors correlates with integration of MMTV provirus in one of the *int* regions mentioned before. However, using restriction enzyme digestions which revealed integration of exogenous MMTV in *int-1* in 16 out of 26 C3H mammary tumors (38), only two out of 25 GR mammary tumors contained an integration of MMTV-provirus in *int-1*. Similarly, using restriction enzyme digestions which detected integration of MMTV-DNA in *int-2* in 17 out of 40 BR6 mammary tumors (39), only one integration in *int-2* was observed in 24 GR mammary tumors (Michalides, unpublished data). One of the integrations in *int-1* was observed in a HD transplant line: as seen in Fig. 4, a fragment of 8.6 Kb contains *int-1* information in an EcoRI digest of GR control liver DNA, using an EcoRI-BamHI *int-1* probe. Transplant tumor passage 11 and later passages of this line show an extra fragment at 14 Kb hybridizing with the *int-1* probe. This coincides with the appearance of an extra MMTV-DNA containing fragment of the same size in transplant 11 and in later

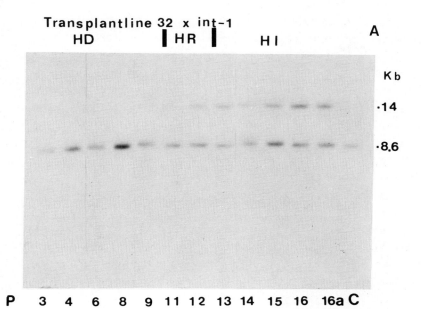
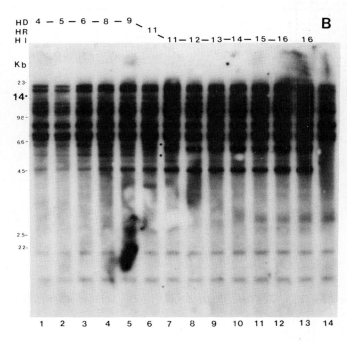

Fig. 4. Insertion of MMTV-DNA in *int-1* domain in a transplant line of GR mammary tumor 32. 20 µg of DNA from mammary tumors of different passages of transplant line 32 were digested with EcoRI and probed for the presence of *int-1 (A)* and MMTV-DNA (B) containing fragments. The numbers below correspond with the passage number. C= control DNA, liver DNA of GR.

passages. The extra *int-1* containing fragment is most likely due to insertion of this MMTV-DNA into the *int-1* region, thereby changing its characteristic EcoRI recognition sites. The emergence of an integration in *int-1* from passage 11 on manifests the prevalence of a particular subclonal population of the tumor mass, which could not be detected in earlier passages and/or was suddenly selected for. Integration of MMTV in *int-1* is not required as an early tumorigenic event in this tumor line, since early passages of this transplant line lack integration of MMTV-DNA in *int-1*. Integration in *int-1* or *int-2* is therefore not likely to be required for mammary tumor development in GR, but may provide one of the multiple steps in the progression of the tumors. Other DNA regions may be involved in mammary tumorigenesis in GR, which are currently being searched for.

The tumor line presented in Fig. 4 shows a conversion of hormone dependency at tumor passage 11 and coinciding loss of two extra MMTV-DNA fragments marked by an asterisk. HD cells withing the original tumor are characterized by these two extra MMTV-DNA bands, other extra MMTV-DNA bands which become more intense

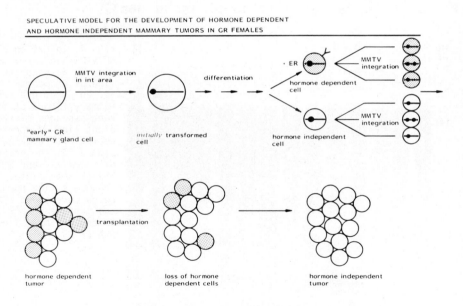

Fig. 5. Speculative model for progression in GR mammary tumors; for description: see text.

in HI passages, appear characteristics of the HI cells in the original tumor. These extra MMTV-DNA fragments serve only as markers for either HD or autonomous cells in this transplant line and are not necessarily causally linked to hormone responsiveness or to the process of tumor induction.

Progression of mammary tumor development in GR may well be depicted in the speculative model presented in Fig. 5. An early mammary gland cell of GR is hit by an integration of MMTV-DNA in a relevant *int* region, which leads to an initially transformed mammary gland cell. This initially transformed cell still retains the capability to differentiate into hormone dependent and hormone independent cells. The final tumor mass then contains hormone dependent as well as independent cells. Hormone dependent cells behave less stable upon transplantation.

GR mice develop hormone dependent mammary tumors, which hardly appear in other strains of mice. The causation of this feature of GR mice may be its characteristic MMTV provirus associated with the Mtv-2 locus. This MMTV provirus is expressed in large amounts early in the life of the animal, whereas the mammary glands of mice infected with milk-borne MMTV become gradually infected with MMTV. The early and massive expression of MMTV in GR could allow transformation of mammary gland cells at early stages of development, at which the initially transformed cells are still capable of differentiation. The speculative parts of this model, i.e. the identification of initially transformed mammary gland cells, and of *int* regions specific for GR mammary tumors, require further study.

These studies can answer the question whether different types of mammary tumors originate by transformation of mammary gland cells in the various stages of mammary gland differentiation, or whether the integration of extra proviruses in different crucial regions of the cellular genome results in different tumor types. The insertional mutagenesis model for MMTV induced mammary tumor formation provides new insights in mammary tumorigenesis. It shows that MMTV functions as a trigger for tumor formation and that the essential genes for mammary tumor development are part of the cellular genome.

ACKNOWLEDGEMENTS

I thank R. Nusse and M. Sluyser for providing materials for the experiment described in the last section of this paper, the latter and P. Emmelot for critical reading of the manuscript; N. van Nuland for preparation of the manuscript.

REFERENCES
1. Duesberg, P.H. (1983) Nature 304, 210.
2. Bishop, J.M. (1982) in: Advances in Cancer Research, Vol. 37, Acad. Press. New York, pp. 1-32.
3. Hilgers, J. and Sluyser, M. (1981) Mammary tumors in the mouse. Elsevier/North Holland. Amsterdam.
4. Cohen, J.C. and Varmus, H.E. (1979) Nature 278, 418.
5. Ringold, G.M. (1979) Bioch. Bioph. Acta 560, 487.
6. Gilboa, E., Mitra, S.W., Goff, S., Baltimore, D. (1979) Cell 18, 93.
7. Robertson, D.L. and Varmus, H.E. (1981) J. Virol. 40, 673.
8. Dudley, J.P. and Varmus, H.E. (1981) J. Virol. 39, 207.
9. Michalides, R. and Nusse, R. (1981) In: Hilgers, J. and Sluyser, M., eds. Mammary tumors in the mouse. Elsevier/North Holland Amsterdam pp. 465-503.
10. Van Ooyen, A.J.J., Michalides, R. and Nusse, R. (1983) J. Virol. 46, 362.
11. Wheeler, D.A., Butel, J.S., Medina, D., Cardiff, R.D. and Hager, G.L. (1983) J. Virol. 46, 42.
12. Donehower, L.A., Huang, A.L., Hager, G.L. (1981). J. Virol. 37, 226.
13. Kennedy, N., Knedlitschek, G., Groner, B., Hynes, N.E., Herrlich, P., Michalides, R., Van Ooyen, A.J.J. (1982). Nature 295, 622.
14. Dickson, C. and Peters, G. (1981) J. Virol. 37, 26.
15. Dickson, C., Smith, R. and Peters, G. (1981) Nature 291, 511.
16. Klemenz, R., Reinhardt, M., Diggelman, H. (1981) Mol. Biol. Rep. 7, 123.
17. Ringold, G.M. Dobson, D.E., Grove, J.R., Hall, H.V., Lee, F., Vannice, J.L. (1983) In: Greep, R.O., ed. Recent progress in hormone research, vol. 39, Acad. Press, New York, pp. 387-421.
18. Groner, B., Kennedy, N., Rahmsdorf, U., Herrlich, P., Van Ooyen, A. and Hynes, N.E. (1982). In: Dumont, J.E. and Nunez, J., eds. Hormones and cell regulation, Vol. 6, pp. 217-228. Elsevier/North Holland, Amsterdam.
19. Hynes, N.E., Kennedy, N., Rahmsdorf, U., Groner, B. (1981) Proc. Natl. Acad. Sci. USA 78, 2038.
20. Diggelman, H., Vessazi, A.L. and Buetti, E. (1982) Virol. 122, 332.
21. Buetti, E. and Diggelman, H. (1981) Cell 23, 335.
22. Owen, P. and Diggelman, H. (1983) J. Virol. 45, 148.
23. Lee, F., Mulligan, R., Berg, P. and Ringold, G., (1981) Nature 294, 228.
24. Huang, A.L. Ostrowski, M.C., Berard, D., Hager, G.L. (1981) Cell 27, 245.
25. Hynes, N.E., Van Ooyen, A.J.J., Kennedy, N., Herrlich, P., Ponta, H. and Groner, B. (1983) Proc. Natl. Acad. Sci. 80, 3637.
26. Buetti, E. and Diggelman, H. (1983). Embo Journ. 2, 1923.
27. Payvar, F., Wrange, O., Carlstedt-Duke, J., Okret, S., Gustafson, J.A. and Yamamoto, K.R. (1981) Proc. Natl. Acad. Sci USA 78, 6628.
28. Govindan, M.V., Spiess, E. and Majors, J. (1982) Proc. Natl. Acad. Sci. 79, 5157.
29. Pfahl, M. (1982) Cell 31, 475.

30. Chandler, V.L., Maler, B.A. and Yamamoto, K.R. (1983) Cell 33, 489.
31. Moreau, P., Hen, R., Wasylyk, B., Everett, R., Gaub, M.P. and Chambon, P. (1981) Nucl. Acids. Res. 9, 6047.
32. Laimins, L.A., Khoury, G., Gorman, C., Howard, B. and Gruss, P. (1982) Proc. Natl. Acad. Sci. USA 79, 6453.
33. Cohen, J.C., Shank, P.R., Morris, V.L., Cardiff, R.D., Varmus, H.E. (1979) Cell 16, 333.
34. Neel, B.G., Hayward, W.s., Robinson, H.L., Fong, J., Astrin, S.M. (1981) Cell 23, 323.
35. Hayward, W.S., Neel, B.G. and Astrin, S.M. (1981) Nature 290, 475.
36. Noori-Daloii, M.,R., Swift, R.A., Kung, H.J., Crittenden, L.B. and Witter, R.L. (1981) Nature 294, 574.
37. Fung, Y.K.T., Lewis, W., Crittenden, L.B. and Kung, H.J. (1983) Cell 33, 357.
38. Nusse, R. and Varmus, H.E. (1982) Cell 31, 99.
39. Peters, G., Brookes, S., Smith, R. and Dickson, C. (1983) Cell 33, 369.
40. Varmus, H.E. (1982) Science 216, 812.
41. Sluyser, M., Van Nie, R. (1974) Cancer Res. 34, 3252.
42. Van Nie, R. and Dux, A. (1971) J. Natl. Cancer Inst. 46, 885.
43. Van Nie, R. and Hilgers, J. (1976) J. Natl. Cancer. Inst. 56, 27.
44. Van Nie, R., (1981) In: Hilgers, J. and Sluyser, M., eds., Mammary tumors in the mouse. Elsevier/North Holland, Amsterdam, pp. 201-266.
45. Bentvelzen, P. (1968) Hollandia, Thesis.
46. Van Nie, R., Verstraeten, A.A. and De Moes, J. (1977) Int. J. Cancer 19,383.
47. Michalides, R. Van Nie, R. Hynes, N.E. and Groner, B. (1981) Cell 23, 165.
48. Van Nie, R., Versteaeten,A.A. (1975) Int. J. Cancer 16, 922.
49. Sluyser, M., Evers, S.G., de Goey, C.C.J. (1976) Nature 263, 386.
50. Michalides, R., Wagenaar, E., Sluyser, M. (1982) Cancer Res. 42, 1154.
51. Sluyser, M., (1981) In: Hilgers, J. and Sluyser, M., eds. Mammary Tumors in the mouse. Elsevier/North Holland, Amsterdam pp. 267-701.
52. Percy, D.H., Morris, V.L. and McInnes, J.(1980). J. Natl. Cancer Inst. 69, 933.

Résumé

Le retrovirus MMTV induit des tumeurs mammaires dans des souches de souris sensibles. Il se réplique via un DNA proviral, qui s'intègre dans le génome des cellules infectées de glande mammaire. L'intégration du DNA de MMTV dans le génome cellulaire se passe dans des sites distribués plus ou moins au hasard. Cependant, des insertions dans deux domaines particuliers et différents sont associées avec une tumorigénèse mammaire dans les souches de souris C3H et BR6. L'insertion du DNA de MMTV en de tels domaines conduit à l'expression d'un gène dans ce domaine, qui est probablement un oncogène cellulaire activé par l'insertion du provirus MMTV. Ces deux domaines d'insertion ne semblent pas impliqués dans l'induction de tumeurs mammaires dans la souche de souris GR. Cette souche de souris développe des grossesses et des tumeurs mammaires hormono-dépendantes et indépendantes. Les tumeurs hormono-dépendantes sont composées d'un mélange de cellules hormono-dépendantes et indépendantes. L'expression du MMTV dans des cellules exprimant le MMTV in vitro, est stimulée par le glucocorticoïde synthétique dexaméthasone. Une petite région définie, d'approximativement 200 nucléotides, située juste en amont du site d'initiation de la transcription est responsable de cet effet de la dexaméthasone. Cette région du génome du MMTV peut agir en tant qu'"amplificateur" (enhancer).

PROLACTIN RECEPTORS AND INTRACELLULAR MEDIATOR FOR PROLACTIN ACTION ON THE MAMMARY CELL

JEAN DJIANE , LOUIS-MARIE HOUDEBINE , PAUL A. KELLY , ISABELLE DUSANTER-FOURT BERTRAND TEYSSOT , MASAO KATOH
Laboratoire de Physiologie de la Lactation, Institut National de la Recherche Agronomique, CNRZ 78350 Jouy-en-Josas, (France)
and Laboratoire d'Endocrinologie Moléculaire, Université McGill, Hôpital Royal Victoria, 687 Avenue des Pins ouest, Montréal, Québec, H3A 1A1, (Canada).

INTRODUCTION

Prolactin is a polypeptide hormone composed of 198 amino-acids associated in an unique chain which contain three disulfide bridges. This hormone, secreted by the anterior pituitary, exists not only in mammals but in all vertebrates, where a total of at least 80 biological functions have been attributed to this hormone (1). In mammals, prolactin is principally implicated in the regulation of the functions of reproduction and lactation, and to a lesser extent in the regulation of water and ions fluxes. The mechanism of action of prolactin, at the cellular level, remains poorly understood. Most of the work has been performed on the mammary gland where prolactin constitutes the main trophic hormone for the induction of cellular differenciation and biosynthetic activity of milk products (proteins, lipids and carbohydrates). In addition, in numerous species, it has been demonstrated that prolactin is essential for the multiplication of the mammary cells. All these actions are focused at the nuclear level, where the transcription of specific genes and DNA synthesis are stimulated (2) as a result of prolactin action. The epithelial mammary cell is one of the few biological system where the expression of specific genes are under the control of a peptide hormone. In this system, prolactin stimulates casein and α-lactalbumin genes transcription, and at the same time increases the half-life of the neosynthetized messenger RNA (3). In addition, prolactin stimulates the translation of milk proteins mRNA (4). All these phenomena converge to increase the expression of milk protein genes in the mammary cell by more than one thousand fold between mid-pregnancy and lactation in the rabbit. These actions of prolactin are modulated by steroid hormones, principally by glucocorticoids, which are potent amplifiers and progesterone which, at least *in vivo*, is a powerful inhibitor (5).

To elicit its actions at the cellular level, prolactin first interacts with well characterized specific receptors, located at the cell periphery (6,7). The mechanisms by which the stimulation is transferred to the nucleus is not clear. None of the classical intracellular relays is applicable to prolactin effects. The aim of the present paper is to review our recent work on the characterization of prolactin receptors at the molecular level and on the identification of an intracellular peptide mediator which can specifically increase the transcription of casein genes in isolated mammary nuclei.

THE PROLACTIN RECEPTOR

1) General characteristics

Prolactin receptors have been identified in many tissues including mammary gland, liver, kidney, adrenals, ovaries, testes, male accessory sexual organs, hypothalamus, choroid plexus. The effects of prolactin in all these organs are not completly known. Most of the work on biochemical characterization and purification has been performed on rabbit mammary gland receptors (8). These receptors are hydrophobic glycoproteins which possess very high affinity for prolactin and related hormones with the same biological activities (human growth hormone, placental lactogenic hormones). The localization of these receptors appears almost exclusively membranous, but we have recently observed that an important part of the receptors become water soluble when the membranes are washed with an hypotonic buffer (P. Berthon et al., unpublished). The receptors are not only located in peripheral membranes (plasma membrane). The concept that peptide hormone receptors are preferentially concentrated on plasmalemma has been progressively attenuated during the last five years, by a number of studies which have demonstrated the binding of labelled hormones to intracellular membranes (9). In the case of prolactin, it has been shown that purified Golgi fractions of rat liver cells are richer in receptors and have a higher affinity for prolactin than plasma membrane fractions (10). In the lactating mammary cell, it appears that about 70% of the prolactin receptors are in fact intracellular receptors. These receptors probably do not participate in the first step of hormonal recognition and their role remains to be established. Prolactin receptors are, at least in the mammary cell, very short lived molecules. When mammary gland explants are cultured in the presence of cycloheximide, 50% of prolactin receptors disappeared from membranes in less than 3 hours, whereas in tissue cultured in the absence of the inhibitor, the level of prolactin receptors remained unchanged up to 24-48 hours. By contrast, the blocage of transcriptionnal activity by actinomycin D, only slightly modified the level of prolactin receptors. This last finding, suggests that the messenger RNA for the receptor protein could have a long half-life, and that rapid modulations in prolactin receptor levels probably occurs at translational or degradative steps. The degradation is effective mainly in lysosomes since the inactivation of these organelles by amines such as chloroquine, ammonium chloride or methylamine, result in a rapid and very important increase in prolactin receptor levels (see Fig.1 and Ref.11).

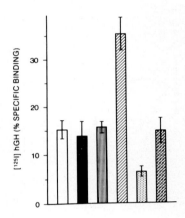

Fig. 1. Effect of insulin (5 µg/ml) alone or in combinaison with chloroquine (100 µM), cycloheximide (1 µg/ml) or a combinaison of both, on prolactin receptor numbers in rabbit mammary gland organ culture.
☐ before culture ; ■ no hormone ; ▥ insulin ; ▨ chloroquine ; ▦ cycloheximide ; ▨ chloroquine + cycloheximide.

Prolactin receptors are subjected to precise regulatory events in mammary cells, where prolactin itself plays the major role. In fact, prolactin can both "up"- and "down"-regulate its own receptors by quite different mechanisms. The down regulation is a rapidly established and easy reversible phenomenon with can occur after an acute stimulation (12). It probably reflects an increase in the rate of the internalization and degradation in lysosomes of the receptors. By contrast, the "up"-regulation of prolactin receptors requires days to be established and remains stable (13,14). In the case of the developing mammary gland, this effect take place at the same time as the increase of intracellular membranous network during cellular differentiation.

2) Relation between prolactin receptor occupancy and the hormonal responses

The interaction of prolactin with its receptor on mammary cells induces a cascade of events which finally results in an increase in milk protein synthesis (caseins) and cell multiplication. When mammary explants are cultured in the presence of varying doses of prolactin, a progressive increase in caseins synthesis and DNA synthesis can be seen and at the same time, a progressive decrease in prolactin receptors (Fig.2) and Ref. 15).

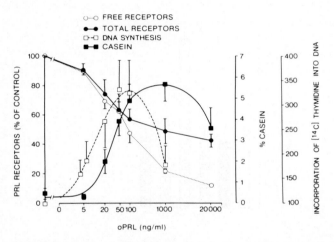

Fig. 2. Effects of increasing concentrations of oPRL on free and total PRL receptors (measured after *in vitro* desaturation with 4M $MgCl_2$), and on caseins and DNA synthesis in explant cultures of rabbit mammary gland. Explants were cultured in the presence of insulin (1 µg/ml) and cortisol (500 ng/ml) for 24 hours.

The maximal effect for casein synthesis as well as for DNA synthesis occurs at about 100 ng oPRL/ml, at high concentration of prolactin a desensitization of the mammary cell to the hormonal stimulus is observed. This phenomenon is more pronounced for DNA synthesis than for casein synthesis. It should be noted that this desensitization is not accompanied by a further decrease in prolactin receptor content. Also, experiments described previously (16,17) demonstrate clearly that this down-regulation of prolactin receptors can be blocked by lysosomotropic agents, which interfere with the degradation of the receptors in lysosomes, without inducing any modification in the intensity of prolactin effects on caseins synthesis. This suggests that the degradation of the receptors does not constitute a crucial step in the transfer of the hormonal message.

3) Molecular characteristics of prolactin receptors

Some characteristics of prolactin receptors have been recently determined by purification of these receptors and labelling the binding unit by photo-affinity cross-linking reagents. The purification procedure summarized in Table 1 involves solubilization of mammary microsomes with the zwiterionic detergent CHAPS (7.5 mM after a pretreatment of the microsomes with CHAPS at 1 mM). The most efficient step of this purification involves an affinity chromatography using oPRL coupled to Affigel-10 according to the method of Shiu and Friesen (8). Only 0.014% of the total protein was recovered after elution with 4M $MgCl_2$ and an average of 650-fold purification from microsomes was achieved resulting in a recovery of 8.9% of the total binding capacity.

	Protein recovery (%)	PRL binding capacity (pmole/mg)	Purification (fold)	Binding sites recovered (%)	Ka (nM)
Crude microsome	100	0.37±0.15	(1)	100	3.8±1.7
CHAPS extract	22.6±3.4	0.72±0.33	(1.87)	43.0±9.5	27.6±9.5
Affinity-purified	0.014±0.002	258±126	(655)	8.9±1.7	16.2±7.2

TABLE I

Summary of purification of PRL receptor from rabbit mammary gland

Crude microsomes from 3 different preparations of lactating rabbit mammary gland (680,1600 and 3300 mg protein, respectively) were solubilized and purified. PRL binding capacity and affinity constants were determined by Scatchard analysis. The values represent the mean ± SEM.

Electrophoresis of these partially purified receptors was performed on 5-15% or 9-15% gradient gels under reducing conditions. Gels were stained by Coomassie brillant blue or by silver staining. Affinity labelling receptors using the photoactive cross-linker HSAB (Hyodroxy-N succinimidyl/azobenzoate) was electrophoresed similarly. One faint band of Mr 32,000 was occasionally detected by Coomassie blue suggesting that this protein constitutes the major component in our preparation. However, silver staining which is much more sensitive, detected at least 7 major bands of Mr 30, 32, 39, 46, 51, 64 and 68 KD. In some preparation a larger 114 KD band was observed. When this partially purified material was affinity labelled using photoactivable cross-linker, one major binding component of MW 31,300 + 400 (n = 9) was specifically labelled by ^{125}I oPRL. In addition, minor components of 63 and 80 KD assuming 1:1 cross-linking between hormone and binding component, are also visible under reducing conditions. A single Mr 65 KD was observed under non reducing conditions (Fig.3).The binding unit of 31 KD is also easily detectable by photoaffinity of crude microsomal preparations, not only from rabbit mammary gland but also from mammary gland of others species and from other tissues (ovaries, liver, adrenals). These findings indicate that the major binding component for prolactin in its targets organs is constituted by a rather low molecular weight protein, in comparison to what is known for

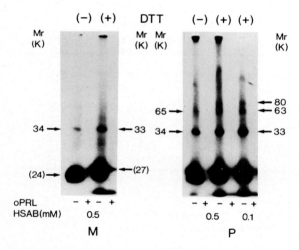

Fig.3. Autoradiogram comparing binding components from microsomal (M) and affinity purified (P) PRL receptors. Purified receptors (1.2 g) characterized in table 1, was labelled by ^{125}I oPRL and photoaffinity cross-linker (HSAB 500 µM) and analysed on a 5-15% gradient acrylamide gel under reducing (DTT) or non reducing conditions. Microsomal receptors were affinity labelled and electrophoresed on the same gel and indicated in the left position of the figure (M).

other receptors of peptide hormones. In addition, the structure of the prolactin receptor does not involve disulfide bridges since, no other binding components were observed, and 32 KD band was not altered, in absence of DTT during electrophoresis. In conclusion, the molecular organization of the prolactin receptor appears completely different from the insulin receptor, which is a macromolecule with several subunits linked together by disulfide bridges. It remains to be established if some of the major 8 proteins which we have purified by affinity chromatography constitute part of the prolactin receptor without having a direct role in the binding activity. One interesting possibility would be, that a proteolysis of a higher molecular weight component take place during the purification procedure or during the affinity labelling, resulting in the appearance of the 32 KD subunit.

4) Antibodies against the prolactin receptor

The partially purified prolactin receptor preparation was injected into guinea pigs, sheep and goats, with the goal of obtaining antibodies against the receptor. Only 50 µg of the antigen injected three time at monthly intervals were sufficient to induce, in goats and sheep, the production of very active antibodies against the receptor. These antibodies have been tested for their capacity to inhibit the binding of prolactin to its receptor and also to immunoprecipitate the hormone receptor complex. As shown in Fig. 4a, sera from sheep, goat and guinea pig were capable of inhibiting the binding of prolactin to its receptor. Significant inhibition was observed at a dilution of 1/10.000 with the sheep antibodies being the most potent. For all the antisera, complete inhibition was obtained at a dilution of 1/100.

The specificity of these antisera has been tested and the results demonstrate that these antibodies are able to inhibit the binding of prolactin on its receptor in all its target tissues (liver, adrenals, ovaries, prostate, kidney, DMBA mammary tumor). Also, only prolactin binding is inhibited and not binding of other hormones to other receptors. The inhibition of binding only estimates the presence of antibodies directed against the part of the receptor molecule which is responsible for the association with the hormone. Results described in Fig. 4b demonstrate that the antibodies can also precipitate the receptor even if it is occupied by the hormone. In other words, there exists also in our polyclonal antisera, antibodies directed against other parts of the prolactin receptor than the binding domain. Interestingly the potency of the various antibodies is quite different in both tests. The sheep antiserum had a lower potency than goat antiserum for the precipitation of occupied receptors whereas the sheep antiserum was the most potent to inhibit binding. This fact could provide some explanation to the finding that sheep antibodies possess a more pronounced agonistic effect of prolactin on the mammary cell (see below).

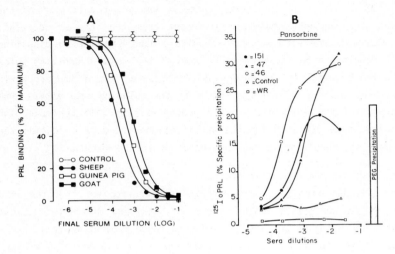

Fig. 4. Panel a : Action of anti-prolactin receptor sera on the binding of prolactin, to mammary membranes. About 100.000 cpm ^{125}I oPRL were incubated for 16 hours at 20°C with rabbit mammary membranes (300 µg protein) in the presence of various concentrations of control serum or anti-prolactin receptor sera from sheep (151) guinea pig or goat.
Panel b : Immunoprecipitation of prolactin receptor complexes by antisera against the prolactin receptor. Rabbit mammary gland membranes were solubilized by Triton X 100 (1%) and diluted. 25 µg of soluble membrane proteins were incubated with ^{125}IoPRL (100.000 cpm) during 16 hours at 20°C. After which, antisera against the prolactin receptor were added at various dilutions. The anti-receptor complexes were precipitated with protein A (Pansorbin). No precipitation of ^{125}IoPRL, occurs when receptors were omitted (WR). All antisera are more efficient than polyethylene glycol (PEG) to precipitate the hormone receptor complexes.

5) Antagonistic and agonistic effects of prolactin receptor antibodies

As expected and as it has been shown previously (18), antibodies against the prolactin receptor can block the effects of prolactin on mammary cells. As a result of the inhibition of binding of prolactin, they prevent casein synthesis, DNA synthesis as well as casein mRNA accumulation (24). However, inhibitory effects were always observed at high immunoglobulin concentrations. At lower concentrations, the anti-prolactin receptor antibodies added to cultured medium of rabbit mammary explants (19) or injected into pseudopregnant rabbit (20) mimicked prolactin action. Interestingly, these antibodies also exhibit an agonistic effect of prolactin for the stimulation of

DNA synthesis. These observations provide strong support for the idea that the prolactin molecule is not required beyond its initial binding to elicit its actions on mammary cell nucleus and that an intracellular relay is generated at the membrane level when the receptor, which by itself is sufficient to transfer the stimulation, is occupied by specific ligands such as prolactin or antireceptor antibodies. These findings also, appear to exclude a crucial role of the direct interaction of prolactin or fragments of this molecule on mammary cell-nucleus to stimulate casein gene transcription.

We have recently analysed the role of the bivalency of the antibodies against prolactin receptor for their capacity to inhibit the binding of prolactin and to mimic prolactin actions on mammary cells. Bivalent fragments (F ab'$_2$) of anti-prolactin receptor antibodies were prepared by pepsin cleavage, whereas the monovalent fragments (Fab') were obtained after a reduction with dithiothreitol of the F(ab'$_2$). When assayed for their ability to inhibit ^{125}IoPRL binding to its receptor, whole anti-receptor serum, bivalent immunoglobulins (F(ab'$_2$)) and monovalent F(ab') have similar potency, half maximal inhibition being reached with 5-10 μg of immunoglobulin/ml. When added to culture medium of mammary explants total immunoglobulins as well as their F(ab'$_2$) and F(ab') fragments, all induce, as prolactin does, a 50% reduction of prolactin receptors. This seems to indicate that the internalization and

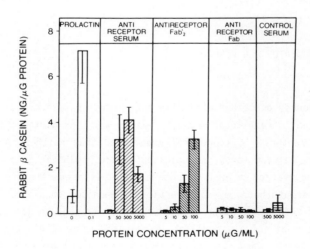

Fig. 5. Effect of prolactin, whole receptor serum and anti-receptor F(ab'$_2$) and Fab' fragments on rabbit casein content (measured by radioimmunoassay) in mammary explants cultured for 24 hours in the presence of insulin (1 μg/ml) and cortisol (500 ng/ml). Results are mean ± SEM of 4 independent cultures.

degradation of the receptors following their occupancy also occur with anti-prolactin antibodies and their di or monovalent fragments. Using a similar culture procedure, the effects of anti-prolactin receptor antibody fragments, on casein and DNA synthesis, were examined. As it can be seen in Fig. 5, whole anti-receptor serum was about 50-60% as effective as optimal doses of prolactin anti-receptor. F(ab') are completely devoid of activity. Also, only bivalent fragments were able to stimulate DNA synthesis, monovalent fragments being completely ineffective (data not shown).

All these data point to the crucial role, in the liberation of a putative intracellular relay of prolactin, of the interaction of at least 2 receptor molecules. This property is reminiscent of what has been shown for insulin receptors (21). In this way, the movement of the receptor molecules within the phospholipid bilayer could constitute an important regulatory parameter in prolactin action. It remains to be established if this association of receptors is also necessary for prolactin, for which at present, there is no evidence of a bivalency. One additional interesting point is that monovalent fragments are able to down-regulate the receptors with the same potency as complete immunoglobulins or prolactin itself. If the down-regulation results from internalization and degradation of the receptors (22), the loss of the biological activity of the monovalent fragments suggest that internalization and degradation are not directly related to the transfer of the hormonal stimulation inside the cell.

AN INTRACELLULAR MEDIATOR FOR PROLACTIN

1) Identification

The work previously described utilizing anti-receptor antibodies, brings the evidence that an intracellular mediator exists which is distinct from the hormone and the receptor. It is generally admitted that neither cAMP, nor cGMP, polyamines, calcium ions, prostaglandins are prolactin intracellular mediator for the activation of gene transcription. It seems therefore that, as for insulin or growth hormone, non classical mechanisms should be considered to account for the transmission of the prolactin message to genes.

The hypothesis that soluble mediator is released after prolactin binding at the membranous level, and that this relay molecule can directly activate casein gene transcription after having been transferred to the nucleus, was studied in the following manner : mammary microsomes were incubated with or without prolactin for one hour at room temperature and the supernatants were saved after pelleting the membranes. This soluble fraction was incubated with mammary nuclei, isolated from rabbits treated for 4 days with the hypoprolactinemic drug, bromocriptine, a treatment which has been demonstrated to provoke a partial deinduction of the transcriptional activity in mammary cell nuclei. These isolated nuclei were incubated in presence of the various nucleotide triphosphates, one of which being mercurated (HgCTP or HgUTP). The neosynthetized mRNA were then selected by an affinity column using SH-sepharose and their quantification was performed after an hybridization with specific cDNA probes. In some experiments, the nuclei were incubated with a ^{32}P nucleotide (^{32}P CTP) and the affinity column was omitted. The hybridization was then performed with cloned cDNAs corresponding to different casein genes inserted into plasmids which were fixed to nitrocellulose filters. The results are summarized in Fig. 6 and referred to a control value of transcriptional activity obtained with nuclei incubated with supernatants from membranes not exposed to hormone. When lactogenic hormones (PRL, hGH, oPL) were incubated with microsomal membranes containing prolactin receptors (mammary gland, ovaries, adrenals) a soluble mediator was released from the membranes which stimulated the transcription of specific genes 3 to 7-fold

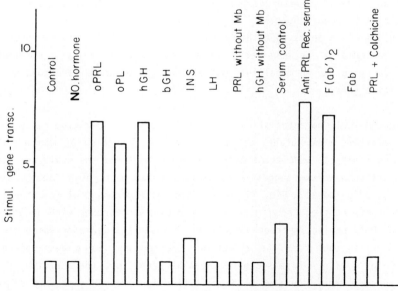

Fig. 6. Effects of various hormones (lactogenic or not), antibodies against prolactin receptor, di and monovalent fragments of these antibodies and of colchicine in combinaison with prolactin, in the generation by mammary gland membranes of the factor stimulating caseins genes transcription in isolated mammary cell nuclei. Results refer to a value 1 obtained with the mammary nuclei of lactating rabbit deinduced in vivo by bromocryptine and incubated in the absence of membrane supernatant.

(αS_1, β caseins, αlactalbumin genes). This mediator was without any effect on rRNA genes or total transcriptional activity (23). Non lactogenic hormones were devoid of activity with the exception of insulin which was slightly active in this respect.

Lactogenic hormones were completely inactive if incubated directly with the nuclei, and finally only membranes which contain prolactin receptors were able to release such a factor in response to lactogenic hormones. Antibodies against the prolactin receptor induced an equivalent response to lactogenic hormones for the release of this mediator, if utilized at moderate concentrations and with a bivalent structure. Fab fragments, which proved to be devoid of activity on caseins and DNA synthesis, although they bind to the receptor (see preceeding section) were also completely inactive in the liberation of the mediator.

It is also interesting that colchicine and other tubulin binding drugs are able to block the liberation of this mediator from membrane, a fact which is in good agreement with what it has been demonstrated in whole cells where these drugs are potent inhibitors of prolactin action (17-24). The additional information is that these drugs act at the membrane level by inhibiting the liberation of the mediator (25).

Therefore, it appears that the mediator released from membranes possesses all the criteria to be considered as an intracellular relay of prolactin action, carrying the hormonal information from the membrane receptors to the PRL sensitive genes.

2) Physicochemical characteristics of the prolactin intracellular relay

When membrane supernatants obtained after an incubation of mammary membranes with prolactin were fractionated on Sephadex G-15, the mediator eluted far from the void volume suggesting that this molecule is of a low molecular weight (< 1000 daltons). Fig. 7a shows a typical experiment of fractionation of supernatant fractions from incubation of mammary membranes with or without prolactin. Only prolactin treated membrane released the mediator which eluted in Fraction III and IV. The effect of the mediator is a dose-dependent process (Fig. 7b). A biphasic effect can be seen, but the activity of the supernatant membrane incubated with prolactin was always higher than control incubation. It is well known that some compounds, and particularly glycopeptides, could have a paradoxical migration on Sephadex gel. We tried to confirm the molecular weight of the mediator using other filtration systems. When an acrylamide-agarose gel was used (Trisacryl GF 05 (IBF)), the mediator was excluded in the void volume of the column, indicating that the value given by sephadex gel might be underestimated. Thus suggests that the mediator might have a molecular weight higher than 3000 d.

Another important characteristic of this mediator is its sensitivity to proteases. When incubated with trypsin (1 µg/ml), the activity of the mediator was completely lost, suggesting that it has a peptidic nature. The mediator appears heat stable since it is resistent to boiling water treatment for 10 min.

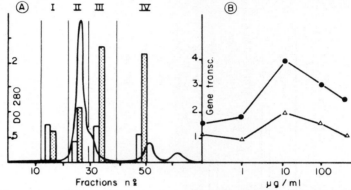

Fig. 7. Panel a. Fractionation of the intracellular mediator of prolactin on Sephadex G-15. Crude microsomes of rabbit mammary gland were incubated (7 mg/ml) with or without prolactin (1 µg/ml) during 1 hour at 20°C. After centrifugation (100.000 g, 1h) the supernatants were fractionated and results in I, II, III, IV fractions. The activity on casein gene transcription of these different fractions was measured with isolated nuclei. ▨ activity with supernatant obtained by incubation of membrane with prolactin ; ▢ incubation without prolactin .
Panel b. Fractions III of the two types of incubation (with prolactin ● without prolactin △) tested at different concentrations for their capacity to stimulate casein gene transcription.

3) Mechanism of action of prolactin mediator on nuclei

It was important to determine if the mediator which has been identified is an activator of transcription at the initiation step, and not simply a factor which may facilitate the elongation of previously (in vivo) initiated RNA chains. For that purpose, we analysed the effect of the mediator on neosynthetized RNA labelled with γ-S ATP, γ-S GTP and ^{32}P CTP. The resulting RNA were isolated using a Hg-cellulose column. Only RNA chains initiated *in vitro* were retained through their thio-phosphate 5'end. Hybridization of the ^{32}P labelled RNA retained and those not retained by the Hg-cellulose column were performed with plasmids containing cloned cDNA of caseins genes bound to nitrocellulose filters. The results of Fig. 8 indicate that the transcription of caseins genes was stimulated by the mediator essentially at the initiation step, since the effect of the mediator was observed only with the RNA fraction labelled by the γ-thionucleotides. Another demonstration relies on the utilization of specific inhibitors of initiation. Compounds such as AF013, heparin, sarkosyl and s-adenosyl homocysteine were added to nuclei with the mediator. In the presence of these inhibitors, the effect of the mediator on the initiation of caseins mRNA synthesis was completely abolished (Fig.8).

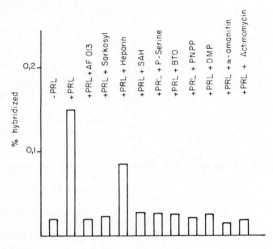

Fig. 8. Effects of various inhibitors on the initiation of the transcription by the prolactin mediator. Initiation inhibitors were (AF013 50 µg/ml ; sarkosyl 0.5% : heparin 50 µg/ml ; S-adenosylbromocysteine 20 mM). Phosphatase inhibitors were (P-serine 10 mM L-bromotetramizole oxalate 10 mM : P-nitrophenyl phosphate 5 mM ; 2,3-dimercaptopropan 1-ol 10 mM). Inhibitors of transcription were antinomycin D 50 µg/ml and α-amanitin 10 µg/ml.

The mechanism through which the initiation of the transcription of genes could be stimulated is an important question, for which few data are available in other biological systems. It is generally admitted that a high transcriptional activity is associated with an hyperphosphorylation of nuclear proteins. We have postulated that the prolactin mediator is acting through a modification of the phosphorylation state of nuclear proteins. For that purpose we have tested in our system, the effects of inhibitor of protein phosphatase. As shown in Fig. 8, several of these compounds prevented the action of the mediator without interfering significantly with the initiation process of the transcription of whole genes. Interestingly, inhibitors of protein phosphatase acting on phosphotyrosine (Zn ion, sodium vanadate) were totally inactive (not shown). These results suggest that the mediator acts on the nucleus probably via an activation of protein phosphatase and that the protein which are dephosphorylated probably contained phosphoserine or phosphothreonine rather than phosphotyrosine.

The properties of the mediator released by insulin as described by several groups (26-27) share many similarities with the prolactin mediator (low molecular weight, trypsin sensitivity...). It has been clearly demonstrated that this mediator acts via a dephosphorylation of regulatory enzymes such as pyruvate dehydrogenase or glycogen synthetase (28). The experiments describs above, on dephosphorylation of nuclear proteins induced by the prolactin mediator extend these similarities to an additional parameter. It is tempting therefore to speculate that a common mechanism could be involved in the transfer of the stimulation of insulin or prolactin in their target cells.

CONCLUSIONS

The Fig. 9 gives a schematical representation of the proposed mechanisms which could be responsible for prolactin recognition at the mammary cell surface and those by which the transfer of the stimulation at the nuclear level is effective.

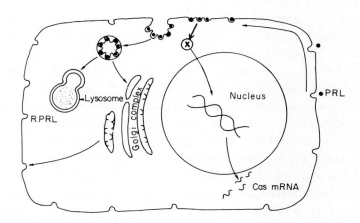

Fig. 9. Schematic representation of the mechanism transferring the prolactin information to milk protein genes.

Prolactin binds to specific receptors at the cell surface. These receptors possess at least one binding unit of low molecular weight (32 KD). Following binding, an aggregation of the receptors probably occurs, a phenomenon which seems determinant for the activation of some enzymatical activities which provoke the release from a membranous precursor of an intracellular soluble mediator of peptide nature and of low molecular weight. This mediator is able to stimulate the transcription of milk proteins gene at the initiation step and via a dephosphorylation of some nuclear proteins. It is not known, however, if all the actions of prolactin are mediated by this molecule. Particularly, the internalization of the hormone and the receptors, followed by their degradation may have specific roles but it appears clearly that these phenomena are unrelated to the mechanism of the transfer of stimulation to caseins genes. Among other things it remains also to determine whether the mediator is also involved in the stimulation of DNA synthesis by prolactin.

It is of course of great interest to study the mechanism by which the mediator is generated from membranes. The experiments performed with anti-receptor antibodies fragments suggest that membrane fluidity, which could regulate the movements of some intramembranous regulatory proteins, may be important. Preliminary experiments have revealed that the release of the mediator is triggered by the addition of small amounts of trypsin, suggesting that an activation of a protease could constitute a key step in the liberation from a membranous precursor, of a fragment constituing the active mediator. An interesting possibility could be that this mediator is a fragment of the prolactin receptor or of a protein closely associated to the receptor. Preliminary experiments indicate that antibodies against the receptor which are polyclonal and produced against a partially purified receptor, are able in concentration much higher than those which provoke the liberation of the mediator, to interact with a preformed mediator and prevent its effect on nuclei. These results suggest that the prolactin mediator or its precursor has been co-purified with the receptor and that antibodies against the mediator are present in the antisera obtained against partially purified receptor. It is now evident that progress in the field of the mechanism of prolactin action will be limited by purification and identification of the comparative structure between prolactin receptor and prolactin intracellular mediator.

REFERENCES

1. Nicoll, C.S. (1974) in "Handbook of Physiology Sect 7 Endocrinology" (E. Knobil and W.H. Sawyer, Eds), Vol.4, Part 2, 253.
2. Teyssot, B. and Houdebine, L.M. (1980) Eur. J. Biochem. 110, 263-272.
3. Guyette, W.A., Matusik, R.J. and Rosen, J.M. (1979) Cell, 17, 1013-1023.
4. Teyssot, B. and Houdebine, L.M. (1981) Eur. J. Biochem., 117, 563-569.
5. Teyssot, B. and Houdebine, L.M. (1981) Eur. J. Biochem., 114, 597-608.
6. Shiu, R.P.C., Kelly, P.A. and Friesen, H.G. (1973) Science, 180, 968-971.
7. Djiane, J., Durand, P. and Kelly, P.A. (1977) Endocrinology, 100, 1348-1356.
8. Shiu, R.P.C. and Friesen, H.G. (1974) J. Biol. Chem., 249, 7902-7911.
9. Posner, B.J., Khan, M.N. and Bergeron, J.J.M. (1982) Endocytosis of peptide hormones and other ligands, in "Endocrine reviews", Vol.3, 280.
10. Kelly, P.A., Djiane, J. and Leblanc, G. (1983) Proc. Soc. Exp. Biol. Med., 172, 219-224.
11. Djiane, J., Delouis, C. and Kelly, P.A. (1982) Mol. Cell. Endocr. 25, 163-170.
12. Djiane, J., Clauser, H. and Kelly, P.A. (1979) Biochem. Biophys. Res. Comm., 90, 1371-1378.
13. Djiane, J. and Durand, P. (1977) Nature,,266,641-643.
14. Posner, B.J., Kelly, P.A. and Friesen, H.G. (1975) Science, 188, 57-59.
15. Djiane, J., Houdebine, L.M. and Kelly, P.A. (1982) Endocrinology, 110, 791-795.
16. Djiane, J., Kelly, P.A. and Houdebine, L.M. (1980) Mol. Cell Endocr., 18, 87-98.
17. Houdebine, L.M. and Djiane, J. (1980) Mol. Cell Endocr., 17, 1-15.
18. Shiu, R.P.C. and Friesen, H.G. (1976) Science, 192, 259-261.
19. Djiane, J., Houdebine, L.M. and Kelly, P.A. (1981) Proc. Nat. Acad. Sci, USA, 78, 7445-7448.
20. Dusanter-Fourt, I., Djiane, J., Houdebine, L.M. and Kelly, P.A. (1982) Life Sciences, 32, 407-412.
21. Kahn, C.R., Baird, K.L., Jarett, D.B. and Flier, J.S. (1978) Proc. Natl. Acad. Sci., USA, 75, 4209.
22. Hizuka, N., Gorden, P., Lesniac, M.A., Van Obberghen, E., Carpentier, J.L. and Orci, L. (1981) J.B.C., 256, 4591-4597.
23. Teyssot, B., Houdebine, L.M. and Djiane, J. (1981) Proc. Natl. Acad. Sci., 78, 6729-6733.
24. Houdebine, L.M. (1980) Eur. J. Cell Biol., 22, 755-760.
25. Teyssot, B., Djiane, J., Kelly, P.A. and Houdebine, L.M. (1982) Biol. Cell, 43, 81-88.
26. Larner, J., Galasko, G., Chang, K., Depaoliroach, A., Huang, L., Daygy, P. and Kellog, J. (1979) Science, 206, 1408-1411.
27. Seals, J.R. and Jarett, L. (1980) Proc. Natl. Acad. Sci, USA, 77, 77-81.

RESUME

La prolactine est reconnue au niveau des cellules cibles par des récepteurs spécifiques. La purification de ces récepteurs a été éffectuée à part des membranes de glande mammaire solubilisées par un détergent (CHAPS), par chromatographie d'affinité utilisant la prolactine liée à un support solide (Affigel 10). Analysé en électrophorèse en conditions dénaturante et réductrice (SDS, DTT), le produit de purification est composé d'au moins 7 unités protéiques de poids moléculaire en 30 et 68 KD. Après marquage covalent utilisant la prolactine radioactive et un agent pontant photoactivable (HSAB), l'unité protéique majeure directement responsable de la liaison de l'hormone est d'un poids moléculaire de 32 KD.

Des anticorps contre le récepteur ont été obtenus. Ils sont capables d'empêcher la prolactine de se lier à son récepteur et ceci dans tous les organes et chez toutes les espèces animales étudiées. Du point de vue biologique, ces anticorps polyclonaux sont capables soit de mimer, soit d'inhiber l'action de la prolactine selon les concentrations utilisées.
Les propriétés mimétiques de la prolactine ne sont observées qu'avec les immunoglobulines natives ou les fragments divalents (F ab'$_2$). Les fragments monovalents (Fab), bien que capables d'interagir avec le récepteur, ont perdu la capacité d'induire le transfert de l'information hormonale à l'intérieur de la cellule mammaire.

L'incubation de membranes possédant le récepteur de la prolactine, avec des hormones lactogènes (PRL, hGH, hormones lactogènes placentaires) provoque la libération d'un médiateur soluble capable de stimuler *in vitro* la transcription des gènes des caséines dans des noyaux isolés de cellules mammaires. Ce médiateur possède toutes les spécificités requises pour être considéré comme un relais intracellulaire de la prolactine. Les anticorps anti-récepteur, uniquement s'ils ont conservé une structure bivalente, sont capables de provoquer la libération de ce relais à partir des membranes. Ce médiateur de nature peptidique (sensibilité à la trypsine) thermostable, est d'un poids moléculaire voisin de 3000 d. La transcription des gènes des caséines est stimulée par ce médiateur, spécifiquement au niveau de l'initiation de nouvelles chaines de mRNA. Les inhibiteurs de phosphatases empêchent le médiateur d'activer la transcription des gènes des caséines, suggérant que ce phénomène est induit par une déphosphorylation de certaines protéines nucléaires et que le médiateur agit probablement en stimulant une phosphatase.

L'ensemble de ces résultats suggère que l'interaction de la prolactine avec son récepteur périphérique au niveau des cellules cibles est suivie d'une agrégation des récepteurs qui provoque la libération d'un relais intracellulaire responsable de l'activation de la transcription des gènes des protéines du lait. Ce relais pourrait être le produit d'une protéolyse limitée du récepteur ou d'une protéine en relation directe avec le récepteur.

TELOMERIC DNA REARRANGEMENTS AND ANTIGENIC VARIATION IN AFRICAN TRYPANOSOMES

ETIENNE PAYS AND MAURICE STEINERT
Département de Biologie Moléculaire, Université Libre de Bruxelles,
67, rue des Chevaux, B1640 Rhode Saint Genèse (Belgium)

INTRODUCTION

Antigenic variation in trypanosomes has probably been first described by Franke in 1905 (1). Yet it still remains a major challenge to health of man and its livestock in tropical Africa, since it has precluded any attempt of vaccination against the parasites. African trypanosomiases are characterized by the occurence of successive waves of parasitemia (2), the trypanosomes of each new wave showing altered antigenic specificity and therefore escaping the immune defense of their host (3-8). Since the first observations by Ritz (9) on cloned trypanosomes, it has been established that such cloned populations may express repertoires of more than a hundred variable antigen types or "VATs"* (10,11). The serological specificity of each variant is associated with a 12-15 nm thick coat of surface antigens which completely covers the plasma membrane of the parasite (12). Although these variant-specific antigens (VSAs)* are glycoproteins (13), the carbohydrate moiety does not seem to be involved in the determination of the antigenic specificity. In infections following either syringe passage or cyclical transmission through the tsetse fly, the different VATs tend to appear in a semi-ordered sequence, the same so-called predominant VATs always appearing in the first parasitemia waves (11, 14-17). This recurrence of the same VATs after transmission has been the first hint that antigenic variation in trypanosomes could not be simply explained by gene mutation. As demonstrated more recently by the isolation of specific VSA mRNAs (18-21), the expression of a variable antigen repertoire is controlled at the transcriptional level by the switching off and on of a collection of genes, each one of these coding for a VSA, and it is generally admitted that, at one and the same time, only one gene of this collection is activated.

This paper reviews the main lines of our present knowledge on the different mechanisms developed by these parasites to selectively activate a single VSA gene among a repertoire, and to allow for a rapid evolution of this repertoire. Data about the transcription of VSA genes and their genomic structure are also

*nomenclature of the variant-specific antigens, used throughout the text, follows the recommendations by Lumsden (24): VAT=variable antigen type; VSA= variant-specific antigen; AnTAR=Antwerp Trypanozoon antigen repertoire; AnTat=Antwerp Trypanozoon antigen type.

presented. For other recent reviews on the molecular basis of antigenic variation in trypanosomes, see also (22, 23).

ACTIVATION OF VSA GENES

A. Duplicative transposition

The expression-linked copy of the gene (ELC) is the expressed one. The activation of several Trypanosoma brucei brucei VSA genes is linked to the synthesis of an additional gene copy, called ELC for "expression-linked copy", which is transposed in a new genetic surrounding (26, 27). This observation has been extended to VSA genes from other trypanosome species and sub-species: T. b. gambiense (28-30), T. b. rhodesiense (29, 31) and T. equiperdum (32). Since in many cases the presence of an ELC was found strictly associated with expression of the gene, and since the control of VSA gene expression is transcriptional (20, 27), it has been inferred that the ELC could be the template used for VSA mRNA synthesis.

The first evidence in support of this hypothesis came from the analysis of chromatin structure around the ELC and its template, or "basic copy" (BC). It is well documented that the chromatin associated with actively transcribed DNA is in such a configuration that it allows for a rapid DNA digestion by low amounts of DNAaseI, whereas the DNA in "inactive" chromatin appears better protected (33). When applied to trypanosome nuclei from the clone expressing the antigen AnTat 1.1, this DNAaseI digestion assay clearly revealed that only the ELC was in an "active" chromatin configuration (34). The hypersensitivity to DNAaseI of actively transcribed sequences in mammalian cells has been attributed to the presence of two chromatin non-histone proteins, HMG 14 and 17, that can be removed by nuclei extraction with 0.35 M NaCl (35). This observation has been confirmed in AnTat 1.1 trypanosomes, as shown in fig. 1. When digested by PstI, the AnTat 1.1 ELC is almost entirely contained in a 2 kb fragment (arrowhead, fig. 1), which is digested by DNAaseI faster than the other AnTat 1.1-specific fragments; however, the hypersensitivity to DNAaseI of the 2 kb PstI fragment is lost after extraction of AnTat 1.1 nuclei by 0.35 M NaCl (fig. 1).

Other indirect evidences, based on the analysis of the VSA cDNA (DNA complementary to the mRNA), also suggested that the ELC is the transcribed VSA gene. Indeed, in variants 117 and 118 from T. b. brucei, only the restriction map of the ELC could match that of the cDNA, whereas the sequence of the 3'-terminal part of the BC was found to differ from that of the cDNA (36); similarly, the restriction map of the AnTat 1.1B cDNA was only found identical with that of the AnTat 1.1B ELC, among all AnTat 1.1-specific fragments of the AnTat 1.1B

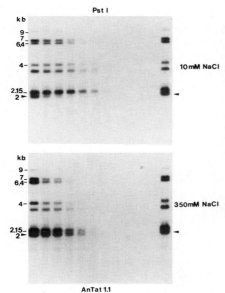

Fig. 1. The AnTat 1.1 ELC is in active chromatin. AnTat 1.1 nuclei were isolated, washed by either 10 mM NaCl (top) or 350 mM NaCl (bottom), then digested for different periods by DNAaseI, as described (34). The DNA was extracted from these nuclei, then digested by PstI, electrophoresed through 0.85% agarose gels, blotted onto nitrocellulose (76) and hybridized with a (32P)labelled AnTat 1.1 cDNA probe (20, 77). The ELC is almost entirely comprised within a 2 kb PstI fragment (arrowhead), which is highly sensitive to DNAaseI provided that HMG proteins are not removed from chromatin, as it is the case after 350 mM NaCl extraction. High sensitivity to DNAaseI is correlated with gene transcription (33, 35).

DNA (28). Therefore, at least in some cases, the VSA mRNA could only have been synthesized on the ELC sequence. A direct proof of this assumption was obtained by the comparison of extensive DNA sequences from the cloned BC, ELC and cDNA of two genes, AnTat 1.10 and 1.1B (37): in both cases, only the ELC sequence was identical with that of the cDNA.

<u>The VSA gene expression site</u>. The restriction map of the ELC environment differs from that of the BC, as shown in figure 2 for several variants of the same antigen repertoire (AnTAR 1). This observation clearly implies that all the ELCs have been inserted in a new genetic surrounding. Since in this location the ELC is transcribed, it shall be described as the VSA gene "expression site". For all ELCs yet examined (in T. b. brucei: 117 and 118 (38), AnTat 1.1, 1.1B,. 1.3, 1.3B, 1.3C, 1.8, 1.10, 1.13, 1.16, 1.30 and 1.45 (30, 39, 40 and E. Pays et al, unpublished); in T. b. gambiense: LiTat 1.6 (30); in T. equiperdum: BoTat 1.1 (32, 41)), this site appears to be a telomeric DNA region, devoid of most restriction sites ("barren" DNA region). The best evidence for this telomeric location has been provided by De Lange and Borst (42), who showed that the 3' neighbourhood of the ELC can be progressively digested by the exonuclease BAL31. This observation has now been extended to all VSA expression sites described so far (39, 40 and H. Eisen, pers. commun.). Absence of restriction

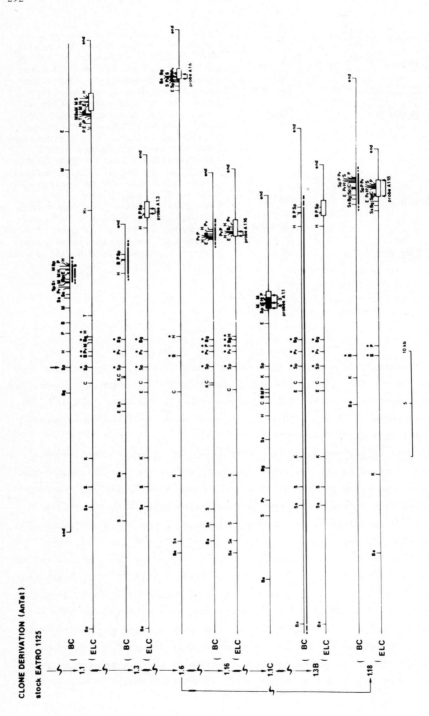

sites is indicative of a peculiar, probably very simple, base sequence. A remarkable feature of this "barren" region is that, as yet, all attempts to clone its DNA have failed (23 and E. Pays et al, unpublished). Indirect evidences suggest that it could contain modified cytidine residues (H. Eisen, pers. commun.). The length of the barren DNA stretch between the ELC and the chromosome end is variable, as illustrated in figure 2 for variants AnTat 1.1, 1.3, 1.16, 1.3B and 1.18; we shall return to this point in the last section of this paper. The restriction map of the 5' environment of the barren region seems identical for several variants (117 and 118 (38); AnTat 1.1, 1.10, 1.1B, 1.3, 1.8, 1.13 and 1.3B (39, 75, and see fig.2); AnTat 1.16, 1.18 and 1.3C (75, and see fig. 2)), suggesting that a chromosome telomere can be used for the expression of different ELCs. This could however not be a general rule (32).

The duplicative transposition is effected by gene conversion. Gene conversion is a DNA recombination leading to the replacement of a sequence by the copy of another one. It has been reported to occur between members of the immunoglobulin gene family (43-48) as well as in the globin gene family (49), in the major histoincompatibility complex sequences (50, 51) and in yeast transposable elements (52). It has been described as a mechanism allowing for both polymorphism and homogeneisation of gene family sequences (48, 53, 54). The yeast mating type interconversion, described as a "cassette" model for sequence replacement (55), is also believed to occur by such a mechanism. Since in several trypanosome variants, different ELCs have been found to be inserted in a seemingly identical genetic environment, it could be hypothesized that at each antigenic switch, they were replacing each other in the same VSA gene expression site, driving out the formerly expressed gene copy. This hypothesis found support in the observation that a sequence in the 3' end of the gene seems to

Fig. 2. Restriction maps of different telomeres carrying sequences involved in VSA expression, in a series of trypanosome clones derived from AnTat 1.1 (⌇⌇⌇=switching of antigenic type; ʃ =trypanosome cloning). The derivation of the clones was obtained by cell cloning after immunological selection from naturally diversifying trypanosome populations; the cloned variants were passaged from mouse to mouse or from mouse to rat at intervals of no more than three days, allowing for a more than 99.9% homogeneity of the antigen type in each clone (10, 25). Isolation and analysis of the genomic DNA from each clone were performed using standard procedures (20, 26). Symbols used for restriction endonuclease sites are: B=BglI; Ba=BamHI; Bg=BglII; C=ClaI; E=EcoRI; H=HindIII; Hi=HinfI; K=KpnI; M=MspI; P=PstI; Pv=PvuII; S=SalI; Sp=SphI; Ss=SstI; T=TaqI. Sites labelled with a dot seem conserved in several maps. The maps have arbitrarily been aligned on a SphI site (arrow). The bars under BC maps represent the known extent of the ELCs, with in most cases length uncertainty at both ends; the ELCa (bar labelled "a") in AnTat 1.1 map is the AnTat 1.1 ELC, whereas ELCb is the transposed copy in AnTat 1.1C (black box in AnTat 1.1C map) (see text). The boxes represent the known extent of the cDNAs.

be shared between different antigen coding sequences (28, 56-58), and could be considered as one of the recognition sequences involved in the recombinational processes of gene copy replacement (30, 36, 38, 59, 60). Similarly, there is indirect evidence that a repeat located 5' to the VSA gene may be required for transposition (31, 61). A proof for this ELC replacement, or gene conversion, hypothesis was provided by the analysis of two variants successively derived from AnTat 1.1, namely AnTat 1.10 and 1.1B (37). The BCs for both the AnTat 1.1 and 1.10 ELCs were found to be members of the same gene family, respectively designated as the "6.4 kb" and "9 kb" sequences, according to the length of the specific PstI fragments each one engenders (see fig. 1). Among the five AnTat 1.1 gene family members, these sequences were the only two located in unstable telomeric regions, similar to the VSA gene expression site; curiously however, they were differently oriented towards the DNA terminus. It was observed that in AnTat 1.10 variant, the "9 kb" ELC has replaced the "6.4 kb" ELC all along the coding sequence. In the next variant (AnTat 1.1B), an incomplete "6.4 kb" ELC has in turn replaced the previous "9 kb" ELC , but only partially. The full AnTat 1.1B message is thus a chimaeric sequence, composed, from 5' to 3', of 876 bp from the "6.4 kb" ELC, 133 bp of unknown origin, then 634 bp from the "9 kb" ELC, conserved in situ from the preceding AnTat 1.10 variant (37). A schematic representation of the AnTat 1.1B "6.4 kb" and "9 kb" sequences, showing the chimaeric AnTat 1.1B gene structure within the expression site, is provided in figure 3. During all the DNA rearrangements leading to AnTat 1.1B expression, the "6.4 kb" and "9 kb" sequences remained unmodified; the partial copy replacements observed here can thus be attributed to segmental gene conversion, and not to unequal crossing over, for instance. Interestingly, a very similar partial gene conversion has been observed in another, independent, AnTat 1.1 expressor clone. Indeed in the AnTat 1.1C variant, a partial copy of the "6.4 kb" sequence has replaced the corresponding stretch from the "9 kb" sequence. In this case however, the target for the gene conversion is the "9 kb" sequence itself, rather than its ELC (fig. 3). From a complete DNA sequence determination of both AnTat 1.1B and 1.1C genes, it could be observed that the junction between the target "9 kb" domain and the incoming block is the same in both cases. In each variant, the converted sequences are the transcribed ones (37, 75).

<u>The extent of gene conversion is highly variable</u>. The extent of the converted region often exceeds the size of the coding sequence. In variants 117 and 118 (38, 61), as well as in AnTat 1.3, 1.8, 1.13, 1.18 and LiTat 1.6 clones (30, 39, 40), a stretch of about 1.5 kb is cotransposed in front of the gene,

to reach an ELC length of about 3 kb. This size corresponds, in the VSA gene, to the distance between two DNA blocks which appear repeated in the genome (30, 38, 60, 61); one of these blocks is present within the 3' end of the VSA gene, and differs from the other one, which is located 1.5 kb upstream from the gene (30, 31, 38, 60, 61), and is made up of repeated A-T rich elements (61). Both repeats seem conserved between different VSA genes, and could thus be used as initiation points for gene conversion between these genes. In several cases however, the extent of the gene conversion leading to antigenic variation is lower than 3 kb, as observed for the AnTat 1.1, 1.10 and 1.1B clones, where the ELC length is only about 2 kb, 1.5 to 2 kb and 1 kb, respectively. In these three variants, a sequence of at least 1 kb, that we called the "companion" sequence, is present in front of the ELC. We found strong evidence that this sequence is a relic from the cotransposed element of a previous ELC (39). In other AnTat 1.1 expressors however, as AnTat 1.1D or 1.1E, this "companion" sequence is absent from the expression site, and in these cases the ELC is of a more usual size, at least 2.5 kb long, up to the 5' repeat (39). It follows that the 5' recombination locus between two ELCs is variable. Such a variability has already been reported for the 3' recombinational point in other clones (59). It most probably depends on the extent of homologies between the target and the incoming sequences.

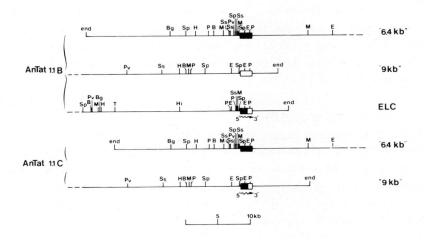

Fig. 3. Restriction maps of the "6.4 kb" and "9 kb" sequences in AnTat 1.1B and AnTat 1.1C DNAs, and comparison with the AnTat 1.1B ELC map. The sequence corresponding to the AnTat 1.1 mRNA is boxed in each map. The specific "6.4 kb" and "9 kb" sequences are represented in black and white, respectively. The wavy arrows represent the transcription of AnTat 1.1B and 1.1C mature mRNAs.

If the two recombinant genes are extensively similar, like in the AnTat 1.1 to 1.10 or AnTat 1.10 to 1.1B switches, the converted stretch can be smaller than the coding sequence, provided that the rearrangement leads to enough antigenic variation to ensure its selection. Some kind of additional constraint seems however to preside to the generation of successful ELCs by gene conversion between closely related sequences, since a virtually identical recombination has been observed between the "6.4 kb" and "9 kb" sequences in two independent AnTat 1.1 expressor clones, AnTat 1.1B and 1.1C (fig. 3). In both cases, the 5' half of the gene is replaced by the corresponding stretch of the other gene family member. As a possible interpretation, the coding sequence would be made up of two halves roughly corresponding to two protein domains, so that only recombinations between these two domains could be selected. Along this line, it is clear that only the 5' half of the gene codes for the surface-exposed epitopes, since the only replacement of the 5' half leads to antigenic variation in the AnTat 1.1B and 1.1C clones.

If, in opposition with the AnTat 1.1, 1.10 and 1.1B cases, the target sequence for the gene conversion is very different from the incoming one, the converted stretch can be very long, as illustrated in the AnTat 1.1C to 1.3B switch (see pedigree in fig. 2). The restriction map of the AnTat 1.1C "active" telomere is indeed very different from the telomere harbouring the AnTat 1.3B gene and from the other known VSA gene expression sites (fig. 2); as a consequence, the ELC of the ensuing variant, namely AnTat 1.3B, is very large, at least 40 kb long (75, and see fig. 2), probably up to a region of homology very distant from the VSA sequence. Such large telomeric conversions could be responsible for the similarities between different telomeres, apparent in the restriction maps in figure 2.

B. Non-duplicative gene activation

A VSA can be synthesized by at least two different gene activation mechanisms. The duplicative transposition is not the only mechanism leading to VSA gene expression. Non-duplicative VSA gene activation, which does not involve the synthesis of an ELC, has indeed also been observed, for different genes of the IlTAR 1 repertoire (19, 62, 63), as well as the MiTAR 1 repertoire (23). In our AnTAR 1 repertoire, two genes, AnTat 1.6 and 1.1C, behave similarly (75 and M. Laurent et al, submitted). This alternative gene activation mechanism may lead to the synthesis of a VAT which can also be expressed following gene conversion, as illustrated in the AnTat 1.1C and 1.1B clones. These two clones are indeed different AnTat 1.1 expressors (AnTat 1.1 homoisotypes), both

synthesizing the same AnTat 1.1 VSA, but by different mechanisms: the AnTat
1.1B gene is activated by duplicative transposition, whereas activation of the
AnTat 1.1C gene does not depend on the transposition of an ELC into the VSA
gene expression site (fig. 3). Despite these differences, activation of both
genes leads to the appearance of trypanosomes harbouring indistinguishable
surface-exposed epitopes (37, 75). On the other hand the same sequence can be
activated by either of the two different mechanisms, and be expressed in
different VATs: this has been observed for the "9 kb" AnTat 1.1 gene family
member, which can, as seen above, either give rise to an ELC (in variant AnTat
1.10) or on the contrary become a target for gene conversion (in AnTat 1.1C).
Similarly, Bernards et al. (74) found that the same 221 antigen gene can be
expressed by either of the two mechanisms.

Relevance to expression in the metacyclic form and to predominance. The
AnTat 1.6 VSA gene is one of the few to be selectively expressed in the tsetse
fly salivary glands (metacyclic antigen types: see ref. 64); moreover, it is
one of the VSA genes to be expressed very early in parasitemia, after either
cyclical or mechanical transmission (predominant antigen types: see ref. 10 and
65). However, the non-duplicative gene activation mechanism used in AnTat 1.6
expression cannot be simply linked, if at all, to these properties, since
variant AnTat 1.1C, whose specific gene is also activated without duplication,
is neither metacyclic nor predominant. Moreover, other predominant VATs, of
either AnTAR 1 (AnTat 1.3: ref. 40) or BoTAR 1 repertoire (BoTat 1.1: ref. 32)
are expressed after duplicative transposition of their respective genes. As yet
the only salient feature that may be related to the non-duplicative mode of
gene activation is the telomeric location of these genes: no example of
non-telomeric gene activation by this mechanism has so far been found (23, 63,
75).

A possible scheme: reciprocal recombination between telomeres. In order to
get a deeper insight into this alternative gene activation mechanism, we cloned
and analyzed the genes coding for the VSAs of the variants from which the AnTat
1.1C or 1.6 clones were derived, and to which they gave rise, namely AnTat 1.16
and 1.3B for AnTat 1.1C, and AnTat 1.3 and 1.16 for AnTat 1.6, respectively
(see pedigree in fig. 2). Strikingly identical observations have been made in
both cases. Indeed, in association with the non-duplicative activation of these
two genes, we observed that the ELC from the preceding variant is conserved in
an inactive form: the AnTat 1.3 ELC in AnTat 1.6 DNA (Laurent et al, submitted)
and the AnTat 1.16 ELC in AnTat 1.1C DNA (75, and fig. 4). Moreover, the
activated AnTat 1.6 and 1.1C genes are both lost from the DNA of the ensuing

variants, AnTat 1.16 and 1.3B, respectively (75, and fig. 4). A possible diagrammatic interpretation, taking all these observations into account without the involvment of multiple recombination steps, is presented in figure 5. According to this model, at least some telomeres carrying VSA sequences, as AnTat 1.6 or 1.1C, can be exchanged with the "active" telomere (or expression site) containing the previously expressed ELC. The crossing-over point should be located between the gene and the putative unique VSA transcription promoter. As a result, inactivated ELCs could be conserved in the DNA of the ensuing variants, whereas telomeric genes activated by reciprocal recombination could be chased from the genome after replacement by the incoming ELC. This mechanism has already been considered regarding the AnTat 1.3 gene (40): the AnTat 1.3 BC is indeed located in a telomeric region whose restriction map is very similar to the VSA gene expression site, so that the possibility of crossing-over between homologous regions from the 5' environment of the gene was raised. As shown in figure 2, the restriction maps of most telomeres share similarities in the 5' environment of the VSA sequences, strengthening this hypothesis. However, the search for the putative crossover point has been unsuccessful up to the present. For instance, the non-duplicative activation of AnTat 1.1C and 1.6 genes did not lead to any restriction map alteration within about 40 kb upstream from the VSA sequence (75, and M. Laurent et al, submitted). Still, there could be room for recombination between the gene and the transcription promoter, since there is evidence that the latter could be located more than 40 kb upstream from the gene (see next section).

Alternation of the two gene activation mechanisms and evolution of the antigen gene repertoire. The modifications of ELCs by gene conversion clearly add to antigenic diversity: an obvious example is provided by the AnTat 1.1B ELC, which codes for a chimaeric protein with both AnTat 1.1 and 1.10-specific domains. The ELC is however generally lost from the genome during the antigenic switch, since it is often the target for the next gene conversion. The conservation, in some cases, of the ELC in the following variants provides the cell with a means to save the rearranged sequence, and to develop VSA multigene families. The new gene family member can evolve by point mutation (78) or by gene conversion, as proposed for instance by Young et al (66) in their analysis of the IlTat 1.3 gene family: these modifications could lead to the appearance of new antigen coding sequences, as observed in the AnTat 1.1 gene family, where the "6.4 kb" sequence, encoding for the AnTat 1.1 antigen, has probably evolved from a copy of the AnTat 1.10-coding "9 kb" sequence (37). On the contrary, alternation of the non-duplicative activation and the duplicative transposition mechanisms implies the loss of VSA sequences from the genome. The antigen gene

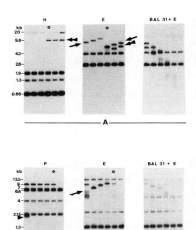

Fig. 4. Behaviour of AnTat 1.16 and 1.1C sequences through the AnTat 1.16 to 1.1C switch. AnTat 1.16 (A) and 1.1C (B) cDNA probes have been hybridized with restriction endonuclease digests of AnTat 1.1, 1.3, 1.6, 1.16, 1.1C and 1.3B DNAs, respectively from left to right lanes in the two first blocks in (A) and (B). The digestions in the first block (HindIII or PstI, for (A) and (B) respectively) allow to look at the 5' environment of both genes; they reveal the conservation of the AnTat 1.16 ELC (double arrowhead in A) and the loss of the AnTat 1.1C gene in the last variant (B). Digestions in the second block (EcoRI) allow to look at the 3' environment of both genes; they reveal the variably sized 3' terminus of the telomeres carrying the respective BCs (arrows) or conserved AnTat 1.16 ELC (double arrowhead). The third pannel illustrates in both cases the progressive shortening of these telomeres in AnTat 1.1C DNA, upon BAL31 digestion (1 unit/µg DNA) for different times: 0, 5, 8, 12, 16 and 21 min., from left to right respectively. Dots on top of some lanes refer to the DNAs from clones actually expressing the gene corresponding to the probe.

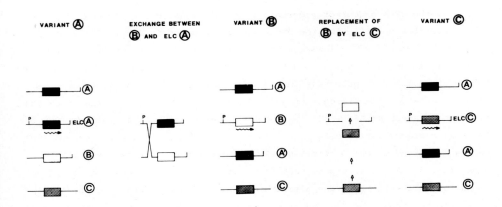

Fig. 5. Possible scheme of gene activation in a series of three successive variants, accounting for the observations reported in fig. 4 if A, B and C are the AnTat 1.16, 1.1C and 1.3B genes, respectively. The wavy arrow represents gene transcription, and P the transcription promoter.

repertoire thus evolves by gain and loss of genes; since the same sequence can be activated by either of the two mechanisms (see, for instance, the AnTat 1.1C and 1.10 clones), a given gene could be amplified in some repertoires or lost from other ones, depending on the way the two activation mechanisms alternate. On the other hand, a succession of variants all activated by the hypothetical telomere exchange would only lead to the translocation of VSA sequences; this could also modulate the expression of a VAT repertoire, since the location of VSA genes seems to influence their ability to be expressed early or late in chronic infection (unpublished results). The rapid evolution of antigen repertoires thus appears to be closely linked to the mechanisms used by the trypanosomes to change their antigen types (75).

TRANSCRIPTION OF VSA GENES

Evidences for processing of a large mRNA precursor

Although none of the VSA genes studied so far appears to be split by introns within its coding sequence, evidence of splicing has been found in VSA mRNAs. A 35 bp stretch, not encoded for by the ELC, was indeed found to be present at the 5' end of the mature mRNA in several variants, including one whose gene is activated without duplication (67-69). This short sequence should therefore be transcribed elsewhere in the genome, on a so-called "mini-exon", and should be linked to the mRNA according to a still undefined processing pathway. This processing could be complex, since several discrete transcripts, perhaps intermediates in the mRNA precursor processing, can be revealed when hybridizing the RNA with probes specific for the sequence cotransposed in front of the gene (30, 39, 61, 63, 67). The location of the "mini-exon" with respect to the VSA gene is still unknown, but should be at least 40 kb upstream (69). Accordingly, the putative VSA mRNA precursor should be at least 40 kb long. In AnTat 1.1B mRNA, we observed the presence of a 120 bp A-T rich sequence upstream from the mini-exon transcript (fig. 6). This sequence is joined to the mini-exon RNA by the AG/A characteristic so far for intron/exon boundaries in trypanosomes (38), as also observed at the 5' limit of the ELC portion transcribed into mature mRNA (fig. 6). At first sight, this A-T rich sequence could thus be considered as an intronic relic from the putative mRNA precursor. If true, the transcription promoter should be located still further upstream from the mini-exon (see summarizing scheme in figure 7). The genomic template for this "A-T rich intron" has been cloned and is presently under study. It appears to be unique (M. Guyaux et al, unpublished), in contrast with the mini-exon which is represented in at least 200 copies in the genome, mostly in 1.4 kb clusters (69, and T. De Lange, pers. commun.).

```
cDNA :  AAGCGACCGCAACTCCCTNAGTTCTAAGAACTTTCTTAATCCATATACATGGGGTA
                       Dde I   Dde I                              50

cDNA :  AATACATTTATAAGGGATAATTGCAAATAATAATAATAATAATTAAAAAAAAACAA
                                              100

cDNA :  GACCAGAACGCTATTATTAGAACAGTTTCTGTACTATATTGAAACAGAAGCCAAAG
          ••      *       *     **       **  RsaI    153
ELC :   CGCGCTCATCACCAAACCGCAATCTCAATACTAAATCATAGAAACAGAAGCCAAAG
                                                •••  1563
```

Fig. 6. Sequence of the 5' extremity of a AnTat 1.1B cDNA clone, and comparison with the AnTat 1.1B ELC sequence. Dots refer to the AG/A consensus sequence for intron/exon boundary in trypanosomes (67). The boxed sequence corresponds to the "mini-exon" transcript.

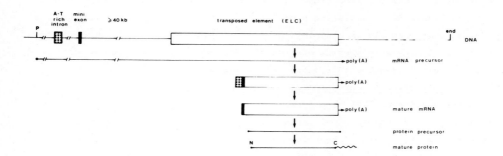

Fig. 7. Hypothetical scheme for VSA gene expression. The putative VSA mRNA precursor could be more than 40 kb long, then spliced in several steps. There is however no evidence yet for colinearity between the gene and mini-exon transcripts. The protein precursor must carry a signal peptide and hydrophobic tail (80, 81), made up of 29 and 23 amino acids, respectively, for the AnTat 1.1B antigen (69). The C-terminus of the mature protein is completed by carbohydrate and fatty acid addition (wavy line)(23).

STRUCTURE OF THE TRYPANOSOME GENOME
Diploidy and the problem of VSA gene alleles

Data from both isoenzyme analysis (70) and comparison between DNA amount and complexity (71) strongly suggest that the trypanosome genome is diploid. Accordingly, the VSA genes should be represented in two allelic forms. Since gene conversion leads to a duplication of the expressed VSA gene (26, 27), this sequence should be present, in the expressor clone, in two chromosome pairs instead of one. In other words, the gene conversion should take place from both alleles simultaneously, on two targets present in two expression sites. These unlikely hypotheses are not sustained by experimental data, neither from measurement of the amount of a given VSA gene per nucleus, nor from the recovery of VSA gene clones per genome: both estimates give only one VSA gene per nucleus (23, 60). Slightly diverging versions of different VSA genes, which could be considered as alleles (39), are more probably due to heterogeneity of the VSA sequence between different cell populations within a clone; this hypothesis would still imply that VSA genes evolve very rapidly, which is in accordance with available data suggesting hypermutagenesis in some VSA sequences (78). If VSA genes are haploid in a diploid genome, chromosome pairing at meiosis should be incomplete. The telomeric location of most VSA genes described so far could perhaps help to resolve this mispairing problem.

A large collection of chromosomes

Using 6 different VSA gene probes, we yet identified 9 different telomeres in a single trypanosome clone (AnTat 1.1). In all cases except one ("6.4 kb" member from the AnTat 1.1 gene family), the VSA sequences are oriented with their 3' end towards the DNA terminus (fig. 2). Among the 14 different clones that we studied, the silent or "basic" copy of the gene of the corresponding surface antigens appeared to be telomeric in 11 cases. These observations are in keeping with the finding, in other stocks of trypanosomes, of an ever increasing number of distinct DNA termini, all involved in antigenic variation (42, 72). Since the total repertoire for surface antigens in a clone is estimated to be above 100 (10, 11), and perhaps up to 1000 (60), the total number of different telomeres, and of course the number of chromosomes, could be quite large. Direct evaluation of this number by microscopic examination is impossible, since the trypanosome chromosomes do not condense at any stage of the cell cycle. Indirect estimations however, for instance based on the extent of genomic DNA digestion by the exonuclease BAL31, suggest a figure of one- to several hundreds chromosomes per nucleus (37). The DNA from these chromosomes could be roughly size-fractionated (79), and it has been shown that VSA sequences (72,

79), as well as highly repetitive DNA (79), could be carried on small chromosomes, or "minichromosomes". A large number of small chromosomes is not exceptional in protozoa, particularly in Euglenida, Dinoflagellida, in Amoeba proteus and in Ciliates (Ciliophora) (73), although in many cases this could be linked to polyploidy, as for instance in the macronucleus of Ciliates.

Size variations of telomeric sequences

In each telomere, the length between the VSA sequence and the corresponding DNA terminus varies between different clones (fig. 2, 4). When followed in a series of successively derived trypanosome clones, this size variation appears to be a regular increase, of about 0.8 kb at each antigenic switch (E. Pays et al, unpublished). This increase is sometimes interrupted by a sharp shortening, often linked to the involvment of the telomere in gene expression (see, for instance, the behaviour of the telomere carrying the AnTat 1.16 gene in figure 4). It occurs independently of the antigenic switch, since a regular telomere length extension has been observed during the clone propagation (74, and E. Pays et al, unpublished). This chromosome "growth" could be an intrinsic property of telomeres in general, depending on the succession of sequence unpairing and foldback pairing within telomeric repeated units after sub-terminal DNA cutting during replication (74). In the telomere carrying the AnTat 1.3 ELC, this size increase is about 10 bp per trypanosome generation (6 h) (E. Pays et al, unpublished), in complete agreement with the previous estimate for another telomeric VSA gene expression site (74). However the rate of extension of other telomeres appears to be different (74, and E. Pays et al, unpublished). The meaning of these observations is still conjectural; at least it precludes any interpretation of the size variations observed near telomeric VSA sequences as due to expression-independent gene recombinations (19, 41, 62).

ACKNOWLEDGEMENTS

Investigations reported herein received support from the FRSM (Brussels), from the ILRAD/Belgian Research Centres Agreement for Collaborative Research (Nairobi), and from the Trypanosomiases component of the UNDP/World Bank/WHO Special Programme for Research and Training in Tropical Diseases (Geneva).

REFERENCES

1. Franke, E. (1905) Munchener Medizinische Wochenschrift, 52, 2059-2060.
2. Fantham, H.B. and Thomson, J.G. (1911) Proc. Roy. Soc. Lond B., 83, 206-211.
3. Gray, A.R. and Luckins, A.G. (1976) in: Biology of the Kinetoplastida,

vol.I (eds Lumsden, W.H.R. and Evans, D.A.; Academic Press, New York) pp. 493-542.

4. Doyle, J.J. (1977) in: Immunity to Blood Parasites in Animals and Man (eds Miller, L., Pino, J. and McKelvey; Plenum Press, New York) pp. 27-63.

5. Cross, G.A.M. (1978) Proc. R. Soc. Lond. B, 202, 55-72.

6. Vickerman, K. (1978) Nature, 273, 613-617.

7. Cross, G.A.M. (1979) J. of Gen. Microbiol., 113, 1-11.

8. Turner, M.J. (1980) in: Molecular Basis of Microbial Pathogenicity (eds Smith, H., Skehel, J.J. and Turner, M.J.; Verlag Chemie, Weinheim) pp. 133-158.

9. Ritz, H. (1916) Arch. fur Schiffs- und Tropen- Hygiene, 20, 397-420.

10. Van Meirvenne, N., Janssens, P.G. and Magnus, E. (1975) Ann. Soc. belge Méd. Trop., 55, 1-23.

11. Capbern, A., Giroud, C., Baltz, T. and Mattern, P. (1977) Experim. Parasitol., 42, 6-13.

12. Vickerman, K. and Luckins, A.G. (1969) Nature, 224, 1125-1126.

13. Cross, G.A.M. (1972) J. Protozool., 19, suppl. 46.

14. Gray, A.R. (1962) Ann. Trop. Med. Parasitol., 56, 4-13.

15. Gray, A.R. (1965) J. Gen. Microbiol., 41, 195-214.

16. McNeillage, G.J.C., Herbert, W.H. and Lumsden, W.H.R. (1969) Exp. Parasitol., 25, 1-7.

17. Miller, E.N. and Turner, M.J. (1981) Parasitology, 82, 63-80.

18. Lheureux, M., Lheureux, M., Vervoort, T., Van Meirvenne, N. and Steinert, M. (1979) Nucl. Acids Res., 7, 595-609.

19. Williams, R.O., Young, J.R. and Majiwa, P.A.O. (1979) Nature, 282, 847-849.

20. Pays, E., Delronche, M., Lheureux, M., Vervoort, T., Bloch, J., Gannon, F. and Steinert, M. (1980) Nucl. Acids Res., 8, 5965-5981.

21. Hoeijmakers, J.H.J., Borst, P., Van den Burg, J., Weissmann, C. and Cross, G.A.M. (1980) Gene, 8, 391-417.

22. Englund, P.T., Hajduk, S.L. and Marini, J.C. (1982) Ann. Rev. Biochem., 51, 695-726.

23. Borst, P. and Cross, G.A.M. (1982) Cell, 29, 291-303.

24. Lumsden, W.H.R. (1982) Systematic Parasitology, 4, 373-376.

25. Van Meirvenne, N., Magnus, E. and Vervoort, T. (1977) Ann. Soc. belge Méd. Trop., 57, 409-423.

26. Pays, E., Van Meirvenne, N., Le Ray, D. and Steinert, M. (1981) Proc. Nat. Acad. Sci. USA, 78, 2673-2677.

27. Hoeijmakers, J.H.J., Frasch, A.C.C., Bernards, A., Borst, P. and Cross, G.A.M. (1980) Nature, 284, 78-80.

28. Pays, E., Lheureux, M. and Steinert, M. (1981) Nucl. Acids Res., 9, 4225-4238.

29. Pays, E., Lheureux, M., Vervoort, T. and Steinert, M. (1981) Molec. Biochem. Parasitol., 4, 349-357.

30. Pays, E., Lheureux, M. and Steinert, M. (1982) Nucl. Acids Res., 10, 3149-3163.

31. Pays, E., Dekerck, P., Van Assel, S., Eldirdiri A. Babiker, Le Ray, D., Van Meirvenne, N. and Steinert, M. (1983) Molec. Biochem. Parasitol., 7, 63-74.

32. Longacre, S., Hibner, U., Raibaud, A., Eisen, H., Baltz, T., Giroud, C. and Baltz, D. (1983) Mol. Cell. Biol., 3, 399-409.

33. Weintraub, H. and Groudine, M. (1976) Science, 193, 848-856.

34. Pays, E., Lheureux, M. and Steinert, M. (1981) Nature, 292, 265-267.

35. Weisbrod, S. and Weintraub, H. (1979) Proc. Nat. Acad. Sci. USA, 76, 630-634.

36. Bernards, A., Van der Ploeg, L.H.T., Frasch, A.C.C., Borst, P., Boothroyd, J.C., Coleman, S. and Cross, G.A.M. (1981) Cell, 27, 497-505.

37. Pays, E., Van Assel, S., Laurent, M., Darville, M., Vervoort, T., Van Meirvenne, N. and Steinert, M. (1983) Cell, 34, 371-381.

38. Van der Ploeg, L.H.T., Bernards, A., Rijsewijk, F.A.M. and Borst, P. (1982) Nucl. Acids Res., 10, 593-609.

39. Pays, E., Van Assel, S., Laurent, M., Dero, B., Michiels, F., Kronenberger, P., Matthyssens, G., Van Meirvenne, N., Le Ray, D. and Steinert, M. (1983) Cell, 34, 359-369.

40. Laurent, M., Pays, E., Magnus, E., Van Meirvenne, N., Matthyssens, G., Williams, R.O. and Steinert, M. (1983) Nature, 302, 263-266.

41. Longacre, S., Raibaud, A., Hibner, U., Buck, G., Eisen, H., Baltz, T., Giroud, C. and Baltz, D. (1983) Mol. Cell. Biol., 3, 410-414.

42. De Lange, T. and Borst, P. (1982) Nature, 299, 451-453.

43. Bothwell, A.L.M., Paskind, M., Roth, M., Imaniski:Kari, T., Rajewsky, K. and Baltimore, D. (1981) Cell, 24, 625-637.

44. Schreier, P.H., Bothwell, A.L., Mueller-Hill, B. and Baltimore, D. (1981) Proc. Nat. Acad. Sci. USA, 78, 4495-4499.

45. Bruggemann, M., Radbruck, A. and Rajewsky, K. (1982) EMBO J., 1, 629-634.

46. Dildrop, R., Bruggemann, M., Radbruck, A., Rajewsky, K. and Bayreuther, K. (1982) EMBO J., 1, 635-640.

47. Bentley, D.L. and Rabbitts, T.H. (1983) Cell, 32, 184-189.

48. Ollo, R. and Rougeon, F. (1983) Cell, 32, 515-523.

49. Sligthom, J.L., Blechl, A.E. and Smithies, O. (1980) Cell, 21, 627-638.

50. Pease, L.R., Schulze, D.H., Pfaffenbach, G.M. and Nathenson, S.G. (1983) Proc. Nat. Acad. Sci. USA, 80, 242-246.

51. Weiss, E.H., Mellor, A., Golden, L., Fahrner, K., Simpson, E., Hurst, J. and Flavell, R.A. (1983) Nature, 301, 671-674.

52. Roeder, G.S. and Fink, G.R. (1982) Proc. Nat. Acad. Sci. USA, 79, 5621-5625.

53. Baltimore, D. (1981) Cell, 24, 592-594.

54. Bregegere, F. (1983) Biochimie 65, 229-237.

55. Hicks, J.B., Strathern, J.N. and Herkowitz, I. (1977) in: DNA Insertion Elements, Plasmids and Episomes (eds Bukhari, A.I. et al; Cold Spring Harbor, New York) p. 457.

56. Matthyssens, G., Michiels, F., Hamers, R., Pays, E. and Steinert, M. (1981) Nature, 293, 230-233.

57. Majumder, M., Boothroyd, J.C. and Weber, H. (1981) Nucl. Acids Res., 9, 4745-4753.

58. Rice-Ficht, A.C., Chen, K.K. and Donelson, J.E. (1981) Nature, 294, 53-57.

59. Michels, P.A.M., Bernards, A., Van der Ploeg, L.H.T. and Borst, P. (1982) Nucl. Acids Res., 10, 2353-2366.

60. Van der Ploeg, L.H.T., Valerio, D., De Lange, T., Bernards, A., Borst, P. and Grosveld, F.G. (1982) Nucl. Acids Res., 10, 5905-5923.

61. Liu, A.Y.C., Van der Ploeg, L.H.T., Rijsewijk, F.A.M. and Borst, P. (1983) J. Mol. Biol., 167, 57-75.

62. Young, J.R., Donelson, J.E., Majiwa, P.A.O., Shapiro, S.Z. and Williams, R.O. (1982) Nucl. Acids Res., 10, 803-819.

63. Majiwa, P.A.O., Young, J.R., Englund, P.T., Shapiro, S.Z. and Williams, R.O. (1982) Nature, 297, 514-516.

64. Hajduck, S.L., Cameron, C.R., Barry, J.D. and Vickerman, K. (1981) Parasitology, 83, 595-607.

65. Hajduck, S.L. and Vickerman, K. (1981) Parasitology, 83, 609-621.

66. Young, J.R., Shah, J.S., Matthyssens, G. and Williams, R.O. (1983) Cell, 32, 1149-1159.

67. Van der Ploeg, L.H.T., Liu, A.Y.C., Michels, P.A.M., De Lange, T., Borst, P., Majumder, H.K., Weber, H., Veeneman, G.H. and Van Boom, J. (1982) Nucl. Acids Res., 10, 3591-3604.

68. Boothroyd, J.C. and Cross, G.A.M. (1982) Gene, 20, 281-289.

69. Michiels, F., Matthyssens, G., Kronenberger, P., Pays, E., Dero, B., Van Assel, S., Darville, M., Cravador, A., Steinert, M. and Hamers, R. (1983) EMBO J., 2, 1185-1192.

70. Tait, A. (1980) Nature, 287, 536-538.

71. Borst, P., Van der Ploeg, M., Van Hoek, J.F.M., Tas, J. and James, J. (1982) Molec. Biochem. Parasitol., 6, 13-23.

72. Williams, R.O., Young, J.R. and Majiwa, P.A.O. (1982) Nature, 299, 417-421.

73. Raikov, I.B. (1982) The Protozoan Nucleus: Morphology and Evolution (Cell Biology Monographs vol. 9; Springer-Verlag, Wien, New York).

74. Bernards, A., Michels, P.A.M., Lincke, C.R. and Borst, P. (1983) Nature, 303, 592-597.

75. Pays, E., Delauw, M.F., Van Assel, S., Laurent, M., Vervoort, T., Van Meirvenne, N. and Steinert, M. (1983) Cell, in press.

76. Southern, E.M. (1975) J. Mol. Biol., 98, 503-517.

77. Rigby, P.W.J., Dieckmann, M., Rhodes, C. and Berg, P. (1977) J. Mol. Biol., 113, 237-251.

78. Frasch, A.C.C., Borst, P. and Van Den Burg, J. (1982) Gene, 17, 197-211.

79. Sloof, P., Menke, H.H., Caspers, M.P.M. and Borst, P. (1983) Nucl. Acids Res., 11, 3889-3901.

80. Boothroyd, J.C., Cross, G.A.M., Hoeijmakers, J.H.J. and Borst, P. (1980) Nature, 288, 624-626.

81. Boothroyd, J.C., Paynter, C.A., Coleman, S.L. and Cross, G.A.M. (1982) J. Mol. Biol., 157, 547-556.

RESUME

Les trypanosomes africains peuvent infester de façon chronique le système sanguin de leurs hôtes vertébrés, grâce à leur capacité à déjouer la réponse immunitaire de ces derniers. Ces parasites sont en effet capables de changer fréquemment d'antigène de surface, par l'expression différentielle d'un répertoire d'au moins cent gènes d'antigènes, dont un seul est exprimé à la fois. Nous avons cloné et caractérisé les cDNAs correspondant aux mRNAs de différents antigènes variables d'un même répertoire de Trypanosoma brucei et utilisé ces séquences comme sondes pour analyser et isoler différents gènes d'antigènes de surface. Plusieurs observations ont été faites:

1. Les séquences spécifiques d'antigènes de surface sont le plus souvent localisées en bout de chromosome (télomères).

2. L'expression de ces gènes passe par des remaniements génomiques. Au moins deux mécanismes différents conduisent à ces remaniements: la conversion génique et, probablement, la recombinaison réciproque. La conversion génique consiste en le remplacement d'une séquence par la copie d'une autre, ce qui se traduit par la duplication d'une séquence et la transposition de cette copie (ELC, pour Expression-Linked Copy) dans un nouvel environnement génomique. La recombinaison réciproque entre différents télomères est le mécanisme le plus plausible qui puisse rendre compte de l'activation de gènes sans duplication. Dans ce cas, le télomère contenant le gène à activer est échangé contre le télomère qui abrite l'ELC précédemment transcrite, la recombinaison hypothétique ayant lieu en aval du promoteur de transcription.

3. Les remaniements génomiques peuvent modifier le réperoire antigénique d'un clone. Le mécanisme de recombinaison réciproque permet en effet de conserver une ELC dans des variants qui n'utilisent plus cette séquence comme matrice de transcription. Cette copie additionnelle constitue donc une nouvelle séquence d'antigène de surface, qui peut évoluer par mutations ou conversions géniques. Inversément, les gènes activés sans duplication peuvent être chassés du génome suite à une conversion génique par une autre ELC : ceci entraîne donc au contraire une perte de certains gènes du répertoire.

4. La longueur de différents télomères subit des variations de taille incessantes, y compris lors de la croissance d'un clone. Nous avons observé une extension régulière de la taille des extrémités de DNA, corrigée par des raccourcissements occasionnels souvent liés à l'intervention du télomère dans les mécanismes d'expression des gènes d'antigènes de surface.

ELECTRON MICROSCOPY AND GENE MORPHOLOGY

Viviane POHL,
Laboratory of Histology, School of Medicine, Free University of Brussels, 2, rue Evers, 1000 Brussels, Belgium

INTRODUCTION

Recently, electron microscopy of nucleic acids has become a tool of increasing usefulness and interest in the fields of molecular genetics and molecular biology. The rapid progress of in vitro recombination and DNA cloning technology has required new methods for characterizing cloned DNA molecules. The visualization of nucleic acid molecules permits direct and easy determination of the size and structure of linear or circular molecules, discrimination between double and single stranded regions and even, in favourable experimental conditions, discrimination between DNA and RNA.

Various hybridization techniques allow the detection and analysis of sequence complementarity between two DNA molecules (heteroduplex, D-loop mapping) or between DNA and RNA molecules (R-loop mapping, R-hybrids). This article does not intend to describe all the possibilities of electron microscopy in molecular genetics and molecular cell biology. It will be restricted to the review of basic techniques for forming, visualizing and analyzing RNA/DNA heteroduplexes (for additional information, see 1,2,3,4) which have played an important part in the elucidation of RNA splicing and what it achieves for the genetic versatility of eucaryotic organisms.

RNA-DNA HYBRIDIZATION : AN OVERVIEW

Two different procedures are used for obtaining DNA-RNA hybrids. In the presence of formamide, RNA-DNA hybrids are much more stable at high temperatures than DNA-DNA duplexes: the difference in stability between DNA/DNA and DNA/RNA hybrids is maximal at high formamide concentrations (70 to

80 %). RNA/DNA association is optimal at conditions where no DNA/DNA hybridization occurs (7); as a consequence, anarchical DNA-DNA hybridization ("bush-like" hybridization) is totally avoided leading to R-hybrids which are easy to interpret.

Formamide solutions for nucleic acid hybridization present an advantage in that they permit specific DNA/DNA and DNA/RNA associations to occur at lower temperatures (8). A linear relationship exists between the formamide concentration and the melting temperature (Tm) of DNA : the Tm of native duplex DNA is reduced by approximately 0.72° C per 1 % formamide (7).

RNA-DNA hybridization :

I. R-loop formation in one step.

White et al. (9) and Thomas et al. (10) have developed the R-loop method which allows the mapping of DNA sequences which are complementary to specific RNA molecules. At temperatures close to the Tm of the DNA and in the presence of formamide a characteristic hybrid structure, commonly called an R-loop, is formed by displacement of one strand of the double helix by the complementary RNA (see Fig. 1). These R-loop bubbles can be easily observed and allow an accurate mapping of the DNA regions homologous to a given RNA.

Thomas et al. (10) have shown that the temperature for the maximum rate of R-loop formation is within one degree of the Tm. This temperature depends on the base composition of the DNA. However the technique presents some disavantages:

- In general, the base composition of a genomic fragment is unknown and it is therefore not easy to define the ideal conditions for R-looping;
- Once formed, the R-loops are considered to be stable but partial displacement of the RNA by branch migration can occur (during sample preparation for spreading,for example)

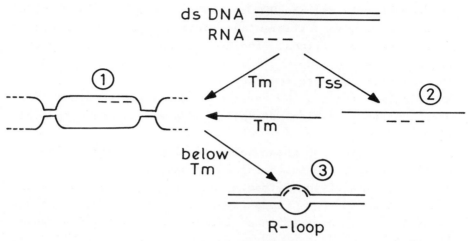

Fig. 1 : <u>RNA-DNA hybrid production</u>
(1) Schematic drawing of partially denatured DNA maintained at near the average melting temperature (Tm). GC rich regions persist as double-stranded structures while AT rich regions are extensively denaturated. In the presence of 70-80% formamide, the Tm of RNA/DNA duplex is 10-15°C higher than that of the DNA/DNA hybrid; consequently, RNA can hybridize with complementary DNA sequences available in denaturation loops.
(2) A RNA/single-stranded DNA heteroduplex formed by incubating RNA with denatured DNA at or immediately above the strand separation temperature (TSS). The hybrid (1) is also an intermediate in the conversion of the heteroduplex (2) to the R-loop structure (3); this conversion occurs upon cooling the hybridization mixture at or slightly below Tm.

II. Single stranded DNA-RNA hybrids and the two-step R-loop method.

Double-stranded DNA is first fully denatured. Lyophilized RNA is added to denatured DNA and allowed to hybridize at a temperature just above the strand separation temperature of the DNA (Tss, some degrees above the Tm) for a time ranging from 2 to 24 hours, depending upon the degree of complementarity between DNA and RNA. Because RNA/DNA heteroduplexes are more stable than comparable DNA/DNA duplexes under these conditions, the RNA can pair with its DNA complement (see Fig. 1). This method conveniently eliminates the need to separate the two DNA strands. The resulting RNA-single stranded DNA heteroduplexes allow straight forward characterization of spliced RNA and genome organization.

Construction and observation of single-stranded DNA-RNA hybrids present the following advantages :
1. Resolution of the method. Very small hybrid regions and single-stranded DNA loops (short exons and short introns respectively), observed in an impressive variety of eucaryotic genes, can be resolved.
2. Hybridization temperature : hybridization may be carried out at higher temperature which permit the unfolding of the secondary structure of G + C-rich RNA.
3. Stability. The displacement of the RNA by the second DNA strand is minimized. This advantage is important : displacement of part of the RNA by branch migration, leads to the appearance of RNA tails, the correct interpretation of which is difficult. The existence of RNA tails is due to two different phenomena : either the RNA and DNA do not hybridize completely because of the heterology in base sequence between RNA tail and the DNA molecule studied or RNA tails are an artefact produced by displacement of the RNA essentially in A+T-rich regions.
4. Finally it can sometime be difficult to localize precisely the end of the RNA hybridized to single-stranded DNA; it is therefore necessary to accentuate single-strand/double-strand junctions. RNA/single-stranded DNA heteroduplexes can be further incubated at slightly lower temperatures to allow the complementary DNA strands, present throughout the hybridization, to reassociate. This double step procedure allows the formation of R-loops in GC-rich as well as in A-T-rich regions (11).

Glyoxal treatment

Kaback et al. (12) and HSU et al. (13) have described methods for stabilizing single-stranded nucleic acids with glyoxal by specifically modifying G residues. Glyoxal has been used for two reasons : first, it has a stabilizing action on the RNA/DNA hybrid molecule (for example, the hybrid can be stored at 4° C for weeks without major degradation); second, it permits RNA molecules to spread in a relaxed measurable form (without glyoxal, RNA appears as aggregates). The dispersion of messenger RNA is necessary in order to localize the sequences which hybridize with a given cloned DNA.

Hybrids mounting for electron microscopy

Following glyoxal treatment, nucleic acids are mounted for electron microscopy. The method routinely used in our laboratory is the basic protein film technique employing cytochrome C as described by Davis, Simon and Davidson (17). In order to discern and measure well extended single - and double-stranded nucleic acids in a particular heteroduplex, it is necessary to introduce a denaturing solvent into the spreading solution (hyperphase) : this melts out most random base-base interactions in single-stranded DNA. We use 3-times cristallized formamide (50% in 0.1 M Tris-HCl, pH 8.5, plus 0.01 M EDTA) as a denaturing solvent (this concentration of formamide gives optimal contrast). Formamide is also advantageous in that it appears to favour the interaction of a greater amount of cytochrome C with DNA and RNA, yielding greater contrast.

HYBRIDIZATION : MATERIALS AND METHODS

Material and methods will be restricted to the basic techniques we have used in our laboratory for forming, visualizing and analyzing RNA/DNA heteroduplexes with reference to the characterization of the structure of thyroglobulin gene which is particularly large (14, 15, 16).

1.1. Glassware and utensils

Glassware and utensils should be very clean, free of grease and detergent, and backed at 180°C for several hours to inactivate RNAses. It is not recommended to siliconize glassware. The small volumes of the experimental mixtures can be handled in Eppendorf tubes which are RNAse free.

1.2. Formamide

The quality of cytochrome C layering (17) with the formamide technique varies to a great extent with different batches of formamide even if pure commercial formamide is used : irregular extension or molecular aggregation of nucleic acid molecules may be observed with some batches. We used recrystalized (2-3 times) formamide (Merck or Fluka, puriss. p.a.). For the hypophase however, pure commercial formamide can be used.

1.3. Hybridization buffer

The hybridization buffer for RNA-DNA hybridization contains 70 % formamide in 0.1 M Pipes (piperazine-N-N'-bis (2-ethanesulfonic acid), 23 mM Tris-HCl (pH 7.8), Na_2 EDTA (5 mM), 0.5 M NaCl and 200 mM KCl. Tridistilled water is required in order to avoid any contaminant.

1.4. DNA

The DNA sample (about 100 µg/ml in 10^{-2} M Tris, 10^{-3} M EDTA, pH 8.5) should be completely protein free to avoid aggregation during the hybridization reactions. The DNA should be a homogeneous population of molecules for example, that of a cloned DNA segment. It is better to linearize circular DNA in order to favour hybridization, preferably using a restriction enzyme to cleave the molecule at single well known site. Molecules linearized in such a way allow good mapping of hybridized sequences and give intelligible hybrids. The DNA should have few, if any, single stranded nick and must be of homogeneous length. Random fragments due to single-stranded nicks in the stock material increase the frequence of multibranched hybridization. On the contrary, when linearization by one site cleavage is not feasible, some nicks should be introduced into the DNA to permit uncoiling and denaturation.

1.5. RNA

RNA preparations must be of high quality, with a low percentage of broken molecules. One of the advantages of RNA/DNA heteroduplex formation and analysis by electron microscopy is that the RNA usually need not be totally prepurified, although enrichment will reduce the background of unhybridized molecules, the hybridization time... and the work of the electron microscopist. We used pure thyroglobulin mRNA (18) concentrated by ethanol precipitation and lyophilized. All ethanol should be removed by vacuum evaporation before attempting to redissolve the pellet in the hybridization mixture.

1.6. Hybridization procedure

2,5 μl of DNA (see 1.4), either small DNA (100 μg/ml) such as pBR 322 with 2-5 Kb inserts, or larger DNA (20-50 μg/ml) such as bacteriophages λ and cosmids with inserts of about 15 and 45 Kb respectively, are diluted in hybridization buffer (see 1.3). 10 μl of the mixture was heated at 67° C for 10 min., added to 150-300 ng mRNA (see 1.5), and incubated for varying time periods (2-24 h) at temperatures which depend upon the DNA GC content (59°C for pBR, 56°C for phages and cosmids). In conditions where DNA is allowed to reanneal, the reaction mixture is cooled at 30-40°C for 10-15 min. Further incubation (30 min. at 30°C) is carried out with glyoxal (Aldrich 40% in water, 3.3% final concentration in 0.01 M sodium phosphate buffer, pH 8.0).

1.7. Layering for electron microscopy and further manipulations

Usually, following glyoxal treatment, the hybridization mixture is diluted ten times with the spreading solution, i.e. hyperphase (deionized 50 % formamide, 0.1 M Tris-HCl (pH 8.5), 0.01 M EDTA, 50-150 μg/ml cytochrome C (Sigma, type V), plus single-stranded φ x 174 DNA and φ x 174 RF DNA (0.1 μg/ml) as the internal standard). The sample (10 μl) is spread onto a hypophase containing 20% formamide (Fluka) and one tenth the electrolyte (i.e. 10^{-2} M Tris; HCl, pH 8.5, 10^{-3} M EDTA) in a 10 cm x 6 cm teflon dish. The method for spreading the hyperphase on the hypophase has been described by Davis et al. (5) but according to the authors, "the only good way to communicate details of technique is visually". The spread hyperphase is allowed to set for 1 minute to allow the cytochrome layer to further relax the hybrids. We pick up the protein-nucleic hybrid mixed film on grids (200 mesh) covered with a freshly spread (2-48 hours) Parlodion film from a 1,5% - 2% w/v solution in n-pentylacetate. High quality Parlodion film is required as Parlodion interacts directly with the cytochrome monolayer and is therefore necessary for obtaining clear images in electron microscopy.

Contrast enhancement is carried out using a two-step procedure : the grids are coloured using freshly-prepared 5.10^{-5}M uranyl acetate - 90% ethanol solution, washed briefly first in 90% ethanol these in 2-methylbutane and air dried. Grids are rotary shadowed at a fixed angle of 8° with 40 mg of platinium/palladium wire in a vaccuum evaporator for additional contrast and then covered with a carbon film. The grids can be immediately observed in the electron microscope (accelerating voltage of 60 KV). Fig. 2 (a and b) are obtained with the same preparation and illustrate the variability of the quality in the spreading.

Fig. : <u>Spreadings obtained with the same preparation</u>.

Fig. 2 (a and b) allows to compare among themselves the five first exons (100-430 base pairs) of human thyroglobulin gene (26) interspersed by four intervening sequences (1.19-1.85 kilobases). Hybrids result from mRNA/single-stranded DNA hybridization. Qualitative differences of the hybrids when observed at the electron microscope may result from very different causes : the quality of formamide, cytochrome C, parlodion film, or more simply, the spreading procedure itself.

Fig. 2a : correct spreading with smooth cytochrome C background, relax aspect of ss and ds DNA, well extended RNA; Fig. 2b : Unsuitable spreading with excessive thickness ("stickly aspect") of the nucleic acid fibers, collapsed shrunken intron (see I.1).

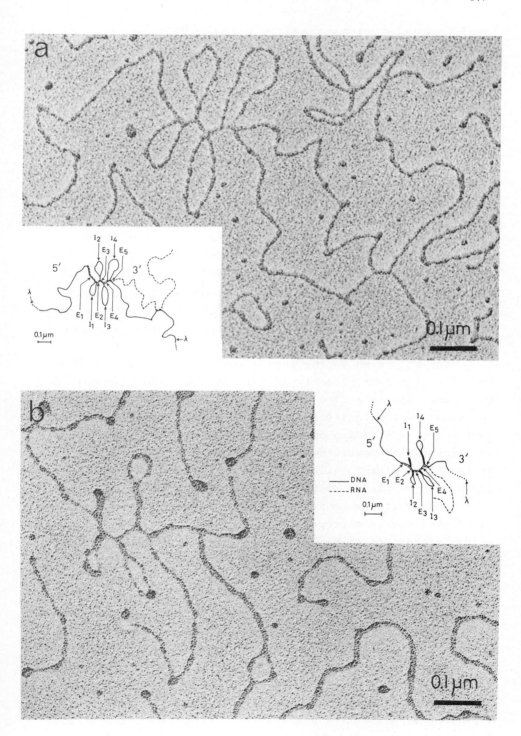

LENGTH MEASUREMENTS AND VALIDITY OF THE MICROGRAPHS

With the formamide - cytochrome monolayer method described above, the length of nucleic acid molecule, whether it is single-(ss) or double-stranded (ds), depends on the conditions under which it is mounted for electron microscopy (19). Absolute measurement of nucleic acid strands does not give an accurate value of their length. However, by measuring lengths relative to appropriate DNA standards, meaningful results can be obtained in quantitative terms.

Circular single- and double-stranded viral DNA molecules, the nucleotide sequences of which have been determined (SV 40, pBR 322, ϕ×174...) are the most frequently used length standards for ss and ds molecules respectively (within ± 4%, RNA/DNA and DNA/DNA segments have approximately the same length/base pairs which does not exceed interpretive uncertainties (9, 10, 20, 21, 22)). For RNA the reference should ideally be a transcription complex at known length.

However, even in the case of homogeneous standard DNA, it is an inherent property of the basic protein film method that the length of the DNA molecule fluctuates around a quantitatively significant mean value with a quantitatively reproducible standard deviation. Despite this, significant and reproducible measurements can be obtained (17) with good preparations giving values with an average deviation of 2-5% for ds and 5-10% for ss-DNA.

Some criteria must be met in order to estimate the value of a hybrid micrograph :
- the appearance of nucleic acid fibers : ds regions (for example hybrid regions) must appear as smooth sinuous lines while ss strands look thinner and kinky (fig. 3a). After glyoxal treatment RNA resembles ds DNA with respect to strand width but it is not as sinuous and it usually presents some small regions with remaining secondary structure. A poorer distinction between ss and ds DNA can be observed as the consequence of these molecules being stretched when the hyperphase is layered onto the hypophase. These images will give unreliable measurements and should not be used.
- general aspect of the preparation : stretched preparations

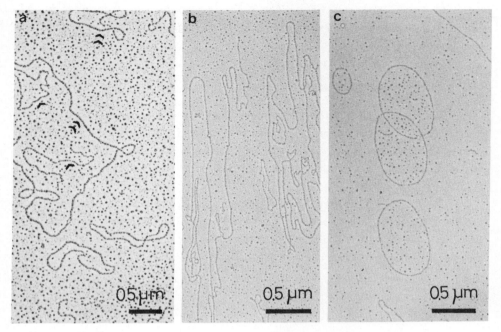

Fig. 3 : <u>Normal and stretched DNAs</u>.
In fig. 3a, smooth linear DNA and circular DNA
(ϕ x 174) ds (⌢) and ss (⌢).
In fig. 3b, stretched linear ds DNA.
In fig. 3c, stretched circular ds DNA (ϕ x 174 phage).

are easily recognizable by the fact that a large proportion of the linear molecules are orientated in the same direction (fig. 3b) and/or the supertightened aspect of circular DNA standards (fig. 3c)
- conversion from measured lengths to number of bases : each hybrid to be measured must be surrounded by several samples of standard ss and ds-DNA; this is necessary in order to be able to convert correctly the measured lengths into number of bases. In fact, significant variations in length may be observed in different regions of a single grid, as a result of physical stretching of the nucleic acids during the surface denaturation of the cytochrome film which cannot be

reproducibly controlled. As a consequence, a conversion factor must be established for each micrograph, and not for the whole set.

LIMIT OF MEASUREMENTS

The shortest RNA-DNA hybrids that have been observed in formamide - cytochrome C preparations are about 40 base pairs long (23, 24). Sequence of 30-50 bases seem to be the extreme limit for accurate measurements of ds nucleic acids (25).

Very short single-stranded loops (intronic, deletion loops) tend to collapse or to contract into "knobs" so that it is difficult to obtain precise measurements for loops below 100 nucleotides in length (26).

CONCLUSION

Electron microscopy (EM) of RNA/DNA heteroduplexes, based on nucleic acid hybridization, has allowed both qualitative and quantitative analysis of gene organization, due to the fact that RNA/DNA heteroduplexes are more stable than DNA/DNA duplexes at high formamide concentration. Mapping resolution can reach approximately 50-100 nucleotides or better, which is sufficient to establish correlations with genomic sequences.

The EM methods present several advantages. Nanogram amounts of RNA and DNA are used. Furthermore the RNA need not be completely pure : heterologous RNA will not hybridize with the DNA and simply appears in the background. Restriction endonuclease maps of the DNA and nucleic acid labeling are not necessary for mapping.

In contrast to methods of mass analysis, complex spliced RNA can be deciphered easily with EM heteroduplex analysis and minor RNA species can even be mapped, with effort, in a background of other RNA species.

However, the EM method has its own limitations. The RNA and DNA preparations must be of higher quality than those necessary for mass analysis, since broken molecules may give confused pictures. RNA shorter than several hundred nucleotides is not easily hybridized nor mapped (but even very short spliced segments can be recognized by deletion looping). For segments present in lesser quantity, EM methods are extremely laborious. The major technical limitation concerns identification and measurement of very small intervening sequences; for loops smaller than 100 nucleotides, the S1 nuclease method (27) will be more sensitive.

Together with other methods (such as nuclease-gel electrophoretic analysis, cloning of cDNA, and nucleotide sequence analysis), EM methods visualizing RNA/DNA heteroduplexes have revealed much of what we know about gene structure in many systems and are unmatched in deciphering the intron-exon organization of eucaryotic genes.

ACKNOWLEDGMENTS

The continuous support and critical interest of Dr. J.E. Dumont, Dr. G. Vassart and their research fellows, H. Targovnik, H. Brocas, D. Christophe, G. de Martynoff, B. Van Heuverswijn, B. Cabrer and L. Mercken are acknowledged. We thank Mrs. G. Pattyn, Mr. G. Vienne and P. Miroir for their excellent technical assistance. This work was supported by grants from the Fonds Defay, the FRSM (n° 3.4530.78) and the University of Brussels. V. Pohl is "Première Assistante" at the School of Medicine, Free University of Brussels, Belgium.

REFERENCES

1. BRACQ, C., 1981. Critical reviews in biochemistry. 10, 113.
2. GRIFFITH, J.D., 1981. Electron microscopy in biology. vol. 1, Wiley Interscience.
3. FERGUSON, J. and DAVIS, R.W., 1978. Advanced techniques in biological electron microscopy II, Koehler, J.K., ed., Springer-Verlag, Berlin-Heidelberg, 123.
4. EVENSON, D.P., 1977. Methods Virol., 6, 219.
5. CHOW, L.T. and BROKER, T.R., 1980. in: Gene structure and expression, P.H. Dean, L.F. Johnson, P.C. Kimball

and P.S. Peulman, eds. Ohio State University Press, Columbus, OH., pp. 175.

6. BROKER, T.R. and CHOW, L.T., 1980. Trends Biochem. Sci., 5, 174.
7. CASEY, J. and DAVIDSON, N., 1977. Nucleic Acid Res., 4, 1539.
8. McCONAUGHY, B.L., LAIRD, C.D. and McCARTHY, B.J., 1969. Biochemistry, 8, 3289.
9. WHITE, R.L. and HOGNESS, D.S., 1977. Cell, 10, 177.
10. THOMAS, M., WHITE, R.L. and DAVIS, R.W., 1976. Proc. Natl. Acad. Sci. USA, 73, 2294.
11. HOLMES, D.S., COHN, R.H., KEDES, L. and DAVIDSON, N., 1977. Biochemistry, 16, 1504.
12. KABACK, D.B., ANGERER, L.M. and DAVIDSON, N., 1979. Nucleic Acid Res., 6, 2499.
13. HSU M.T., KUNG, H.J. and DAVIDSON, N., 1972. Cold Spring Harbor Symp. Quant. Biol., 38, 943.
14. VASSART, G. and DUMONT, J.E., 1973. Europ. J. Biochem., 32, 322.
15. VASSART, G., BROCAS, H., LECOCQ, R. et al., 1975. Europ. J. Biochem., 55, 15.
16. VASSART, G., VERSTREKEN, L., DINSART, C., 1977. FEBS Letters, 79, 15.
17. DAVIS, R.W., SIMON, M. and DAVIDSON, N., 1971. Methods in Enzymology, 21, 413.
18. VAN HERLE, A., VASSART, G. and DUMONT, J.E., 1979. New Engl. J. Med., 301, 239, 307.
19. INMAN, R.B., 1967. J. Mol. Biol., 25, 209.
20. CHOW, L.T., ROBERTS, J.M., LEWIS, J.B. and BROKER, T.R., 1977. Cell, 11, 819.
21. HYMAN, R.W., 1971. J. Mol. Biol., 61, 369.
22. GRIFFITH, J.D., 1978. Science, 201, 525.
23. TONEGAWA, S., 1978. Cell, 15, 1.
24. SAKANO, H., ROGERS, J.H., HUPPI, K., BRACK, C., TRAUNECKER, A., MAKI, R., WALL, R. and TONEGAWA, S., 1979. Nature, 277, 627.
25. CHRISTOPHE, D., POHL, V., VAN HEUVERSWIJN, B., et al., 1982. Biochem. Biophys. Res. Comm., 105, 1166.

26. TARGOVNIK, H., POHL, V., CHRISTOPHE, D., BROCAS, H., CABRER, B. and VASSART, G., in preparation.
27. PETTERSSON, U., TIBBETTS, C. and PHILIPSON, L., 1976. J. Mol. Biol., 101, 479.

RESUME

L'étude au microscope électronique des hétéroduplex RNA/DNA obtenus par hybridation moléculaire est un moyen de choix pour analyser la structure d'un gène. L'obtention de ces hybrides a été permise grâce à la mise au point de conditions expérimentales qui stabilisent préférentiellement les hybrides RNA/DNA par rapport aux duplex DNA/DNA.

Les techniques les plus couramment utilisées pour réaliser l'hybridation entre du RNA messager et son DNA complémentaire génomique ont été décrites; leurs avantages et leurs inconvénients respectifs ont été comparés.

Enfin la validité des images obtenues au microscope électronique a été discutée ainsi que les limites inhérentes à la méthode.

AUTHOR INDEX

Abens, J. 195
Ailhaud, G. 53
Amri, E.-Z. 53
Arrang, J.-M. 213

Bartfai, T. 195
Bauer, K. 231
Belfrage, P. 153
Boheim, G. 99
Butlen, D. 69

Chabre, M. 87

De Block, M. 245
Deblaere, R. 245
Deterre, P. 87
Djian, P. 53
Djiane, J. 269
Docter, R. 179
Dusanter-Fourt, I. 269

Fekkes, D. 179
Foecking, M.K. 37
Forest, C. 53
Fredrikson, G. 153
Fuller, S.J. 139

Garbarg, M. 213
Gierlich, D. 99
Grimaldi, P. 53
Guillon, G. 69

Hallermayer, K. 231
Hamprecht, B. 231
Hanke, W. 99
Hardie, D.G. 117
Heldin, C.-H. 9
Hennemann, G. 179
Hernalsteens, J.-P. 245
Hirschhausen, R. von 99
Holland, R. 117
Houdebine, L.-M. 269
Humbel, R.E. 3

Inze, D. 245

Jimenez de Asua, L. 37

Katoh, M. 269
Kelly, P.A. 269
Kerbey, A.L. 139

Kleefeld, G. 99
Kleinkauf, H. 231
Klumpp, S. 99
Kühn, H. 87
Kunze, N. 231

Merlevede, W. 163
Michalides, R. 255
Mol, J.A. 179
Munday, M.R. 117

Négrel, R. 53

Olsson, H. 153
Otten, M.H. 179
Otto, A. 37
Otto, M.K. 99

Pays, E. 289
Pfister, C. 87
Pohl, V. 309
Poskocil, S. 37
Prochiantz, A. 223

Randle, P.J. 139
Rozengurt, E. 17

Sale, G.J. 139
Schell, J. 245
Schönefeld, U. 99
Schröder, G. 245
Schröder, J. 245
Schultz, J.E. 99
Schulz, M. 231
Schwartz, J.-C. 213
Steinert, M. 289
Strålfors, P. 153

Teyssot, B. 269

Van Montagu, M. 245
Vandenheede, J.R. 163
Vannier, C. 53
Vary, T.C. 139
Visser, T.J. 179

Wasteson, Å. 9
Westermark, B. 9
Westlind, A. 195

Yang, S.-D. 163

SUBJECT INDEX

A

Acetyl CoA carboxylase
- allosteric regulation of, 118-119
- hormones and, 126-135
- phosphorylation, 119-125; 133-135

Adenosine
- mitogenic activity of, 27-28

Adenylate cyclase
- histamine sensitive, 215
- paramecium cilia, 100; 107
- solubilization of, 78
- vasopressin and, 75

ADP ribosylation
- transducin and, 95-96

Agrobacterium tumefaciens
- plant tumor and, 245

Angiotensin II
- acetyl CoA carboxylase and, 132-134

Astroblast
- carnosine synthesis and, 235
- cyclic AMP and, 235

Auxines
- growth control and, 248-249

B

Bombesin
- mitogenic activity of, 23; 25; 31

C

Calcineurin
- in paramecium cilia, 105

Calcium
- histamine receptors and, 216-217
- mitogenic agents and, 20-22
- paramecium cilia and, 101-107
- vasopressin and, 69; 77

Calmodulin
- in paramecium cilia, 105; 106

Cancer
- in plants, 245

Carnosine
- biosynthesis of, 232; 234; 235
- in brain cells, 231-239
- carnosinase, 233
- degradation, 233
- synthetase, 232

Cassette model
- in yeast, 293

Catecholamines
- acetyl CoA carboxylase and, 132-134

Cell differentiation
- insulin and, 53-64
- in plants, 245-252

Cell proliferation
- bombesin and, 23; 25; 31
- cyclic AMP and, 26-31
- glycosylated proteins and, 37-48
- insulin and, 22; 23; 31; 53-64
- ion fluxes and, 18-26
- melittin and, 25-26; 31
- neoplastic growth inplants and, 245
- phorbol esters and, 23-25; 31
- platelet-derived growth factor and, 10

Cell transformation
- in mammary gland, 255
- in plants, 245-246; 249

Cyclic AMP
- carnosine synthesis and, 235
- growth factors and, 18; 29-30
- histamine and, 215
- hormone sensitive lipase and, 157; 159

- ion fluxes and, 30-31
- mitogenic activity of, 26-31
- paramecium cilia and, 111; 112
- vasopressin and, 69

Cyclic AMP dependent protein kinase
- hormone sensitive lipase and, 154-155; 159-160

Cyclic GMP
- histamine and, 217
- in paramecium cilia, 111; 112

Cyclic nucleotide phosphodiesterase
- in paramecium cilia, 105; 110
- rhodsopsin and, 91; 93

Cytokinin
- growth control and, 248-249

D

Deiodinase activity
- low T_3 syndrome and, 180
- in rat liver, 181-184

Diabetes
- pyruvate dehydrogenase and, 139; 141-149
- pyruvate dehydrogenase kinase and, 143-149

DNA
- rearrangements in Trypanosomes, 289

Dopaminergic neurons
- development, 223-228
- maturation, 223-225

G

Gene
- activation of VSA, 290-291; 296-300
- cloning in plants, 250
- conversion, 293-296
- expression site of VSA, 291-293
- human thyroglobulin, 316
- morphology, 309
- Ti plasmid, 246-251

Glial cells
- carnosine synthesis and, 234; 235

Glucagon
- acetyl CoA carboxylase and, 126-132

Glucose oxidation
- in insulin deficient animals, 139-149

Growth factors
- cyclic AMP and, 18; 29-30
- ion fluxes and, 18-26

GTP
- cyclic GMP phosphodiesterase and, 87
- transducin and, 90-91; 93-94

Guanylate cyclase
- in paramecium cilia, 100; 105-107; 109

H

Histamine
- biological response to, 215-217
- in brain, 213-220
- new class of receptors for, 217-220

Hormone sensitive lipase
- phosphorylation, 154-160
- properties of, 153-154

Hybridization
- RNA and DNA, 309

I

Insulin
- acetyl CoA carboxylase and, 53-64; 134-135
- anti-lipolytic effect of, 157-160
- glucose oxydation and, 139; 141-149
- IGF receptors and, 4
- mitogenic activity of, 22; 23; 31
- preadipocyte cells and, 53-64
- protein phosphatase and, 173-174

Insulin-like growth factors
- growth hormone and, 4
- insulin receptor and, 4
- preadipocyte cell proliferation and, 60
- structure and function of, 3-7

Ion fluxes
- cyclic AMP and, 30-31
- growth factors and, 18-26; 31
- melittin and, 25-26

K

Kidney
- vasopressin receptors in, 75; 78-81

Kinase Fa
- protein phosphatase and, 164-171; 173-174

L

Liver
- vasopressin receptors in, 75-78; 81

Low T_3 syndrome
- thyroid hormone deiodination and, 180; 181

M

Mellitin
- cell proliferation and, 25-26; 31

Mouse mammary tumor virus (MMTV)
- genome of, 256-257
- integration and tumorigenesis, 258-261
- transcription and steroid, 257-258

Mutants
- of paramecium, 101

N

Neuronal cells
- culture of, 237-238

Neurotransmitters
- coexistence of classical and peptide, 195-207
- release of, 200; 203-205
- sites of synthesis, 198-199

Noradrenaline
- hormone-sensitive lipase and, 157-159

Nucleic acid
- electron microscopy of, 315-317
- length measurements, 318-320

O

Oncogenes
- mammary tumors and, 255; 259
- in plants, 245

Opines
- in plant tumors, 248

P

Paramecium
- cyclic nucleotides and calcium in, 99-112

Pertussis toxin
- transducin and, 96

Phorbol esters
- acetyl CoA carboxylase and, 133-134
- mitogenic activity of, 23-25; 31

Phosphatidylinositol
- vasopressin and, 77

Phosphorylase a
- vasopressin and, 77

Plasmid
- Ti of A. Tumefaciens, 246

Platelet-derived growth factor (PDGF)
- mitogenic activity of, 10; 18; 37
- $p28^{sis}$ and, 12-13
- purification, 9
- receptor, 11-12
- structure, 10
- synthesis in tumor cells, 13

Preadipocyte cell line OB17
- differentiation of, 53-64
- proliferation of, 53-64

Proinsulin
- insulin-like growth factors and, 3; 5
- preadipocyte cell proliferation and, 60; 62

Prolactin
- colchicine and, 280
- intracellular mediator for, 278-284
- protein phosphorylation and, 282-283
- receptors, 270-278

Prostaglandins
- cell proliferation and, 28; 29; 37

Protein glycosylation
- cell proliferation and, 37-48
- process of, 38-39

Protein kinase
- in paramecium cilia, 105; 109
- rhodopsin and, 91; 94

Protein phosphatase
- acetyl CoA carboxylase and, 125-126
- in glycogen metabolism, 172-175
- hormone-sensitive lipase and, 155; 160
- kinase Fa and, 165-171

Protein phosphorylation
- cell proliferation and, 24-25
- in paramecium cilia, 110
- prolactin and, 282-283
- rhodopsin and, 94

Pyruvate dehydrogenase
- chemistry, 139-140
- fatty acid oxydation and, 141-143
- in insulin deficient animals, 139; 141-149
- phosphorylation, 139-141; 143-148

Pyruvate dehydrogenase kinase
- in diabetes and starvation, 143-149

R

Receptors
- to dexamethasone and MMTV, 255
- H_1 histamine, 213-215
- insulin-like growth factors, 4
- new class of histamine, 217-220
- platelet-derived growth factor, 11-12
- prolactin, 270-278
- vasopressin, 69-82

Retinal cell
- cyclic GMP phosphodiesterase in, 91

Retroviruses
- mouse mammary tumor, 255-265

Reverse T_3
- production of, 179; 180

Rhodopsin
- as photon receptor, 88-90

RNA
- hybridization with DNA, 310-315

S

Sodium-potassium pump
- cyclic AMP and, 30-31
- in 3T3 cells, 18-19; 30-31

Sodium-proton antiport
- protein kinase C and, 25
- in 3T3 cells, 19-20; 25

Somatomedins
- insulin-like growth factors and, 3; 4

Starvation
- pyruvate dehydrogenase and, 139; 141-149
- pyruvate dehydrogenase kinase and, 143-149

T

Thyroid hormone
- conjugation with sulfate, 180; 181; 186-188
- deiodination of, 179-188

Transcription
- hormone stimulated, 256
- of VSA genes, 300-301

Transducin
- in retinal cells, 90-91

Trypanosome
- antigenic variation and, 289
- DNA rearrangement and, 289
- genome of, 302-303

Tumors
- hormone dependent, 255
- mice mammary, 255-265
- in plants, 245-252
- transplanted mammary, 262

Tunicamycin
- cell proliferation and, 43-47
- protein glycosylation and, 39
- structure of, 40

V

Variant-specific antigens (VSA)
- gene activation, 290-291; 296-300
- gene expression site, 291-293

Vasopressin
- acetyl CoA carboxylase and, 132-134
- analogues, 72
- binding, 71-75
- cell proliferation and, 20; 23-25; 31; 37
- effects in mammals, 70; 75-77
- isoreceptors, 71-73
- phosphatidylinositol and, 77
- receptors, 69-82

**THE LIBRARY
UNIVERSITY OF CALIFORNIA**
San Francisco
666-2334

THIS BOOK IS DUE ON THE LAST DATE STAMPER BELOW

Books not returned on time are subject to fines according to the Library Lending Code. A renewal may be made on certain materials. For details consult Lending Code.

14 DAY

OCT - 3 1984

RETURNED

OCT - 9 1984

Series 4128